PUENING

orthogonal Latin squares   pg 210

Aircraft Propulsion Systems Technology and Design

Edited by
Gordon C. Oates
University of Washington
Seattle, Washington

AIAA EDUCATION SERIES
J. S. Przemieniecki
Series Editor-in-Chief
Air Force Institute of Technology
Wright-Patterson Air Force Base, Ohio

Published by
American Institute of Aeronautics and Astronautics, Inc.
370 L'Enfant Promenade, SW, Washington, DC 20024-2518

American Institute of Aeronautics and Astronautics, Inc.
Washington, DC

Library of Congress Cataloging in Publication Data

Aircraft propulsion systems technology and design/edited by Gordon C. Oates.

   p. cm. – (AIAA education series)
   The last of three texts on aircraft propulsion technology planned by Gordon C. Oates. Other titles: Aerothermodynamics of gas turbine and rocket propulsion (c1988); Aerothermodynamics of aircraft engine components (c1985).
Includes bibliographical references.
1. Aircraft gas-turbines. I. Oates, Gordon C.   II. Series.
TL709.A44   629.134'353 – dc20   89–17834
ISBN 0–930403–24–X

FOREWORD

Aircraft Propulsion Systems Technology and Design is the third volume originally proposed and coordinated by the late professor Gordon C. Oates to form a sequence of three texts that represent a comprehensive exposition of the existing knowledge and the state of the art in aircraft propulsion technology today. In completing the editorial work for this text, I would like to dedicate this last volume in the sequence to the memory of Professor Oates who today is recognized as a pioneering giant in aircraft propulsion theory and design whose influence on the propulsion community, both in industry and academe, is sorely missed.

The first volume in the sequence, *Aerothermodynamics of Gas Turbine and Rocket Propulsion,* addresses propulsion system cycle analysis and its use for establishing the design and off-design behavior of propulsion gas turbines. It includes pertinent elements of the aerodynamic and thermodynamic theory applicable to gas turbines as well as a review of the basic concepts of rocket propulsion. The second volume, *Aerothermodynamics of Aircraft Engine Components,* is primarily directed to appropriate aerodynamic and thermodynamic phenomena associated with the components of propulsion systems. The major topics covered include turbine cooling, boundary-layer analysis in rotating machinery, engine noise, fuel combustion, and afterburners.

The present text and third volume in the sequence, *Aircraft Propulsion Systems Technology and Design,* is aimed essentially at demonstrating how the design and behavior of propulsion systems are influenced by the aircraft performance requirements. It also includes an extensive review of how the often conflicting requirements of the propulsion engines and aircraft aerodynamic characteristics are reconciled to obtain an optimum match between the propulsion and airframe systems. The key element in the text is the introductory chapter that integrates technical materials in the sequence by outlining the various steps involved in the selection, design, and development of an aircraft engine. This chapter provides an excellent exposition of the key physical concepts governing gas turbine propulsion systems. The remaining six chapters cover combustion technology, engine/airplane performance matching, inlets and inlet/engine integration, variable convergent/divergent nozzle aerodynamics, engine instability, aeroelasticity, and unsteady aerodynamics.

This text is intended to provide the most recent information on the design of aircraft propulsion systems that should be particularly useful in aircraft engine design work as well as in any advanced courses on aircraft propulsion.

J. S. PRZEMIENIECKI
Editor-in-Chief
AIAA Education Series

Texts Published in the AIAA Education Series

Re-Entry Vehicle Dynamics
  Frank J. Regan, 1984
Aerothermodynamics of Gas Turbine and Rocket Propulsion
  Gordon C. Oates, 1984
Aerothermodynamics of Aircraft Engine Components
  Gordon C. Oates, Editor, 1985
Fundamentals of Aircraft Combat Survivability Analysis and Design
  Robert E. Ball, 1985
Intake Aerodynamics
  J. Seddon and E. L. Goldsmith, 1985
Composite Materials for Aircraft Structures
  Brian C. Hoskins and Alan A. Baker, Editors, 1986
Gasdynamics: Theory and Applications
  George Emanuel, 1986
Aircraft Engine Design
  Jack D. Mattingly, William Heiser, and Daniel H. Daley, 1987
An Introduction to the Mathematics and Methods of Astrodynamics
  Richard H. Battin, 1987
Radar Electronic Warfare
  August Golden Jr., 1988
Advanced Classical Thermodynamics
  George Emanuel, 1988
Aerothermodynamics of Gas Turbine and Rocket Propulsion,
  Revised and Enlarged
  Gordon C. Oates, 1988
Re-Entry Aerodynamics
  Wilbur L. Hankey, 1988
Mechanical Reliability: Theory, Models and Applications
  B. S. Dhillon, 1988
Aircraft Landing Gear Design: Principles and Practices
  Norman S. Currey, 1988
Gust Loads on Aircraft: Concepts and Applications
  Frederic M. Hoblit, 1988
Aircraft Design: A Conceptual Approach
  Daniel P. Raymer, 1989
Aircraft Propulsion Systems Technology and Design
  Gordon C. Oates, Editor, 1989

TABLE
OF
CONTENTS

CHAPTER 1. DESIGN AND DEVELOPMENT OF AIRCRAFT PROPULSION ENGINES

Robert O. Bullock

Gas Turbine and Turbomachine Consultant, Scottsdale, Arizona

DESIGN AND DEVELOPMENT OF AIRCRAFT PROPULSION SYSTEMS

NOMENCLATURE

A	= area
ΔA	= increment of area
a_T	= speed of sound at stagnation temperature
B	= bypass ratio of fan engine
b	= passage height of radial blades
C_p	= specific heat at constant pressure
C_v	= specific heat at constant volume
E_{EJ}	= specific mechanical energy delivered to main engine jet, Eq. (1.18)
E_F	= specific mechanical energy delivered to fan, Eq. (1.19)
E_G	= specific mechanical energy developed by engine, total, Eq. (1.3b)
E_N	= specific mechanical energy of engine, net, Eq. (1.11)
E_P	= specific mechanical energy available from/to propeller, Eq. (1.15)
E_{PT}	= specific mechanical energy produced by power turbine, Eq. (1.17)
F_G	= specific thrust derived from E_G, Eq. (1.12)
F_{PJ}	= specific thrust derived from propeller slipstream, Eq. (1.15)
F_S	= total specific thrust of propulsion system, Eq. (1.19)
g	= acceleration of gravity
H_F	= lower heating value of fuel
ΔH	= change in specific energy
J	= ratio, unit of mechanical work per unit of thermal energy
$(L/D)_{AV}$	= average value of lift-to-drag ratio over a given range of flight
l	= a distance between adjacent radial blades
M_0	= flight Mach number
N	= rotating speed
N_S	= specific speed
n	= length perpendicular to streamline, in plane of flow or projection of flow
Δn	= incremental distance between streamlines
P	= total pressure
P_{SL}	= standard sea level pressure
p	= static pressure
Q	= volume flow rate
R	= gas constant
R_c	= radius of curvature of streamline
r	= radius
ΔS	= increase in entropy

3

T	= total temperature relative to stationary coordinates
T_{SL}	= standard sea level temperature
T'	= total temperature relative to rotating coordinates
T^*	= reduced temperature, $T - [(\gamma - 1)/2](V^2/\gamma g R)$
ΔT	= change in temperature
t	= static temperature
U	= tangential speed of rotor, ω_r
V	= fluid velocity
V_j	= jet speed
V_{FJ}	= effective speed of fan jet
V_{PJ}	= effective speed of propeller slipstream
V_0	= speed of flight
ΔV	= increment in velocity
W_{fuel}	= weight of fuel
W_{to}	= weight at takeoff
$\dot{W}$	= mass flow rate
$\dot{W}_E$	= mass flow rate of engine airflow
$\dot{W}_F$	= mass flow rate of fuel
$\dot{W}_{PJ}$	= mass flow rate in propeller slipstream
$\Delta\dot{W}$	= increment of mass flow rate through incremental area
γ	= ratio of specific heats, C_p/C_v
δ	= P/P_{SL}
ζ	= loss in available specific mechanical energy
θ	= T/T_{SL}
η	= compression efficiency, Eq. (1.4)
η_{ad}	= compressor adiabatic efficiency
η_E	= engine efficiency
η_{EN}	= efficiency of engine exhaust nozzle
η_{ex}	= efficiency of expansion process
η_F	= fan efficiency
η_{FN}	= efficiency of fan nozzle
η_{FT}	= efficiency of turbine driving fan
η_I	= efficiency of inlet
η_P	= efficiency of propeller
η_{PT}	= efficiency of power turbine, including any transmission losses
μ	= momentum function [Eq. (1.26c)] or viscosity
ρ	= gas density
ψ	= mass flow rate function, Eq. (1.26a)
ω	= angular velocity
ω_a	= angular velocity of gas
ω_o	= angular velocity of rotor

Subscripts

0 to 9	= see Fig. 1.1
a	= pertaining to fluid
C	= compressor
C,I	= compressor inlet
C,O	= compressor outlet
IN	= inner, or high-pressure, unit

M	= mean
OUT	= outer, or low-pressure, unit
r	= radial component
T,I	= turbine inlet
T,O	= turbine outlet
Z	= axial component
θ	= tangential component

1.1 INTRODUCTION

This is the third of three volumes reviewing the application of aerodynamic and thermodynamic principles to the gas turbines used in aircraft propulsion. The first volume[1] addresses cycle analysis and its use for establishing the design and off-design behavior of engines. Pertinent elements of the aerodynamic and thermodynamic theory applicable to engines are also presented. It also reviews the basic concepts of rocket engines.

The second volume[2] is primarily directed to the appropriate aerodynamic and thermodynamic phenomena associated with the components of propulsion engines. Important information about turbine cooling is included. This volume also reveals the status of the continuingly developing techniques for necessary boundary-layer analysis and includes a chapter about the aerodynamic development of engine noise. In addition, two chapters are devoted to the combustion of fuel: one on the fundamentals and the other on their application to afterburners.

The subject of combustion is continued in the second chapter of the present volume. It presents a detailed account of the important items involved in the design of primary combustion chambers. The rest of this volume is aimed at demonstrating how the design and behavior of engines is influenced by the goals of various aircraft and the environments provided to the engine by the aircraft. It examines how freedom in the design of aircraft is restricted by the inherent characteristics of engines. This volume also offers an extensive review of how the often conflicting requirements of engines and airplane aerodynamics are reconciled to the needs of the user to obtain a viable product.

This introductory chapter aims at integrating the technical material presented in three volumes by outlining some of the steps involved in the selection, design, and development of an engine. This integration begins with the considerations of the role played by engines. A discussion of factors affecting the selection of the engine cycle variables follows. The considerations governing the selection of rotative speed and the size of the gas turbine components, including the care that must be taken in the arrangement of the components, are reviewed. Special note is made of the causes of losses and their effect on engine performance. The chapter ends with a prognosis about future improvements and opportunities and a description of typical problems encountered during engine development. An appendix offers a brief description of the distinguishing features of centrifugal compressors. They are used in many propulsion engines and no other reference is made to them in this series.

We recognize that many computer programs have been created for treating these and related topics. Some of the programs are proprietary;

others can be purchased on the open market. Since specific calculating procedures are available, this chapter is written to enhance the reader's awareness of the governing physical concepts rather than to present working equations.

1.2 ENGINE DESIGN OBJECTIVES

All aircraft represent an investment of money for an anticipated profit. It is evident that commercial aircraft are built and bought to earn a monetary return on an investment. Although the revenue derived from investment is less tangible for many military aircraft, these too are expected to provide some kind of an identifiable payoff that can be expressed in terms of money. An appropriate economic analysis, therefore, always precedes a decision to design, develop, and manufacture a new or a modified type of aircraft engine. The costs considered include all of the expenses of acquiring and servicing the aircraft involved as well as the engine.

A necessary part of the economic study is the execution of a preliminary design of the proposed vehicle and its powerplant. Chapter 3 of this volume as well as Ref. 3 review the many details that must be considered during such a preliminary design.

Of special interest is the impact that an engine and its fuel requirements have on the results of such economic studies. In long-range airplanes, for example, the required fuel weight may be four times the weight of the payload. The value of the low fuel consumption by these engines is obvious. At the same time, the engines in these airplanes themselves may weigh more than 40% of the payload. The additional airplane structure required to support both the engines and fuel is thus over four and one-half times that of the payload for which the airplane is intended. We now note that the weight of fuel demanded for any given flight is indicated by the following form of the Breguet range equation (see Chapter 3 of this volume):

$$\frac{W_{\text{fuel}}}{W_{\text{to}}} = 1 - \exp\left[-\frac{\text{range}}{JH_F[(\eta_E\eta_P)(L/D)_{\text{AV}}]}\right] \tag{1.1}$$

This ratio can be as high at 0.4 for long-range airplanes. Notice that $(L/D)_{\text{AV}}$ is optimized by aerodynamic considerations. Propulsive efficiency can be expressed as (see Chapter 5 of Ref. 1)

$$\eta_P = \frac{1}{(F_s/2\dot{W}_E V_0) + 1} \tag{1.2}$$

This efficiency depends upon $F/\dot{W}$, which varies with the type of engine (turbojet, turbofan, or propeller). It also depends upon the characteristics of the nozzles or propellers selected by the airplane manufacturer.

Chapter 7 of Ref. 1 examines some of the factors controlling engine efficiency, and these will be mentioned again in the next section. At the moment, it is sufficient to note that this efficiency is almost completely determined by the engine design and the operating regime of the airplane.

Prolonged and expensive research has yielded impressive improvements in engine efficiency and more gains are expected in the future. It can be argued that the savings in fuel alone have made the costs of research a profitable investment. This saving is not the only reward, however, because reducing fuel weight also lessens the required size of the airplane and engines. An additional reduction in acquisition costs is effected. Of course, the payload is often increased by the amount of fuel saved. The overall cost per pound of payload is thus lowered and this is really the basic objective. In practice, an airplane designer may exploit improvements in engine efficiency in various ways to optimize the value of his design.

We cannot forget that benefits also accrue from decreasing the size and weight of the engines required to produce a given thrust. Engines represent a concentration of weight, inertia, and gyroscopic forces that require additional structural weight in the airplane. Even though the engine weight in long-range aircraft may represent only about 5% of the takeoff gross weight, the actual weight associated with engine installations is greater, because extra airplane weight is required to support and restrain the engines. We should also observe that many engine installations are an aerodynamic liability. They produce unwanted local velocity levels and velocity gradients in the surrounding airflow; this phenomenon effectively reduces the value of L/D in Eq. (1.1).

Since engines are a concentrated weight, they can further adversely affect L/D because the center of gravity of the airplane must be controlled in order to maintain inherent stability. Note, however, that the complication of active controls may relieve this problem.

There are many types of aircraft where the engine weight is greater than that needed for long-range airplanes. More than 15% of the takeoff gross weight must often be dedicated to engine weight when aircraft are expected to execute strenuous maneuvers or achieve high acceleration rates. Helicopters that lift heavy loads vertically also belong to this category. In these instances, engines that provide a given output with reduced weight offer substantial payoffs.

Reference 4 presents an informative overview of the value of improved engines. It quantifies incentives for increasing the efficiency and the output per unit weight of engines. This reference also calls attention to the fact that the ultimate costs of aircraft are notably influenced by mechanical features of engines that control the service life, reliability, and costs of manufacture, maintenance, and repair. These features must be recognized from the start by the designers of thermodynamic and aerodynamic components, lest savings in one area be nullified by expenses in another.

In summary, the important design objectives are good thermodynamic efficiency and large power output per unit of engine size and weight. These objectives are often in conflict and the fluctuating cost of fuel obscures their reconciliation. The designs, in any event, must not impose costly fabrication difficulties that negate the economic advantages offered by improved performance. All the critical parts of the engine should be readily accessible to minimize maintenance and repair. The aerodynamic parts should also be as rugged as possible to avert the possibility of accidental damage. This precaution not only lessens repair costs, but it also reduces the amount of standby equipment needed to provide good service.

We close this section with the observation that the design and development of a new engine is very expensive, that it requires the vigorous and rigorous application of many technologies and resources, and that it is not without risk. The development of an engine alone has cost over $1 billion. Competitive designs must approach the limits of technology. This statement is not confined to the aerodynamic and thermodynamic designs, but it applies also to such areas as the structural design and the exploitation of material properties. The design of the control system and actuators and the methods of fabrication and assembly also confront designers. Many intelligent compromises must be made both within disciplines and between disciplines, with each group anticipating how its decisions might make life difficult for a different group. Integrating these efforts requires outstanding talents in both technology and management.

The element of risk arises because engines are complicated and because there is, on the frontiers of technology, the interaction of many phenomena that preclude the possibility of completely anticipating how engines will react to the wide variety of the operating conditions demanded of them. Another risk is associated with predicting the reliability of a design. The ability of 99 out of 100 parts to function perfectly in a hostile environment over a period of say 1000 operating hours is often a factor in making the decision to create a new engine. If a part has a new shape, is made of new materials, or is exposed to new rigors, its ability to survive is always suspect until a number of engines have actually run for 1000 h or more. (This is true even though improved accelerated testing techniques have increased the confidence in abbreviated tests.) If premature failure becomes a problem, the engine manufacturer must spend a lot of money to eliminate the cause. Otherwise, the value of his engine will depreciate.

Incidentally, if the life of a critical part greatly exceeds this target value of 1000 h, the designer can be branded as "too conservative"—the part is deemed to be too heavy or too expensive!

1.3 EFFECT OF THERMODYNAMIC VARIABLES ON ENGINE PERFORMANCE

Figure 1.1 presents a sketch of an elementary gas turbine engine. A compressor converts mechanical energy into pneumatic energy and raises the total pressure of the air between stations 2 and 3. (The station numbers conform to a standard form, see Fig. 5.1 of Ref. 1.) When the engine is in forward motion, additional pneumatic energy converts the kinetic energy of the relative motion into pressure. Combustion of fuel in the burner adds heat and raises the air temperature between stations 3 and 4. The turbines between stations 4 and 5 convert part or nearly all of the available energy at station 4 into mechanical energy. Part of this mechanical energy is transferred to the compressor to effect the compression between stations 2 and 3. Additional mechanical energy may be transferred through a propulsion device such as propeller or fan. Any pneumatic energy remaining at station 5 is used to accelerate the gas to the velocity V_j and the kinetic energy of $\frac{1}{2}V_j^2$ per unit mass of air represents additional power output from the engine. Heat may be added in an afterburner to further increase V_j and the output power.

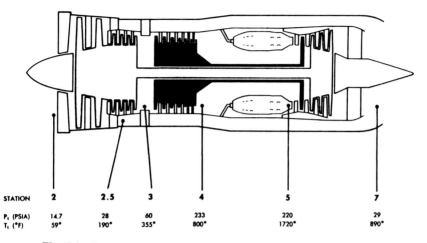

STATION	2	2.5	3	4	5	7
P_t (PSIA)	14.7	28	60	233	220	29
T_t (°F)	59°	190°	355°	800°	1720°	890°

Fig. 1.1   Basic arrangement of components in a gas turbine engine.

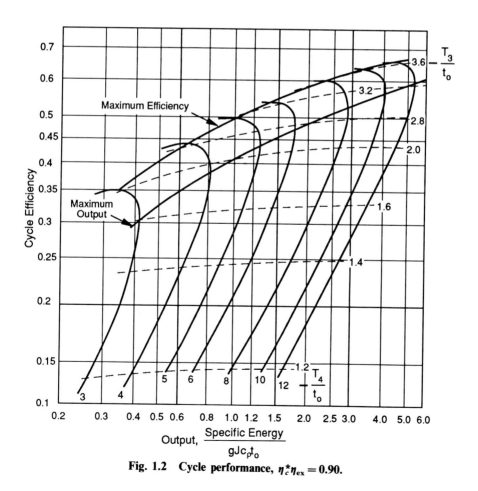

Fig. 1.2   Cycle performance, $\eta_c^* \eta_{\text{ex}} = 0.90$.

The thermodynamic variables currently available to the engine designers are the mass flowrate, energy transferred to the compressor, and gas temperatures at both the turbine inlet, and afterburner outlet. The values selected for a given design must produce the desired mechanical power and efficiency within the objective size and weight.

The engine of an aircraft produces power, which may be in the form of shaft power, kinetic energy in an exhaust jet, or both. The power and efficiency of the engine are the primary concern of the engine manufacturer. Power is then converted into thrust by means of a propeller or jet nozzle. In this section, we first discuss the factors determining engine power and efficiency. Gases with constant specific heats are studied first because the results are easily acquired and generalized. The effects of real gases are then examined. Finally, the development of thrust from available power is studied and the variables affecting the conversion of power into thrust are enumerated.

Analysis with Ideal Gases

Chapter 7 of Ref. 1 identifies the thermodynamic manipulations needed for relating the cycle variables to engine performance. The following review is based on using an ideal gas and the physical principles described in Ref. 1. The review is simplified by consolidating both the compression and expansion processes. The energy delivered to the complete compression process is $C_p(T_3 - t_0)$. That delivered by the expansion process is $C_p(T_4 - t_{9'})$. (Recall that we are assuming an ideal gas with constant values of C_p and γ). The output of the engine is

$$\text{Power} = \dot{W}_E J C_p [(T_4 - t_{9'}) - (T_3 - t_0)] \tag{1.3a}$$

For convenience, the pressure ratio of the compression process is defined as P_4/p_0. That of the expansion process is then $P_4/p_{9'} = P_4/p_0$.

Also for convenience, we define compression efficiency as

$$\eta_c^* = \frac{1 - \left(\dfrac{p_0}{P_4}\right)^{\frac{\gamma-1}{\gamma}}}{1 - \dfrac{t_0}{T_3}} \tag{1.4}$$

This expression represents the ratio of the energy available after compression to that actually invested in the compression process. It is to be distinguished from the commonly used compressor adiabatic efficiency, which is

$$\eta_{ad} = \frac{\left(\dfrac{P_4}{p_0}\right)^{\frac{\gamma-1}{\gamma}} - 1}{\dfrac{T_3}{t_0} - 1}$$

The expansion efficiency is the same as the conventional turbine efficiency

$$\eta_{ex} = \frac{1 - t_9/T_4}{1 - (p_0/P_4)^{(\gamma - 1)/\gamma}} \tag{1.5}$$

The concepts for η_c^* and η_{ex} also define efficiencies of other engine components.

When Eqs. (1.3a) and (1.4) are used with Eq. (1.5), we find that engine specific work is

$$E_G = \frac{\text{power}}{\dot{W}_E} g = gJC_p t_0 \left(\eta_c^* \eta_{ex} \frac{T_4}{t_0} - \frac{T_3}{t_0} \right) \left(1 - \frac{t_0}{T_3} \right) \tag{1.3b}$$

and the efficiency is

$$\eta_E = \frac{\left(\eta_c^* \eta_{ex} \dfrac{T_4}{t_0} - \dfrac{T_3}{t_0} \right)\left(1 - \dfrac{t_0}{T_3} \right)}{\dfrac{T_4}{t_0} - \dfrac{T_3}{t_0}} \tag{1.6}$$

Figure 1.2 illustrates the trends of these equations when $\eta_c^* \eta_{ex}$ has the constant value of 0.9. Engine efficiency is plotted against the quantity $E_G/gJC_p t_0$, which is the ratio of work output to inlet enthalpy per unit of mass flow of the engine inlet. The parameters are T_4/t_0 and T_3/t_0. One may determine that $\eta_c^* \eta_{ex}$ is primarily a function of engine type, turbine inlet temperature, and technical excellence than it is of T_3/t_0. The value 0.9 for $\eta_c^* \eta_{ex}$ is appropriate for a modern supersonic airplane engine without afterburning. The range of options between an engine design for maximum output and one for maximum efficiency is self-evident in this figure.

It is clear that both the output and efficiency of an engine benefit from raising T_4/t_0 and T_3/t_0. Maximum engine efficiencies require the values of T_3/t_0 to be higher than those for maximum output; maximum engine efficiency is thus attained at the cost of larger compressors and turbines and a heavier engine. If the subsequent saving in fuel weight is too small, an increase in airplane weight and drag is the result.

It is worth noting that we can also express Eq. (1.3a) by

$$E_G = gJC_p T_5 \left[1 - \left(\frac{p_{9'}}{P_5} \right)^{(\gamma - 1)/\gamma} \right] \tag{1.3b}$$

The procedures outlined in Chapter 7 of Ref. 1 show that if T_3/t_0 is increased from unity, with T_4 fixed, the magnitude of T_5 continuously decreases. The quantity $P_5/p_{9'}$, however, increases—rapidly at first and then more and more slowly until it too decreases.

The rise in $P_5/p_{9'}$ overcomes the fall in T_5 until maximum power is reached. At higher values of T_3/t_0, the reducing value of T_5 controls the trend and power decreases. The rate of change of power is slow until $P_5/p_{9'}$ also begins to decrease; output then falls rapidly. Efficiency continues to improve because the input decreases faster than the output; these balance out at the point of maximum efficiency.

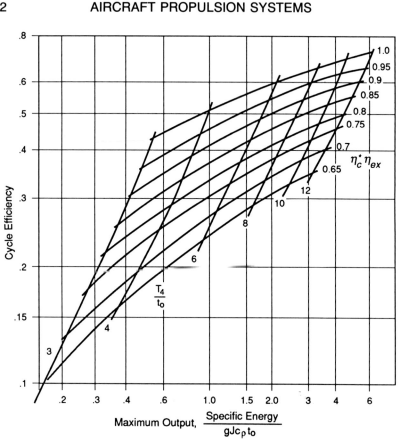

Fig. 1.3   Maximum output and corresponding cycle efficiency, $\eta_c^*\eta_{ex}$ = parameter.

Changes in component efficiencies can have dramatic effects on engine performance. These are illustrated in Fig. 1.3 where the engine efficiency associated with the maximum engine output obtainable at a given T_4/t_0 is plotted against this maximum output and in Fig. 1.4 where the maximum efficiency obtainable is plotted against the corresponding power. The equations for these curves were derived by differentiating Eqs. (1.3) and (1.6) with respect to T_3/t_0 and equating the result to zero. The results are, for maximum power,

$$\frac{T_3}{t_0} = \sqrt{\eta_c^*\eta_{ex}\frac{T_4}{t_0}} \qquad (1.7)$$

and for maximum efficiency,

$$\frac{T_3}{t_0} = \left(\frac{\sqrt{1+\phi}-1}{\phi}\right)\frac{T_4}{t_0} \qquad (1.8)$$

where

$$\phi = \frac{1}{\eta_c^*\eta_{ex}}\left[\frac{T_4}{t_0}(1-\eta_c^*\eta_{ex})-1\right] \qquad (1.9)$$

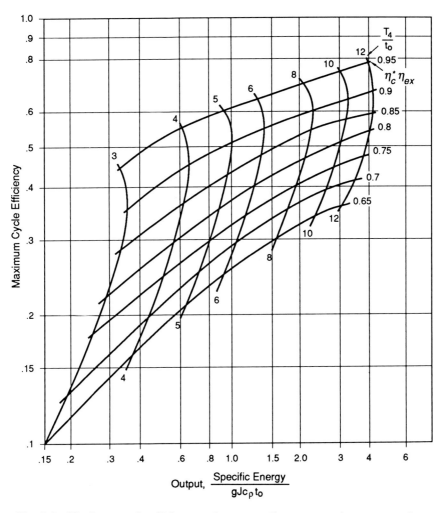

Fig. 1.4 Maximum cycle efficiency and corresponding output, $\eta_c^* \eta_{ex}$ = parameter.

These relations are reliable only to the extent that $\eta_c^* \eta_{ex}$ does not vary greatly with T_3/t_0 in the ranges of interest, but this seems to be a minor problem.

The value of $\eta_c^* \eta_{ex}$ varies from about 0.7 for subsonic turboprops to over 0.90 for supersonic flight. The high values of η_c^* and η_{ex} for the inlet compression and the nozzle expansion explain the large value of their product at high flight speed.

We note from Figs. 1.3 and 1.4 that both engine output and efficiency benefit from high component efficiency, although engine efficiency is the more sensitive. We also note that the power at the maximum efficiency point falls when $\eta_c^* \eta_{ex}$ rises above about 0.9. Observe that T_3/t_0 approaches unity in Eq. (1.8) at efficiencies of unity and that the corresponding value of E_G in Eq. (1.3) is zero. There is, therefore, a limit to the practical engine efficiencies that may be sought.

When high work output is the principal goal of a design, every effort is placed on making the product $\eta_c^* \eta_{ex} T_4$ as large as possible. It is worthwhile to sacrifice η_{ex} to increase T_4 only as long as this product is increased.

The interrelations of turbine inlet temperature and component efficiency are more intricate when high engine efficiency is sought. The overview provided by Fig. 1.4 shows that increases in T_4/t_0 are indeed welcome as long as any accompanying efficiency penalties due to turbine cooling are small.

This section was written to provide some insight into the factors determining engine power and efficiency. We have found that high turbine inlet temperatures are indeed desirable. They cannot be raised indiscriminately, however, if component efficiencies are too adversely affected. This is a danger when aircooling is carelessly applied to turbine blades; see Chapters 4 and 5 of Ref. 2 for details about this problem. We may appreciate that a desire for the potential uninhibited benefits for high turbine inlet temperatures has prompted research on materials that can withstand high temperatures and stresses with little or no cooling. Ceramics and materials derived from carbon are examples of current effort.

Effects of Real Gases on Calculated Performance

Recall that we have assumed C_p and γ to be constants. This concept is true when the average square of the linear speeds of the molecules accounts for all of their energy. When molecules consist of two or more atoms, however, the atoms rotate about each other and vibrate; this energy is absorbed by the gas. More and more energy is diverted this way as the temperature is raised; thus, the value of C_p continually increases while γ decreases. We note that air consists almost entirely of the diatomic gases oxygen and nitrogen and that C_p increases and γ therefore decreases with increasing air temperatures. After combustion, we also find important quantities of triatomic elements—carbon dioxide and water vapor. Combustion thus causes a further increases in C_p and reduction in γ. The gas constant is also lowered. We now need to determine how these changes affect our thoughts about engine thermodynamic design.

The result of a spot check of the effect of real-gas properties is presented in Table 1.1. These results are typical. The assumed fixed conditions for the cycle are noted. The calculated gas properties are: the enthalpy increase for compression, the energy added by fuel to obtain the stipulated turbine inlet temperature, the enthalpy converted into mechanical energy during the expansion process, the energy produced by the engine, and the engine efficiency. The first column shows the results of assuming C_p to be 0.24 and γ to be 1.4.

The second column shows the results when calculations use the real-gas properties for air and the products of combustion. The gas property data are found in Ref. 5. (This reference provides all the material necessary for making such cycle computations. Pertinent data from Ref. 5 is reprinted in many handbooks and is used by many members of the gas turbine

Table 1.1   Effect of Real Gas Properties

$V_0 = 0$, $t_0 = 520°R$, $P_3/p_0 = 20$, turbine efficiency = 0.88

$\left(\dfrac{P_2}{P_1}\right)\left(\dfrac{P_4}{P_3}\right)\left(\dfrac{P_8}{P_5}\right) = 0.92$, compressor adiabatic efficiency = 0.85, compressor efficiency = 1.00

Property	$C_p = 0.24$ $\gamma = 1.4$	Real gases	$C_p = 0.24$ (comp), 0.274 (turb) $\gamma = 1.4$ (comp), 1.333 (turb)
ΔH comp	1.99	1.98	1.99
ΔH fuel	387	475	488
ΔH turb	353	377	369
Output, ΔH eng	154	179	171
η, eng	0.40	0.38	0.35

industry.) Note that performance predictions based on the previous simplifying assumption have noticeable inaccuracies. Experience shows, however, that the superiority of one set of specified cycle conditions over another is usually correctly indicated by the simplifying assumptions. The third column shows the result of assuming one set of constant gas properties for the gas during the compression of air and another set for expansion of the hot gases. These results are in somewhat better agreement with reality and this technique is often used when preliminary estimates are made with a hand-held calculator.

Be aware that even the tables of Ref. 5 do not take all possibilities of error into account. Accounting for humidity is often necessary. Moreover a slight but finite time is required for the vibration and rotation of atoms within a molecule to reach equilibrium when the temperature is changed, particularly when the temperature falls and the internal molecular energies of rotation and vibration are converted to the kinetic energy of linear motion. When the rate of change in temperature with time is large, the calculated results using these tables are in error because there is not enough time for equilibrium to be achieved within the turbine. Most inaccuracies from this source are not detected, however, because the errors are usually less than the unavoidable measurement errors in gas turbines.

A more serious problem is the phenomenon of dissociation. This subject is introduced in Chapter 1 of Ref. 2. Essentially, small but noticeable amounts of CO and OH separate from the CO_2 and H_2O molecules at temperatures above about 1800 K. This dissociated represents incomplete combustion and it thus prevents the total fuel energy from being used. Dissociation also changes the values of specific heats and thus introduces further errors when some gas tables are used. We note that other tables, e.g., Ref. 6, claim to provide data for dissociated states. However, another time delay between the temperature and the attainment of equilibrium specific heats can be experienced when the changes in temperature are rapid, causing the inaccuracies to recur.

Standard textbooks on thermodynamics (e.g., Ref. 7) are suggested for some further study of dissociation and other variable gas properties. Understanding these phenomena is essential for the proper interpretation of some engine data.

1.4 DEVELOPMENT OF THRUST

Turbojet Engines

In a turbojet (Fig. 1.5), all the energy developed by the engine appears as kinetic energy of the gases expelled through a nozzle into the atmosphere. The resulting high-speed jet produces the thrust. If the mass flow does not vary throughout the engine and nozzle, then

$$E_G = \tfrac{1}{2}V_j^2 \tag{1.10}$$

In this section all velocities are assumed to be uniform. This is the energy represented by the abscissa of Fig. 1.2—it is the relative output specific energy of an engine. This is an ideal relation; we should multiply E_G by the efficiency η_{EN} for real cases.

The net specific energy available is

$$E_N = \tfrac{1}{2}(V_j^2 - V_0^2) \tag{1.11}$$

or
$$\frac{E_N}{gJC_pt_0} = \frac{E_G}{gJC_pt_0} - \frac{\gamma-1}{2}M_0^2$$

The quantity $[(\gamma-1)/2]M_0^2$ is subtracted from the abscissa of Fig. 1.2 to get the available propulsive energy.

The thrust per unit of mass flow is calculated from

$$F_G = (V_j - V_0) = V_0\left(\sqrt{\frac{2E_G}{V_0^2}} - 1\right) \tag{1.12}$$

(Unvarying mass flow is again assumed.)

The propulsive specific energy actually realized is $V_0(V_j - V_0)$. In view of Eq. (1.11), the propulsive efficiency is

$$\eta_p = \frac{2V_0}{V_j + V_0} \tag{1.13a}$$

This agrees with Eq. (1.2), which becomes

$$\eta_p = \frac{1}{(F_s/2V_0)+1} = \frac{2}{\sqrt{2E_G/V_0^2}+1} \tag{1.13b}$$

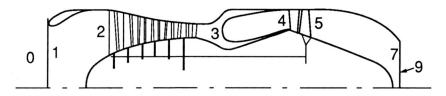

Fig. 1.5   Schematic of turbojet.

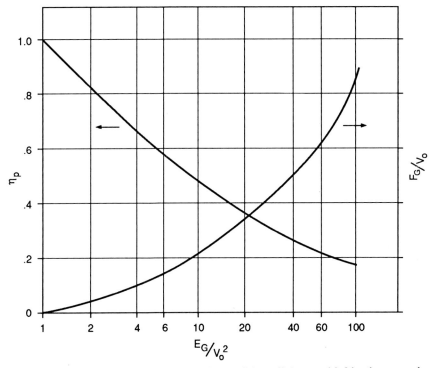

Fig. 1.6   Variation of specific thrust and propulsion efficiency with kinetic energy in propulsive jet.

The quantity E_G controls both F_s/V_0 and η_p. Figure 1.6 displays these relations. We see that high values of propulsion efficiency are inimical to large values of either F_s/V_0^2 or E_G/V_0^2.

This behavior is certainly not a blessing and it did indeed discourage the financial support of initial jet engine development. Some early proponents of these engines foresaw, however, that such designs could inherently handle much larger weight flows of air than reciprocating engines and could then produce more thrust per unit engine weight. The thrust required for a fighter aircraft could be obtained with engines that were much smaller and lighter than the reciprocating engines they would replace, even though the available technology limited $\eta_c^* \eta_{ex}$ to values between 0.80 and 0.85 and T_s/t_0 and T_4/t_0 to values of about 1.6 and 4.0, respectively. Moreover, the resulting airplanes required no propeller. These assets could compensate for a portion of the range lost because of low efficiency; also, some range could be waived to develop airplanes that would overtake and outclimb the competition.

Dr. Hans von Ohain of Germany was the first to demonstrate the reality of this belief. His engine was installed in a Heinkel He 178, which made its first successful flight on Aug. 27, 1939. Meanwhile, Sir Frank Whittle of England was also developing a turbojet engine to power a Gloster Aircraft E28/39; the first flight was on May 14, 1941.

As their designers anticipated, these fighter airplanes had a decisive tactical advantage over aircraft powered by piston engines. The high thrust-to-weight ratio of the jet engines did indeed enable the airplanes to outmaneuver and outclimb conventional designs. Since they were not encumbered by propellers they could also fly much faster. Because the pioneering engines suffered from low efficiency and consequent high fuel consumption, their range of operations could not extend far beyond their fuel supply.

The impact of these aircraft, however, had far-reaching effects. Governments were quick to support the research and development of the technologies for gas turbine engines and high-speed aircraft. Maintaining competitive strength compelled industry also to make investments in this area. Notable improvements in gas turbine engines were inevitable as a result of increased component efficiencies and elevated turbine inlet temperatures. Further improvements came from reductions in the engine weight. Reduced weight was derived from advances in stress and vibration analysis that, in turn, provided better structural efficiency. The discovery of materials having superior stress-to-weight ratios afforded further weight reduction.

In a short period of time, the turbojet engine evolved to the point where it could be used for intercontinental flights. Operators discovered that jet-powered aircraft vibrated far less than those with piston engines—thus noticeably reducing maintenance costs and increasing airplane availability. Fewer airplanes were required for a given service because of the high speed attainable. Higher rates of revenue generation and more passenger comfort were offered. A revolution in commercial air service began and many areas of the world became accessible by overnight flights. This brief discussion calls attention to the fact that intangible and unforseen factors also control the success of a venture with a new airplane or engine.

At the present time, production of turbojet engines for commercial airplanes has practically ceased, although many old designs are still in service after 15 or more years. More efficient turbofans offering the virtues of turbojets have replaced them. In addition, turboprops provide more competitive service in some areas. This is also applicable in many military requirements.

We see from Fig. 1.6 that very-high-speed flight in the stratosphere is the ideal milieu for turbojets, provided the needed value of F_G can be controlled to increase of a lower rate than V_0^2. Adequate thrust, as well as high propulsive efficiencies, can be provided at low ratios of F_G/V_0.

At the same time, high values of $\eta_c^*\eta_{ex}$ are possible because of the comparatively good efficiencies of the inlet and thrust nozzle, even when additive drag at the inlet and boattail drag at the outlet are included in η_c^* and η_{ex}. (See Chapters 3 and 4 of this volume for the discussion of additive drag and boattail losses.)

Turbojets are also the natural choice for short-range missiles, where the fuel weight is small and the payoff for high efficiency is small. This application still attracts interest in turbojets.

Afterburning Turbojet Engines

The term afterburning, or reheat, applies to engines in which a second combustor is placed between stations 5 and 7 (see Fig. 1.7) to increase the temperature of the gas downstream of the turbines that drive the compressors. Afterburners are stationary and are consequently subject to far less stress than turbine rotors. The metal parts are much more easily cooled than turbine airfoils and stoichiometric temperatures can thus be approached unless unwanted disassociation intervenes. Engine thrust can be augmented by about 50% at takeoff and by over 100% at high speeds. If we again assume constant specific heat, the output specific energy may be expressed as

$$E_G = gJC_pT_5[1 - (p_4/P_7)^{(\gamma - 1)/\gamma}] \tag{1.14}$$

This is the same as Eq. (1.3b), except for the subscripts. We observe that T_7 is deliberately greater than T_5, while pressure losses in the afterburner make P_7 slightly less than P_5. The result can be a large increase in E_G, and an accompanying increase in V_j augments the thrust. However, a thorough study of performance (e.g., Chapter 7 of Ref. 1) reveals that the fuel consumption rate is raised to a point that greatly reduces the efficiency.

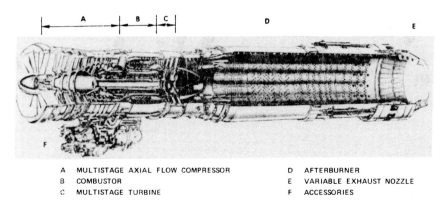

A	MULTISTAGE AXIAL FLOW COMPRESSOR	D	AFTERBURNER
B	COMBUSTOR	E	VARIABLE EXHAUST NOZZLE
C	MULTISTAGE TURBINE	F	ACCESSORIES

Fig. 1.7 Jet engine with afterburner.

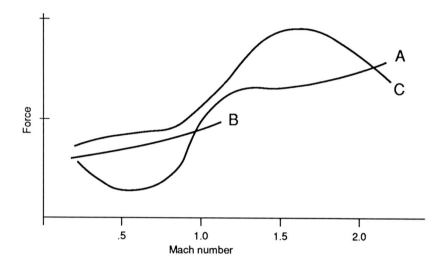

Fig. 1.8 Drag and thrust available for prescribed flight path: a) drag (equals thrust required), b) available thrust in turbojet, c) available thrust in turbojet with afterburning.

For a stipulated level of T_7, we find from Eq. (1.14) that P_7/p_0 should be as high as feasible to derive the most thrust by afterburning. This ratio is dependent on the individual values of T_2/t_0, T_3/T_2, compressor efficiency, turbine efficiency, and total pressure loss at the engine inlet. These inter-relations cannot be portrayed in a simple graph. In most cases however, conditions for maximum values of P_7/p_0 are found near the upper right-hand corner of Fig. 1.2. The associated values of T_3/t_0 lie between those for maximum power and maximum efficiency. Since the engine weight and the resulting airplane drag increase with T_3/T_2, the desirable magnitudes of T_3/t_0 lie below those for maximum P_7/p_0.

Although afterburning provides significant increases in the engine thrust-to-weight ratio, it is always at the expense of engine efficiency. Even so, augmenting thrust by afterburning for short periods of time can reduce the total weight of fuel consumed during a mission. For example, a given change in altitude can be realized in far less time with afterburning. The total fuel required to change the altitude by a given amount may thus be less with afterburning, even though the fuel consumption per unit of time is much greater. Similarly, the short bursts of power required for acceleration, vigorous maneuvers, or high speeds are often most economically made with afterburning.

Figure 1.8 illustrates this point. The abscissa indicates a flight Mach numbers schedule for a hypothetical airplane as it climbs from sea level to the stratosphere and simultaneously accelerates to the ultimate value shown. Curve A indicates the thrust demanded by the airplane to realize these requirements, while curve C indicates the thrust available from an afterburning engine. The difference between these curves is the thrust available for acceleration. Observe that, even with afterburning, the difference between these curves is small when supersonic Mach numbers are first encountered. This situation is often called the "thrust pinch." The intersection of the two curves at the highest Mach number is the condition for level flight at the indicated flight speed and given altitude.

Curve B in Fig. 1.8 represents the thrust available from a similar engine, but without an afterburner. The available forces for acceleration are inadequate and this engine could not push an airplane through the thrust pinch regime. Of course, a larger and heavier nonafterburning engine might have done the job, but it would have increased both the thrust required and decreased the total fuel weight in the airplane—to the detriment of the value of the airplane. Accurate analysis of problems of this sort is a vital necessity and Chapter 3 of this volume is recommended for further study. Texts similar to Ref. 3 provide additional information.

We have noted the weight of afterburners and the pressure loss that they always suffer between stations 5 and 7. See Fig. 1.7. This may be a determinant of whether or not to utilize an afterburner. We must also keep in mind that afterburners occupy a generous portion of the overall length and thereby further increase airplane drag. (Chapter 2 of Ref. 2 justifies the necessity for the length and other features of afterburner design.) Although relatively small, this drag detracts from the thrust whether the afterburner is operating or not. Moreover, at least part of the pressure dissipation $P_5 - P_7$ is independent of engine operation and some loss in thrust is always present.

These two penalties limit the value of afterburning for long-range flights. Even so, a small amount of afterburning has been found to be useful for commercial supersonic transports. The best configuration seems to require a thrust augmentation of about 10%, which reduces the disadvantages. Chapter 7 of Ref. 2 provides insight into the noise problem.

Turboprop Engines

A typical arrangement of a turboprop engine is illustrated in Fig. 1.9. Instead of converting all the available energy at station 5 into kinetic energy, the gas partially expands through additional turbine stages that drive a propeller through the gears. The propeller provides the thrust by imparting momentum to a large mass of air. The specific kinetic energy in the resulting airstream is low and good propulsive efficiencies are realized, as indicated in Fig. 1.6.

Equations (1.11–1.13a) still apply. Specifically, we have for specific thrust

$$F_{PJ} = V_{PJ} - V_0 = V_0 \left[\sqrt{\frac{2E_P \eta_P}{V_0^2}} - 1 \right] \qquad (1.15)$$

where V_{PJ} is the absolute mean velocity behind the propeller and E_P the effective specific energy in the slipstream behind the propeller.

We also observe that

$$F_{PJ} V_0 = E_P \eta_P \qquad (1.16)$$

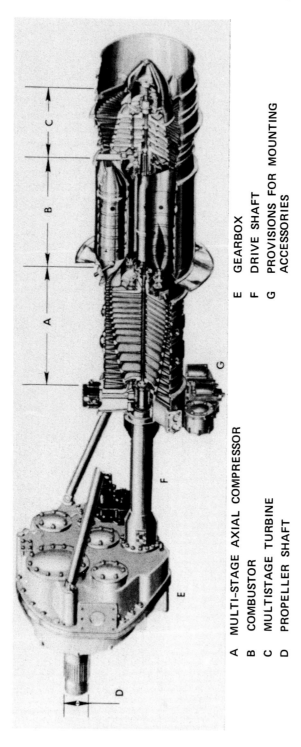

A   MULTI-STAGE AXIAL COMPRESSOR   E   GEARBOX

B   COMBUSTOR   F   DRIVE SHAFT

C   MULTISTAGE TURBINE   G   PROVISIONS FOR MOUNTING

D   PROPELLER SHAFT   ACCESSORIES

Fig. 1.9   Turboprop engine.

The available energy taken from the engine to drive the propeller is

$$\dot{W}_E E_{PT} = \frac{E_P \dot{W}_P}{\eta_{PT}} \qquad (1.17)$$

The efficiency η_{PT} comprehends all the losses in useful energy brought about by the installation of the propeller and its drive system. It thus includes the mechanical efficiencies of the gearing and bearings as well as the aerodynamic efficiency of the power turbine. Also included are the losses in useful energy incurred by the additional airplane drag, such as that due to the added weight of the turbine, gear box, and propeller. We must also include any drag resulting from the disturbed flow in the propeller slipstream. Notice that the efficiency of the propeller itself is excluded.

Inasmuch as gases are expelled from the engine, they must assume the velocity V_{jE} and provide additional thrust. The total thrust per unit of engine airflow is

$$F_S = \frac{\dot{W}_{PJ}}{\dot{W}_E} V_0 \left[\sqrt{\frac{2E_{PT}}{V_0^2} \eta_{PT} \eta_P \frac{\dot{W}_E}{\dot{W}_{PJ}}} - 1 \right] + V_0 \left[\sqrt{\frac{2(E_G - E_{PT})}{V_0^2} \eta_{EN}} - 1 \right]$$

since $EG = E_{PT} + E_{EJ}$.

This thrust has a maximum value when

$$\frac{E_{EJ}}{V_0^2} = \frac{E_G}{V_0^2} \frac{1}{1 + \frac{\eta_{PT} \eta_P}{\eta_{EN}} \frac{\dot{W}_{PJ}}{\dot{W}_E}} \qquad (1.18)$$

Equation (1.18) is useful for identifying the factors involved in determining an optimum division of engine power between the propeller and the jet. It also tests the sensitivity of a given arrangement. In actually pursuing solutions to this problem, the concepts outlined in Chapter 3 are used. This is a trial-and-error technique that finds the optimum propeller and jet combination, not for one flight condition alone but for the entire range of conditions encountered during a typical flight, or mission, by the subject airplane. Observe that propeller vendors select the value of N_{PJ}.

We observe that helicopters require engines similar to those of turboprops. The speed of most helicopters is relatively low and the propelling blades must be able to produce a thrust equal to or greater than the gross weight of the aircraft. The exhaust gas duct is therefore usually designed to maximize the shaft power produced by the engine. In fact, the exhaust duct is an efficient diffuser in many of these installations.

At one time, turboprop engines were considered to be a fruitful use of gas turbines for the propulsion of commercial and long-range airplanes. High values of η_P and η_{PT} were available, together with high levels of thrust. The gas turbines were lighter than reciprocating engines and offered much better propulsive efficiencies than the turbojet engines we have discussed. The attainable speed of efficient turboprops was limited by the propeller and proved to be too low, however. The revenue suffers in comparison to turbojets because the higher speeds of turbojet engines attracted and carried more passengers and freight per unit time. A further detraction of the turboprop was propeller noise, which can be uncomfortable to passengers. Turboprops, however, do find wide application in aircraft that carry a relatively small number of passengers on short intercity flights. The total elapsed time for a short flight can be less with a turboprop than that of a jet. Turboprops have also been necessary when available landing strips are too short for jets.

As previously noted, the lower speed of turboprops has been the fault of propellers. The vector sum of high forward speed and the needed propeller blade speed causes supersonic relative Mach numbers to be inevitable near the tips of propeller blades. Acceptable values of propeller efficiency could not be maintained with available technology. Overcoming this problem has been difficult. NASA has guided and funded programs in recent years to extend the range of good propeller efficiency into the regime of high subsonic flight Mach numbers. Encouraging results have been obtained and prototypes have been tested on airplanes to determine their installation drag, endurance and reliability, and cost of maintenance. All these factors must be explored before any new technology can be exploited. However, we may expect useful designs to emerge in the near future.

Turbofan Engines

The objective of modern turbofans, see Fig. 1.10a, is to achieve acceptable thrusts and efficiencies at higher flight speeds than a turboprop can. As with a turboprop, the main engine can operate in any regime of Fig. 1.2 without disturbing the effectiveness of the thrust-producing components. Similar to a propeller, a fan imparts energy to a large quantity of air: the quantity is less than that of a propeller, however, being limited by factors affecting the installation weight and drag, as well as fan efficiency.

Figure 1.10a differs from Fig. 1.9 only by the fact that the turbine located between stations 5 and 7 drives an axial flow fan or compressor. The fan raises the pressure of air passing through it so that a portion of this flow, $\dot{W}_F$, can be discharged with the velocity V_{FJ}, which is lower than that prevalent in turbojets, but higher than that behind propellers. The rest, of the flow into the fan blades, $\dot{W}_E$, passes through the engine. We call the ratio $\dot{W}_F/W_E$ the bypass ratio.

The tolerable Mach numbers of the tips of the fan rotors are much higher than those of conventional propeller blades. Available compressor technology can be applied to produce the desired fan pressure ratios with high efficiencies at elevated Mach numbers (the relative close spacing of the fan blades and the presence of a casing surrounding the fan are the principal reasons for this difference).

The designer of fan engines has the responsibility, however, of selecting the energy delivered to the fan and the mass flow through it. The ensuing discussion calls attention to the principal elements governing that decision.

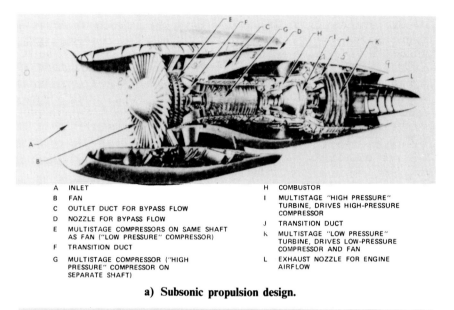

A   INLET
B   FAN
C   OUTLET DUCT FOR BYPASS FLOW
D   NOZZLE FOR BYPASS FLOW
E   MULTISTAGE COMPRESSORS ON SAME SHAFT
     AS FAN ("LOW PRESSURE" COMPRESSOR)
F   TRANSITION DUCT
G   MULTISTAGE COMPRESSOR ("HIGH
     PRESSURE" COMPRESSOR ON
     SEPARATE SHAFT)

H   COMBUSTOR
I   MULTISTAGE "HIGH PRESSURE"
     TURBINE, DRIVES HIGH-PRESSURE
     COMPRESSOR
J   TRANSITION DUCT
K   MULTISTAGE "LOW PRESSURE"
     TURBINE, DRIVES LOW-PRESSURE
     COMPRESSOR AND FAN
L   EXHAUST NOZZLE FOR ENGINE
     AIRFLOW

a) Subsonic propulsion design.

THE GREATER NUMBER OF STAGES IN THE FAN PROVIDES THE HIGHER PRESSURE RATIOS NEEDED FOR
EFFICIENT OPERATION AT SUPERSONIC FLIGHT. AFTERBURNING IS APPLIED TO BOTH THE FAN FLOW AND
THE ENGINE FLOW. THE DESIGN PRESSURE MUST BE CAREFULLY BALANCED AT A, WHERE THE TWO FLOWS
UNITE.

b) Supersonic propulsion design.

Fig. 1.10   Turbofan engines.

By using convenient efficiency terms, we can express the total thrust
developed by the fan and engine exhaust as

$$F_s = B[\sqrt{\eta_{FN}(V_0^2\eta_I - 2E_F\eta_F)}V_0] + [\sqrt{2\eta_{EN}(E_G - BE_F/\eta_{FT})} - V_0] \quad (1.19)$$

The first set of brackets contains the fan thrust, while the second set
denotes the thrust resulting in the engine exhaust jet. Observe that E_G
represents the specific energy produced by the engine, while the amount
BE_F/η_{FT} is diverted to drive $\dot{W}_F$ through the fan. The difference is the ideal
energy of the engine jet. We use terms η_{FN} and η_{EN} to represent the effects
of the degradation of energy by the internal and external drag of the fan
nacelle and nozzles, as well as that due to the weight of the added parts.
For our purpose, it is necessary to note only that something like these
efficiencies must exist.

It has been mentioned that it is the designer's task to select the values of E_F and B that make F'_s a maximum at a specified flight condition. This is another task to be accomplished by trial and error in actual designs. We can gain insight, however, by assuming the efficiency terms to vary only slightly with E_F and B. Equating the partial derivatives of F_s with respect to E_F and B to zero then yields a short expression for the desired optimum value of E_F:

$$\frac{E_F}{V_0^2} = \frac{(1 - \eta_I\eta_{FN}) + \sqrt{(1 - \eta_I\eta_{FN})}}{\eta_F\eta_{FN}} \qquad (1.20a)$$

The partial derivatives also provide an optimum value for B for any E_F.

$$B = \eta_{FT}\left[\frac{E_G}{E_F} - \frac{(E_F/V_0^2)}{2\eta_{EN}Z}\right] \qquad (1.20b)$$

where

$$Z = \sqrt{\eta_{EN}\eta_I + 2\eta_F\left(\frac{E_F}{V_0^2}\right) - 1}$$

These results are revealing. They show that the optimum energy added by the fan depends only on the flight velocity, on the efficiency of the fan itself, and on the parts of the fan installation that degregates energy. It is independent of the rest of the engine. The magnitude of E_F is particularly sensitive to η_{FN}, which reflects the aerodynamic losses due to the fan installation, as well as those in the exhaust nozzle of the fan duct. The entire fan assembly should thus be designed in close cooperation with airplane designers. The fan energy should increase as higher losses are encountered or as flight velocity becomes higher. Fan pressure ratio is determined from E_F and η_F.

These equations emphasize that all the specific energies, the energies per unit mass of gas, should increase with V_0^2. All elements must be more highly loaded as flight Mach numbers increase.

We also note that selection of the bypass ratio and the energy in the engine exhaust gas are strongly dependent on the aerodynamics of the whole airplane. An incentive for designing engines and airplanes together is clearly demonstrated. These remarks can be extended to installations that include afterburning and the employment of afterburning in a fan engine for supersonic propulsion is illustrated in Fig. 1.10b.

1.5 OFF-DESIGN PERFORMANCE
OF GAS TURBINE PROPULSION ENGINES

So far we have been concerned with engine power and efficiency at one particular operating condition as determined by the ambient inlet pressure and temperature, turbine inlet temperature, and assigned component properties. Engines are, however, expected to start from a standstill, accelerate, and run over a range of flight speeds with a variety of compressor and turbine inlet temperatures. They must eventually decelerate and stop. Smooth changes from one operating condition to another are essential. Providing stipulated thrusts with reliability and with the required or higher efficiencies is expected at all specified conditions.

As Chapter 3 of this volume emphasizes, we must ultimately be concerned with both engine and airframe performance for a given mission. We can tentatively assume that an airplane, flight path, and thermodynamic cycle have been selected and that components have been chosen to provide a desired efficiency at a given value of thrust. What an engine does at other levels of thrust (or at off-design) needs to be evaluated. The result is influenced by the characteristics of the compressors and turbines and their arrangement in an engine.

Any change in specific power causes the operating point of each component to move. The principal variable causing this change is the ratio of turbine inlet to compressor inlet temperature, which we shall call the engine temperature ratio or ETR. Flight altitude, fuel flow, and flight Mach number are the determinants of this ratio. Note, however, that power and thrust are directly proportional to the inlet pressure unless low pressures reduce the Reynolds number beyond the point where component performance deteriorates.

The changes in the component operating points with the ETR have a crucial influence on the effectiveness and reliability of a given engine design. Anticipating and understanding the physical events governing these changes are valuable assets for both design and development. This subject is reviewed in this section. Instead of pursuing the formal attacks described in Chapter 8 of Ref. 1, we shall use a few approximations that are adequate for locating possible problem areas on component maps. The relative behaviors of several designs are then compared.

Two pictures of a turboprop engine provide some background for engine configuration. Figure 1.9 is a cutaway view showing the compressor, turbine, gears, and propeller shaft. The provisions for accessories, which usually consist of an electric generator, oil pump, and hydraulic fluid pumps, are also shown. The relative sizes of these particular accessories are indicated by a photograph of the complete engine in Fig. 1.11. Engines often provide additional power in the form of compressed air, which is bled from the compressor. In this section, however, we shall ignore the small power requirements for accessories and other parasitic demands and deal only with those of the propellers and jets.

Compressor and Turbine Characteristics

This section presents some background information for the discussions that follow.

Compressor maps. The characteristics of compressors are described by performance maps similar to Fig. 1.12. The abscissa is the corrected flow $W(\sqrt{\theta}/\delta)_I$ and the ordinate the pressure ratio. The parameters are corrected speed $N/\sqrt{\theta_I}$ and adiabatic efficiency. Figure 1.12a represents a class

Fig. 1.11   Complete turboprop engine.

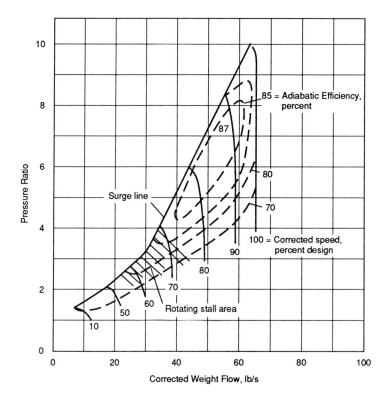

a) **High pressure ratio.**

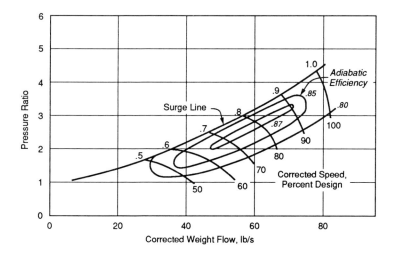

b) **Low pressure ratio.**

Fig. 1.12 Compressor maps.

of axial flow compressors having pressure ratios of the order of 10, while Fig. 1.12b is more representative of units with design pressure ratios of about 4. (The abscissa of Fig. 1.12b has been scaled to agree with that of Fig. 1.12a.) Notice that in comparison to the latter figure, most of the $N/\sqrt{\theta_l}$ curves in Fig. 1.12a extend over a relatively small range of airflow and that the surge line has an almost discontinuous slope. The prevalence of rotating stall at the lower speeds is also noted. The difference between the curves is a consequence of stage matching problems that are intensified by increases in compressor pressure ratio. See Chapter 3 of Ref. 2.

Every effort should be made to stay to the right of the surge line. At high speeds, surging can be explosive. Running at or beyond the surge line at any speed can cause the compressor efficiency to deteriorate so badly that an engine cannot function. Special features, such as adjustable compressor stator vanes or interstage bleed, have been used to cope with this problem when it arises. These modifications are also discussed in Chapter 3 of Ref. 2. Remember though that these devices increase the manufacturing costs as well as the costs of control and maintenance. They should be used reluctantly.

Rotating stalls are induced in the low-pressure stages at low corrected speeds. Although blade stalling is a prerequisite for surge, this phenomenom is not necessarily the inevitable result of a stall. When stalls occur, they often excite blade vibrations that may cause failure. If compressor blades must be excessively thickened to forestall damage, their aerodynamic performance is impaired. Prolonged operation with rotating stalls is therefore avoided whenever possible. Adjustable vanes and bleed can also provide relief from rotating stalls.

Turbine maps.   Two curve sheets are usually used to describe turbine performance. A typical set is shown in Fig. 1.13. The first graph shows how the corrected turbine weight flow varies with the pressure ratio across the turbine. Observe that this flow is practically constant when the turbine pressure ratio exceeds 2.0. This curve is similar to that of a nozzle, which is to be expected since turbine blades are really curved nozzles. In this example, the rotating speed has little effect on the gas flow rate—a result that is expected in turbines where the rotor blades are unchoked and the mean radius of the flow path is nearly constant. In other designs, however, the weight flow is slightly reduced at the lower pressure ratios when the rotor speed increases.

The second curve shows how turbine efficiency depends on the pressure ratio and rotating speed. Observe that there is an appreciable interval at the higher speeds in which the efficiency varies only slightly with the pressure ratio and that the pressure ratio for peak efficiency increases with speed.

Turbines are sometimes equipped with adjustable nozzles—the throat area of which can be varied to raise or lower the maximum gas flow through them. Reasons for their use will be discussed later. We should be aware that they are effective over a limited range of adjustment. If the area is closed too far, the pressure ratio across the turbine rotors is lowered; any tendency for adverse pressure gradients within them becomes aggravated, thus penalizing turbine efficiencies. On the other hand, nozzle performance is penalized if they are opened too far; moreover, the throat area of the rotors then controls the maximum flow rate and Fig. 1.13a becomes a strong function of rotor speed as well as pressure ratio. Variable nozzles are another item that increases the costs of the controls, manufacturing, and maintenance. They lose effectiveness when they are an afterthought in a developed turbine.

Single-Shaft Turbojet

Previous remarks have noted that this engine is now used only for special purposes in modern aircraft because of the conflicting requirements of engine and propulsive efficiencies. The component arrangement is, however, a building block for many engines. Studying a simple jet engine, which is sketched in Fig. 1.14, provides an understanding of the behavior of more complicated ones. A picture of a production engine, which includes an afterburner, is shown in Fig. 1.15.

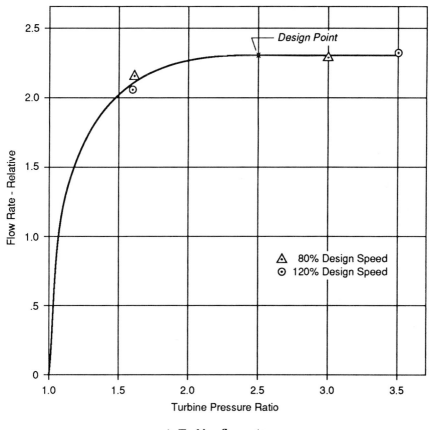

a) Turbine flow rate.

Fig. 1.13 Turbine performance maps.

Continued.

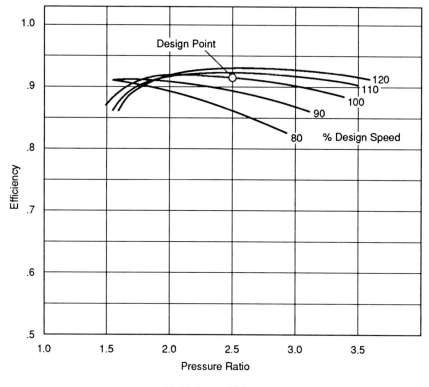

b) Turbine efficiency.

Fig. 1.13   (Continued) Turbine performance maps.

A definite design point is needed and a representative one is noted on the compressor and turbine maps of Figs. 1.12 and 1.13. This is where these components are supposed to operate when the design value of the ETR is imposed. Performance of the components is assumed and conventional cycle analysis enables the design point performance to be estimated. An approximate technique can be used for anticipating the way that the compressor and turbine operating points change when the ETR is raised or lowered. Such a procedure is now described. It ignores Reynolds number effects, but one should recognize that anything that changes the Reynolds number has the potential of producing other noticeable changes in performance.

Method of approximate analysis.   We shall assume an ideal gas having the properties expressed in Sec. 1.3. This allows the major trends to be easily revealed. To establish these trends, it is also convenient to make the reasonable assumption that the gas flow rate in the turbine just behind the combustor is about the same as that in the compressor immediately in front of it. Moreover, if we confine our observations to the cases where the turbine is choked (see Fig. 1.13a), we can plot straight lines on the coordinates of a compressor map to show the constant corrected turbine

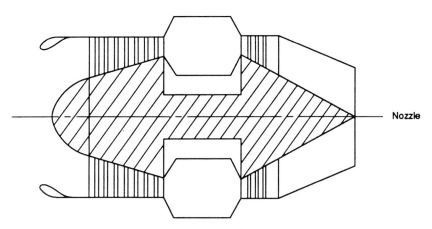

Fig. 1.14   Jet engine with single shaft.

flow rate. The following equation is needed:

$$\frac{P_{C,O}}{P_{C,I}} = \frac{(\dot{W}\sqrt{\theta}/\delta)_{C,I}\sqrt{T_{T,I}/T_{C,I}}}{(P_{T,I}/P_{C,o})(\dot{W}\sqrt{\theta}/\delta)_{T,I}} \tag{1.21}$$

We may use a constant value (e.g., 0.95) for $P_{T,I}/P_{C,O}$ without obscuring any important trends. The equation then indicates that compressor pressure ratio varies linearly with the corrected weight flow when the turbine is choked. The slope of the line is proportional to the square root of ETR. Lines A-A and B-B of Fig. 1.16 thus represent the requirements of continuity for two values of ETR. An operating point would be indicated by the intersection of a line representing the given ETR with the curve for an appropriate corrected speed.

The needed corrected speed is estimated by equating the work of the compressor to that of the turbine. This procedure is consistent with our neglect of the work required by the accessories and friction. Compressor energy is proportional to $\Delta T_C/T_{C,I}$, which in turn is evaluated from compressor maps by using the definition of adiabatic efficiency. A typical variation of $\Delta T_C/T_{C,I}$ with the compressor pressure ratio is shown on the right side of Fig. 1.17. The left side shows the usual pressure ratio curve. Line C-D corresponds to the choked turbine flow when ETR has its design point value; line A-B represents the choked turbine flow when ETR is an arbitrary 64% of its design value—a condition where the power is much lower than at design. The intersection of this line with each of the constant $N/\sqrt{\theta}_{C,I}$ lines is projected on the right-hand side of Fig. 1.17 to obtain curve A-B. Compressor work is thus determined as a function of $N/\sqrt{\theta}_{C,I}$ for the selected value of ETR.

Before one can consider turbine work, the turbine pressure ratio must be estimated. Note again from Fig. 1.13a that turbine corrected flow is ostensibly constant when the pressure ratio across it is 2.0 or more. In the

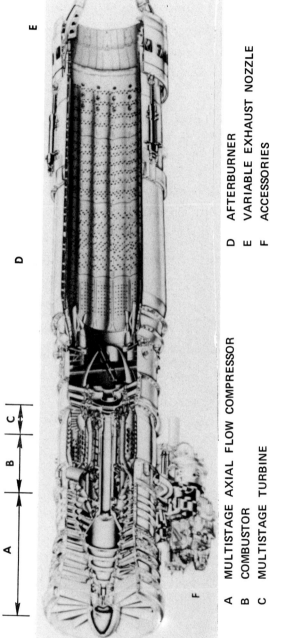

A   MULTISTAGE AXIAL FLOW COMPRESSOR

B   COMBUSTOR

C   MULTISTAGE TURBINE

D   AFTERBURNER

E   VARIABLE EXHAUST NOZZLE

F   ACCESSORIES

Fig. 1.15   Cutaway view of turbojet engine with afterburner.

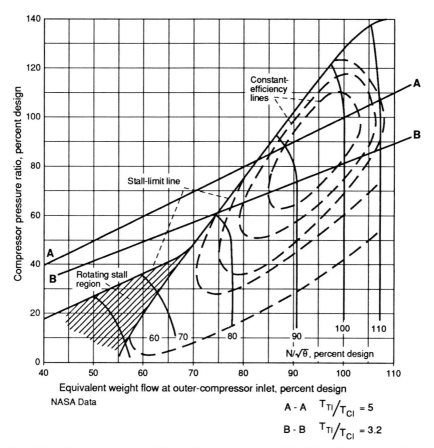

Fig. 1.16 Compressor map with two lines of constant $T_{T,I}/T_{C,I}$: line A-A = 5, line B-B = 3.2.

same way, a nozzle is choked when its pressure ratio exceeds 1.85. Combining the nozzle flow data with Fig. 1.13a yields the results of Fig. 1.18. The choked or nearly choked flow in the engine nozzle forces the turbine pressure ratio to be constant at overall pressure ratios higher than about 3.7. This condition is satisfied in most engines in regimes where specific output and efficiency are important.

Let us now calculate turbine work for this constant pressure ratio. Curves such as those of Fig. 1.13b are used; efficiency and pressure ratio determine $\Delta T_T/T_{T,I}$, which is plotted in Fig. 1.19 vs the turbine corrected speed for turbine pressure ratios of 2.0, 2.5, and 3.0. Curve A'-B' of Fig. 1.18 is the same as curve A-B of Fig. 1.17. The necessary change of coordinates uses the equations

$$\left(\frac{\Delta T}{T}\right)_{T,I} = \left(\frac{\Delta T}{T}\right)_{C,I} \frac{T_{C,I}}{T_{T,I}} \qquad (1.22a)$$

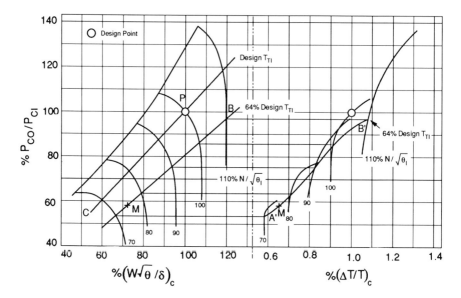

Fig. 1.17   Compressor pressure ratio and temperature rise.

$$\left(\frac{N}{\sqrt{\theta}}\right)_{T,I} = \left(\frac{N}{\sqrt{\theta}}\right)_{C,I} \sqrt{\frac{T_{C,I}}{T_{T,I}}} \qquad (1.22b)$$

The value of 2.5 is selected for the design pressure ratio of the turbine. We tentatively assume this to be unchanged even when the ETR is 64% of its design value. The intersection of curve A'-B' with the curve for the 2.5 pressure ratio fixes match point M. This point is transferred to Fig. 1.17 by Eqs. (1.22) in order to determine from the compressor pressure ratio whether the assumption of an unchanging turbine pressure ratio of 2.5 is reasonable. In this case, we may say that it is; we implied a design point compressor pressure ratio higher than 6.0; therefore, the estimated pressure at the new match point is close enough to 3.7 for our estimate to be valid.

Conclusions from analysis.   The important result of this example is that the pressure ratio across the turbine and the turbine operating point are practically unchanged over the interval in which the overall pressure ratio is above about 3.7. This condition embraces a wide range of ETR. The value of $\Delta T_I / T_{T,I}$ may be considered constant over the same interval. A quick appraisal of $\Delta T_C / T_{C,I}$ and the compressor match point is thus provided by the compressor map alone.

It is also useful to observe that $(N/\sqrt{\theta})_{T,I}$ is almost constant at the same time. One may say that the percent change in the rotating speed is a little greater than the percent change in the ETR.

These observations enable us to predict the off-design trends of arrangements similar to Fig. 1.14. Curve C-C of Fig. 1.16 was calculated from an engine model using exact relations and, at 80% speed and higher, these match points are consistent with those estimated by assuming a constant value for $\Delta T_T / T_{T,I}$. The corresponding values of $(N/\sqrt{\theta})_{C,I}$ are those expected.

Typical performance trends. Curve C-C of Fig. 1.16 outlines the important off-design features of the basic turbojet engine. This curve, or operating line, is the locus of match points for various values of the ETR at sea-level static conditions. The pressure ratio across this turbine is virtually constant. Decreasing the ETR moves the points to the left. The curve does not stray too far from the regions of the best available compressor efficiency, as long as the corrected speeds are above about 75% of design, although the best obtainable compressor efficiency drops as the ETR is lowered. The decreasing pressure ratios accompanying lower values of ETR adversely affect engine efficiency, particularly at subsonic flight Mach numbers where nearly the entire compression temperature ratio is provided by the compressor.

Substantially lower values of ETR push the compressor into the area of rotating stall and the operating line flirts with the surge line. This condition often occurs while an engine idles. If rotor speed must be rapidly raised from, say, 70% speed, the ETR must exceed its equilibrium value. Equation (1.21) or lines A-A and B-B in Fig. 1.16 show that the transient operating point then moves closer to or beyond the surge line. This occurrence has caused turbines to overheat or fail.

Except for possible Reynolds number effects, Fig. 1.16 also applies to various flight conditions. If the design point designates sea-level static conditions, the extreme right of curve C-C designates subsonic flights in the stratosphere, where the inlet temperatures are low. This operating point provides the highest specific power and best engine efficiency available in this example. Compressor stall and possibly surge is encountered again at supersonic flight speeds, where ETR is lowered at a given turbine inlet temperature because compressor inlet temperature is elevated. ETR is less than one-half of its value at takeoff when the flight Mach number is 3.0—the corrected speed is forced to be less than 70% of its value at takeoff. Chapter 6 shows that inlet flow distortions require the compressors to run with a considerable margin for the pressure ratio, or flow, between the operating point and the surge line. The situation shown in Fig. 1.16 would require remedial action.

The previously mentioned variable compressor stators and turbine nozzles, as well as interstage compressor bleed, can be employed to alleviate problems at the lower corrected speeds. Other aerodynamic concessions have been made, but an effective solution that does not deteriorate performance has yet to be found. (Modest reductions in compressor and turbine efficiency might be tolerated at supersonic speeds because most of the compression and expansion takes place in the engine inlet and nozzle.)

Note that power can be reduced at any time by lowering the inlet pressure and thus reducing the engine airflow. The specific power and engine efficiency can be maintained over a wide operating spectrum if the entropy is not increased. Increasing altitude accomplishes this result.

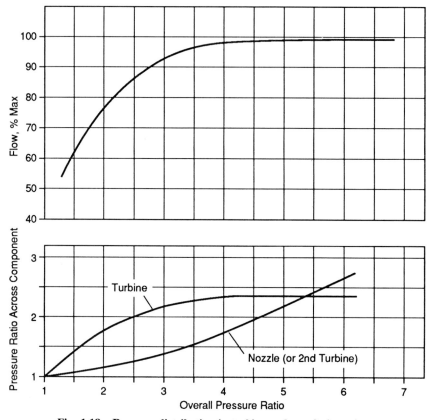

Fig. 1.18   Pressure distribution in turbine and nozzle in series.

However, a prior mission analysis should already have accounted for this.

A word about variable jet nozzles is in order at this point. Opening the nozzle increases the turbine pressure ratio when the ETR is fixed. Turbine speed increases and the intersection of curve A'-B' with the 3.0 pressure ratio curve in Fig. 1.19 indicates the trend of the new match point. The higher speed causes the compressor to develop a higher pressure ratio with a higher weight flow. Closing the exhaust nozzle lowers the turbine pressure ratio; a sample result is shown by the intersection of curve A'-B' with the 2.0 curve in Fig. 1.19. These adjustments, of course, affect the thrust, which can be estimated. Variable nozzles are essential in engines with afterburning.

Single-Shaft Turboprop

Simplified analysis.   This type of engine is represented by the sketch of Fig. 1.20. The design differs from the turbojet in that the engine speed can be independently set by adjusting the propeller pitch. The magnitude of

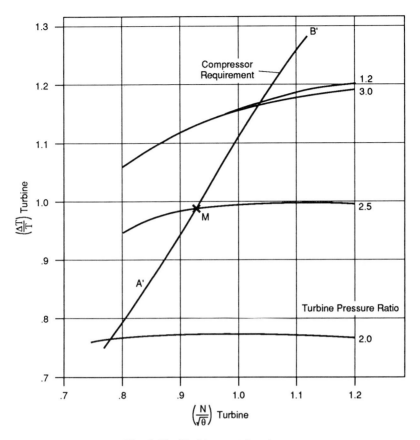

Fig. 1.19 Turbine match point.

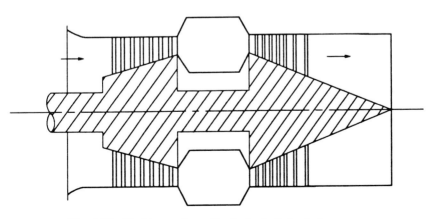

Fig. 1.20 Turbine engine with single shaft for turboprop.

$(N/\sqrt{\theta})_{C,I}$ is then known when the compressor inlet temperature is given. The speed parameter of a compressor map (i.e., Fig. 1.12) is thus established as an independent variable. A match point is defined on a compressor map by the intersection of the given corrected speed line with a line for a given ETR such as curve A-A in Fig. 1.16. The corrected speed of the turbine is then determined. If we assume that the pressure ratio through the engine inlet to be about equal to that through the exhaust nozzle, we can say that turbine pressure ratio is approximately equal to that of the compressor. A match point on the turbine map is thus located and the estimate of the powers developed by the turbine, compressor, and engine are routine.

Performance trends. Adjustable propeller pitch provides for good propulsion efficiency at a variety of flight speeds and engine powers. Constant engine speed is the preferred mode of operation, however, even at reduced power levels. The engine then quickly responds to demands for higher power. (If rapid increases in engine speed were necessary at lower speeds, the large polar moment of inertia of the propeller and gears would delay the response. This can be dangerous.) Engine efficiency at reduced power with constant speed is inferior to that obtainable along curve C-C of Fig 1.16, but this situation is seldom necessary and usually has only a small effect on most missions. Optimum altitudes are usually selected for cruise to get good efficiency.

A single-shaft turboprop engine has a characteristic that is useful when steep descents into tight airports are made. The propeller can be used as a windmill to absorb power from the velocity of the airstream to drive the engine. In this way, the flight speed is reduced to low levels in spite of a steep angle of descent. This drag force can also cause problems in a two-engine airplane when one engine fails.

Free Turbine Turboprop

Figure 1.21 differs from Fig. 1.20 in that the turbine functions are separated—one turbine drives a compressor and another the propeller. The turbine that drives the compressor behaves as it does in a single-shaft turbojet and the propeller turbine has the pressure ratio and flow characteristics of a nozzle. Curve C-C of Fig. 1.16 again represents an operating line.

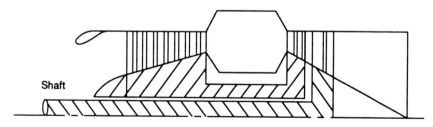

Fig. 1.21   Free turbine engine for shaft power.

This engine offers better efficiency at reduced specific power than the single-shaft engine running at constant speed. The speed of the gas generator is reduced when the ETR is lowered; however, since there is no propeller or large gear train on this shaft, the inertia is comparatively low. The finite time required to respond to a demand for more power is small and usually not objectionable.

The free turbine engine is the preferred choice for helicopters, which require the main rotor to spin freely if the engine should stop. The load imposed by the turbine, gears, and accessories is small enough to allow this to happen without the use of a clutch or a free-wheeling device.

When the engine is idling, both the engine and propeller speeds are low. This is a desirable quality for ground operations because of the low noise level. Low engine speeds at part power are also useful when extended periods of time for loitering are necessary.

Two-Shaft Jet Engine

Simplified analysis. The maximum practical pressure, or temperature, ratio for the compressor of a single-shaft jet engine is limited. The volume flow rate at the inlet determines the diameter and the rotating speed is limited by the highest allowable Mach number at the inlet stage. As the pressure increases within a compressor, the volume flow rate reduces and the air temperature rises. The hydraulic diameters of the flow passages continuously decrease from inlet to outlet and the corresponding pressure ratio produced by each stage becomes smaller and smaller. Eventually, the unit becomes heavy, inefficient, and impractical.

The increase in air temperature through a compressor lowers the Mach numbers of the flow about the blades. It is thus reasonable to use two compressors; then, after a moderate pressure ratio is developed, the flow is passed to another compressor having a smaller diameter and a greater rotating speed. In the same way, the use of two turbines also offers advantages. Designs with two separate shafts, as shown in Fig. 1.22, are thus recommended. The reduced size of the inner spool offers the additional benefit of a reduction in weight. Although the pressure ratio of each compressor is smaller than that used in Fig. 1.12a, the overall pressure ratio and efficiency can be increased markedly. Figure 1.23 presents typical maps

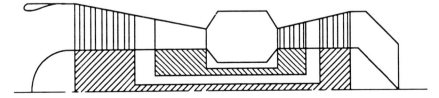

Fig. 1.22   Two-shaft jet engine.

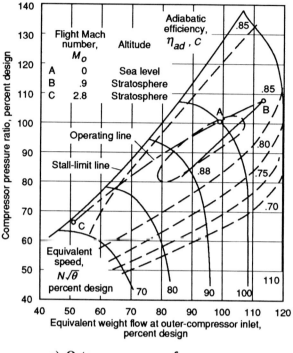

a) Outer-compressor performance map.

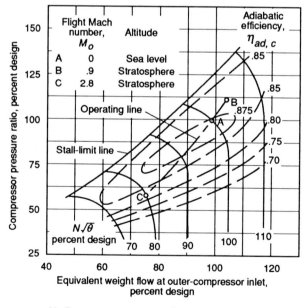

b) Inner-compressor performance map.

Fig. 1.23 Typical compressor maps for two-shaft jet engines.

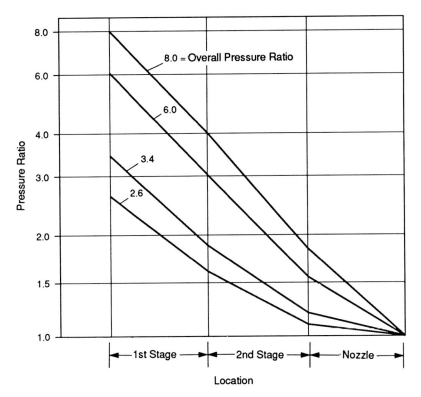

Fig. 1.24 Pressure distributions through two turbines and a nozzle in series.

of the compressors. The trends of the turbine maps of Fig. 1.13 are still appropriate.

The hierarchy of pressure ratios across the turbine is similar to that of the gas generators previously described. An example is shown in Fig. 1.24 for several overall pressure ratios up to 8 across the turbines and engine nozzle. A logarithmic scale is used for the ordinate so that the constant-pressure ratios across components appear as parallel lines. All elements were assumed to be choked at the arbitrary overall pressure ratio of 8. The low-pressure turbine begins to unchoke when the pressure ratio is 6, while the high-pressure turbine is practically choked down to an overall pressure ratio of 3.4.

The effect of changes in the ETR on the inner spool is similar to that of the single-shaft engine. However, one must use the ratio of the turbine inlet temperature to that at the inlet of the inner compressor (IETR) instead of ETR. The conclusions reached above about single-shaft turboprops then apply here. The match points thus estimated anticipate the trends of the formally computed operating line of the inner rotor in Fig. 1.23b.

An operating point of the outer spool is partly determined by the fact that $(\Delta T_T/T_{T,I})_{\mathrm{OUT}}$ is practically constant K over the operating range of interest when the engine nozzle area is not varied. A value for $(T_{C,O}/T_{C,I})_{\mathrm{IN}}$

is selected from the inner-spool compressor map, similar to Fig. 1.23b, and the corresponding value of $(\Delta T_C/T_{C,I})_{\text{OUT}}$ is calculated using

$$\left(\frac{\Delta T_C}{T_{C,I}}\right)_{\text{OUT}} = K\frac{\text{ETR} - (\Delta T_C/T_{C,I})_{\text{IN}}}{1 + K(\Delta T_C/T_{C,I})_{\text{IN}}} \qquad (1.23)$$

The value of K is found at the design point, where the other variables are known.

The outer compressor flow is corrected to its outlet conditions with

$$\left(\frac{W\sqrt{\theta}}{\delta}\right)_{\text{OUT}} = \left(\frac{W\sqrt{\theta}}{\delta}\right)_{\text{IN}} \left(\frac{\sqrt{T_{C,o}/T_{C,I}}}{P_{C,o}/P_{C,I}}\right) \qquad (1.24)$$

The variation of $(\Delta T_C/T_{C,I})_{\text{OUT}}$ with this flow and corrected speed are calculated from the data of Fig. 1.23a and is presented in Fig. 1.25. We could also plot the pressure ratio against these variables, but this is not shown. The magnitude of the corrected outlet flow is known, since it must equal the flow given in Fig. 1.23b for the selected match point. The value of $(\Delta T_C/T_{C,I})_{\text{OUT}}$ has been found with Eq. (1.23) and this match point is spotted on Fig. 1.25 in order to estimate the corrected speed of the outer unit. A curve similar to Fig. 1.25 for the pressure ratio enables this quantity to be estimated. If a check of the turbine operation is desired, the variation of $(\Delta T_T/T_{T,I})_{\text{OUT}}$ with the corrected turbine speed can be found by using the scheme applied to Fig. 1.19.

Performance trends. From estimates of $\Delta T_C/T_{C,I}$ calculated from the data at the design point, we can again estimate where the compressors operate during various off-design conditions. The curves CAB in Fig. 1.23 are somewhat representative, although this example (which happens to be the only one readily available) represents the case where the engine nozzle area is varied to maintain the outer unit at a constant mechanical speed. The general trends are valid, although the mechanical speed of the outer unit would vary over a much wider range for the same ETR if the engine nozzle were fixed.

The speed of the outer shaft is nearly proportional to the speed of the inner one as long as the corrected flow through the engine nozzle is fixed—when the ETR is lowered from its design value, the speeds of both shafts reduce at about the same percent rate. In the example of Fig. 1.22, this would be true at overall pressure ratios above about 6. At lower overall pressure ratios, the percent rate of change in speed is greater for the outer unit until the inner turbine is also unchoked. The variation in turbine power with the overall pressure ratio is seen to be dependent on where the turbine is located in the gas stream.

Figure 1.23 indicates that the outer compressor of a two-shaft engine operates near its surge line at 70% of the design corrected speed. Recall that this phenomenon was also observed in Fig. 1.16. It does not follow, however, that this event will cause the compressor system of a two-shaft unit to surge. The surge line of the outer compressor was observed when it

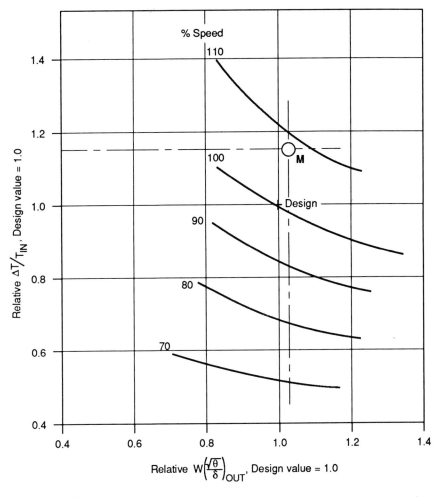

Fig. 1.25   Compressor temperature rise ratio vs flow, corrected to compressor outlet.

was tested as a single compressor. As part of a two-stage unit, surge is improbable at this point because the pressure ratio characteristics of the inner unit cause the overall pressure ratio to increase as the flow rate is reduced. This usually satisfies the criterion of compressor stability when the compressors are closely coupled aerodynamically.

The outer unit may still be exposed to the fluctuating airflow of rotating stall, just as the single-stage compressor was. These stalls are usually less severe and the problems are evaded with fewer performance penalties.

This is an opportune point at which to mention flutter—the subject of Chapter 7. In contrast to the vibrations forced by rotating stall, flutter is a self-excited phenomenon of aeroelasticity. The stalling of an airfoil can precipitate flutter; other causes of stall are discussed in Chapter 7. Flutter due to stall can be found in the outer compressor of Fig. 1.23 at low speeds

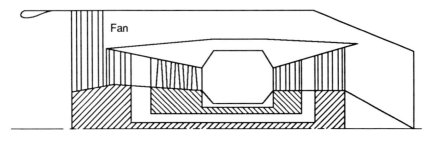

Fig. 1.26   Two-shaft turbofan engine.

and in the inner compressor at high speeds. Blade vibrations due to "choke flutter" can exist in the inner compressor at low speeds, while the outer compressor may experience "unstalled supersonic flutter" at high speeds.

Multishaft Turbofan Engines

An arrangement of this type of engine is suggested by the sketch of Fig. 1.26. A photograph of a production version for subsonic flight is shown in Fig. 1.27. Recall that the fan is nothing more than a compressor producing a relatively low pressure ratio. The off-design behavior of the components of this engine is similar to the two-shaft turbojet, since the turbines and jet divide the pressure ratio in about the same way. Observe that when the ETR is lowered, the power to the jet diminishes faster than that to the fan. The discussion of turbofan engines in Sec. 1.3 demonstrated that there is an optimum division of these powers, which depends on flight speed and internal efficiencies. Prolonged operation at two or more flight conditions requires the power division to be compromised unless some variable feature is provided.

Another engine currently in production uses three concentric shafts: there are two sets of compressor and turbine rotors and a turbine on the third shaft drives the fan. The thrust nozzle is downstream of the third turbine. Reducing the ETR first causes the nozzle to unchoke; the fan turbine becomes unchoked at a slightly lower ETR. The changes in the ratio of fan power to jet power is different for this design and a different off-design efficiency is thus expected.

Still another concept is shown in Fig. 1.28. The objective of the design is to put the turbine driving the fan immediately downstream of the high-pressure turbine; the tortuous flow path was accepted in order to achieve this objective. This arrangement was a deliberate attempt to provide a favorable balance of fan and jet power at low levels of specific engine power, while still maintaining acceptable engine efficiencies. The aerodynamic aims were almost realized in spite of the intricate flow path. There was an unfortunate mechanical problem, however. The regions of high pressure within the engine are on the right side of the photograph for both the compressors and turbines. A large pneumatic force, which could not be accommodated by rolling element thrust bearings, is exerted toward the left

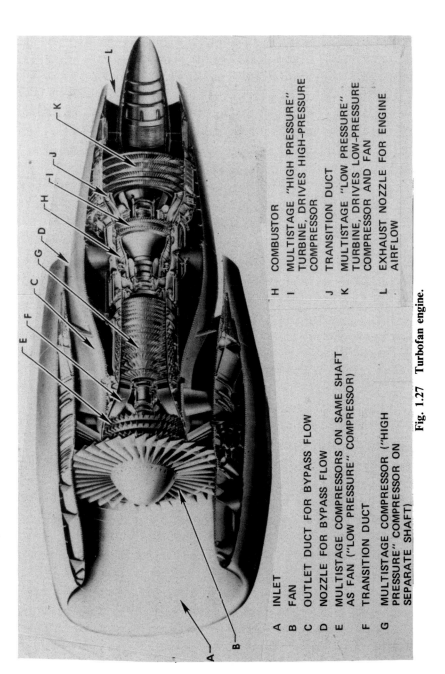

A INLET
B FAN
C OUTLET DUCT FOR BYPASS FLOW
D NOZZLE FOR BYPASS FLOW
E MULTISTAGE COMPRESSORS ON SAME SHAFT
   AS FAN ("LOW PRESSURE" COMPRESSOR)
F TRANSITION DUCT
G MULTISTAGE COMPRESSOR ("HIGH
   PRESSURE" COMPRESSOR ON
   SEPARATE SHAFT)
H COMBUSTOR
I MULTISTAGE "HIGH PRESSURE"
   TURBINE, DRIVES HIGH-PRESSURE
   COMPRESSOR
J TRANSITION DUCT
K MULTISTAGE "LOW PRESSURE"
   TURBINE, DRIVES LOW-PRESSURE
   COMPRESSOR AND FAN
L EXHAUST NOZZLE FOR ENGINE
   AIRFLOW

Fig. 1.27 Turbofan engine.

on the rotors. Pressurized air bearings requiring rotating air seals having large diameters are necessary. Precise control of clearances is mandatory to prevent the associated leakage losses from overwhelming the gains; thus, the costs of manufacturing and maintenance control the value of this design. This is a good illustration of a need for close cooperation among aerodynamic, mechanical design, and manufacturing specialists early in a program.

Multishaft Turboprop Engine

It is obvious that the multishaft concepts can be extended to a turboprop engine with the propeller turbine replacing the role of the fan and most of the role of the thrust nozzle. The off-design behavior would be similar to that discussed in the previous section on the two-shaft jet engine.

Afterburning

A photograph of a single-shaft turbojet with afterburning was presented in Fig. 1.15. Two designs of two-shaft turbofan engines with afterburners are shown in Fig. 1.29. Notice that the gas stream from the engine is mixed with that of the fan ahead of the afterburner. The static pressures of both streams must be the same where they merge, e.g., at point A in the upper figure. This requirement affects the off-design performance of both the fan and gas generator, but the general principles previously noted still apply. Multistage fans and afterburners typify configurations needed by supersonic aircraft. Except for their proportions, the rotating structures resemble those of Fig. 1.27. The function of the mixers just ahead of the afterburners is explained in Chapter 2 of Ref. 2.

Igniting an afterburner poses a special off-design and control problem that is treated in Chapter 2 of Ref. 2. In brief, the effective Mach number of the flow in the nozzle throat is unity during normal operation. When the

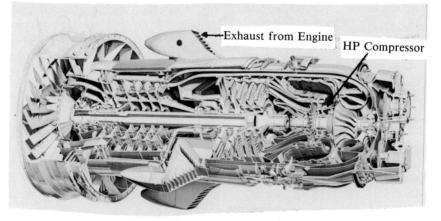

Fig. 1.28   Arrangement of a turbofan engine.

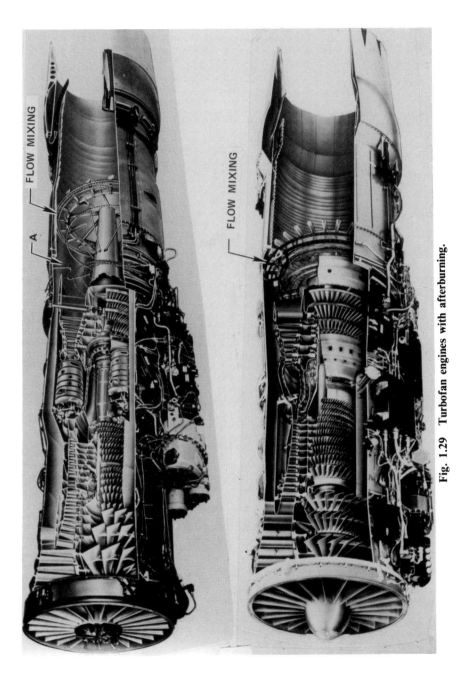

Fig. 1.29   Turbofan engines with afterburning.

gas temperature is suddenly raised, the nozzle throat area must rapidly increase to pass the mass flow without an excessive pressure rise. The rate of change of the fuel flow must be accurately synchronized with the rate of change of the throat area. If the area increases too rapidly, thrust is momentarily lost just when an increase is demanded. If it is opened too slowly, the fan and possibly the compressor are pushed into surge. The magnitude of the control problem varies inversely with the permissible surge margins of the compressors.

Summary

Changes in the ETR, the ratio of turbine inlet temperature to compressor inlet temperature, cause changes in the physical variables affecting the performance of all components. A compressor is the component most affected. Lowering the ETR invariably reduces the compressor flow rate, speed, and pressure ratio. Engine specific power is thus reduced at the expense of efficiency. Changes in the ETR may also move the compressor operating point to areas where its stability is threatened by surge or where stalls or flutter induce harmful blade vibrations.

The performance of a turbine is affected far less. In jet engine designs, the pressure ratio across a turbine is nearly constant for many of the important operating regimes. Reductions of speed with ETR almost maintain the turbine at a uniform corrected speed. Its dimensionless operating point changes only slightly. In fact, we may assume this behavior and then estimate a compressure match point with reasonable accuracy when only trends are required.

The other components have not been mentioned in this section, but we should be aware from the other chapters of these volumes that they too have problems. The engine inlet needs the careful attention described in Chapter 4, particularly if supersonic flight is included in the mission. Transient flows at the inlet is given special attention in Chapter 6. Many of the points made in these chapters may be profitably used in the design of compressors and fans. Similarly, the discussion of nozzle aerodynamics in Chapter 5 has many features that are applicable to turbines.

The ignition and stability of combustors are not always predictable and satisfactory results must be demonstrated under the conditions they would experience during the given flight envelope. Chapter 2 of this volume as well as Chapters 1 and 2 of Ref. 2 survey the problems involved.

There is need for mutual understanding and cooperation among all designers and developers of these components. One aspect of this need is well stated in Chapter 4 of Ref. 2.

1.6 LOSSES IN AVAILABLE ENERGY

The term losses describes the difference between the energy ideally needed to propel an airplane a given distance and the latent energy of the fuel actually consumed. The use of efficiencies has implied the existence of losses without identifying their cause. Since minimizing losses is a major function of engine design, it is now appropriate to review the sources of various losses in engines. This is the purpose of this section.

Note first that the thermodynamic cycle used in propulsion engines has an inherent loss mechanism that is not included under aerodynamics—the fuel energy rendered unavailable because adiabatic combustion is necessary. In comparison to a reversible isothermal process, this type of combustion wastes about half of the potential energy of the fuel. This loss is not recognized in Eq. (1.3).

The losses implied by the "ideal cycle efficiency," the quantity $(1 - t_0/T_3)$ in Eq. (1.3), are not aerodynamic either. Neither are they included as part of the compression or expansion efficiencies. It is the losses causing the component efficiencies to be less than unity that are the subject of the rest of this section, which first examines the requirements for a consistent definition of losses and then discusses their sources.

Losses in Engine Performance

Chapter 3 of Ref. 2 presents an excellent discussion about keeping track of losses. The authors use increases in entropy as the natural quantity for evaluating various losses of available energy in turbines. This idea is equally true for the other components and for engines. We note first that Eq. (1.3) can be written as

$$\frac{E_G}{gJC_p t_0} = \left[\frac{T_4}{t_0} - \frac{T_3}{t_0}\right]\left[1 - \frac{t_0}{T_3}\right] - \frac{T_4}{t_0}[1 - \eta_c^* \eta_T]\left[1 - \frac{t_0}{T_3}\right]$$

We then have the following explicit expression for the losses in terms of compression and expansion efficiencies:

$$\zeta = \frac{T_4}{t_0}(1 - \eta_c^* \eta_T)\left(1 - \frac{t_0}{T_3}\right) \qquad (1.25a)$$

Entropy may be defined by

$$\frac{T_{\text{OUT}}}{T_{\text{IN}}} = \left(\frac{P_{\text{OUT}}}{P_{\text{IN}}}\right)^{(\gamma - 1)/\gamma} \exp\left(\sum \frac{\Delta S}{C_p}\right) \qquad (1.25b)$$

The losses of Eq. (1.25a) are then expressed by

$$\zeta = \frac{t_0}{T_3}\left[\exp\left(\sum \frac{\Delta S}{C_p}\right) - 1\right]$$

The efficiencies of Eq. (1.3) are seen to be related to the sum of the entropy increases in the individual components. Most loss data, however, are expressed in terms of total pressure and Eq. (1.25b) enables them to be interpreted as increases in entropy as well.

Requirements of Significant Expressions for Losses

Care must be taken in either assigning or using an expression for total pressure loss. Such data are often presented in a conventional form that is useful for the analysis of piping systems, but inadequate for propulsion engines or turbomachinery where momentum is important. Note that we often use one-dimensional flow calculations to estimate the flow rate and momentum, even in situations where there are significant gradients normal to the gas velocity vectors. (See Fig. 1.30.) The early designers of steam turbines approached this problem by using two coefficients—one for flow rate and another for thrust. The same idea is currently used for engine thrust nozzles. (See Chapter 5.)

Two coefficients are usually used in compressor design also—one for blockage and another for losses in total pressure. The total pressure coefficient is too often an arbitrary average that may or may not be related to either blockage or momentum, which has resulted in some errors. Moreover, the total pressure loss coefficient may not account for all of the loss—every nonuniform flow represents a potential loss because mixing always increases entropy. This mixing loss is also determined by the laws for the conservation of flow rate, energy, and momentum and the principles involved are most easily discussed with the aid of the equations for these

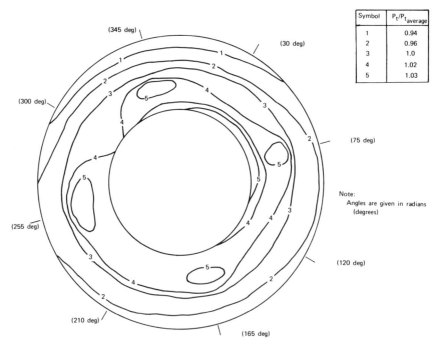

Symbol	$P_t/P_{t_{average}}$
1	0.94
2	0.96
3	1.0
4	1.02
5	1.03

Note:
Angles are given in radians
(degrees)

Fig. 1.30 High pressure compressor, inlet total pressure map, no engine inlet distortion.

quantities, as

Flow rate:
$$\dot{W} = \frac{PA}{\sqrt{T}}\psi = \sum \frac{P}{\sqrt{T}}\psi\, \Delta A \qquad (1.26\text{a})$$

$$\dot{H} = \dot{W}C_pT = \sum C_pT\, \Delta \dot{W} \qquad (1.26\text{b})$$

Momentum:
$$f = pA\mu = \sum p\mu\, \Delta A \qquad (1.26\text{c})$$

The definition of ψ and μ are given in Fig. 1.31. Observe that both are functions of Mach number with the trends depicted in the figure. Their ratio determines ϕ, which is again a function of the Mach number only. (This function is also indicated in the figure.)

In application, we would use these equations to determine the magnitudes of $\dot{W}$, $\dot{H}$, and f, which are the same before and after the mixing that makes the flow uniform. The ratio $f/\dot{W}$ yields the value of ϕ for the uniform flow, so the corresponding value of either ψ or μ can then be calculated. The effective total pressure is then found from Eq. (1.26a) or (1.26c). Using this pressure with the known temperature, flow area, and flow rate correctly estimates mutually consistent values for all other flow variables. This procedure applies also to calculating shock losses in supersonic flow and total pressure losses during heat transfer as in Chapter 2 of Ref. 2.

Note that angular momentum is conserved when the flow is also rotating. This requirement is expressed by

$$\sum V_\theta\, \Delta \dot{W} = \text{const}$$

The value of T and P in Eqs. (1.26a) and (1.26b) must be changed also. We can subtract the amount $[(\gamma - 1)/(\gamma g R)] \cdot (V_\theta^2/2)$ from T to create a reduced total temperature T^*. The associate reduced total pressure is

$$P^* = P\left(\frac{T^*}{T}\right)^{\gamma/(\gamma - 1)}$$

We also have

$$\sum T\, \Delta \dot{W} \sum T^*\, \Delta \dot{W} = \frac{\gamma - 1}{\gamma g R} \sum \left(\frac{V^2}{2}\right)\Delta \dot{W}$$

The procedures outlined above are then also valid for rotating flows. This brief treatment indicates how loss data can be either presented or used in the design of gas turbine engines. This example was presented only to illustrate what is involved; other consistent systems can also be developed.

Losses are often divided by a quantity associated with kinetic energy to form a loss coefficient. Quantities such as $\frac{1}{2}V^2$ or $(P - p)$ have been used

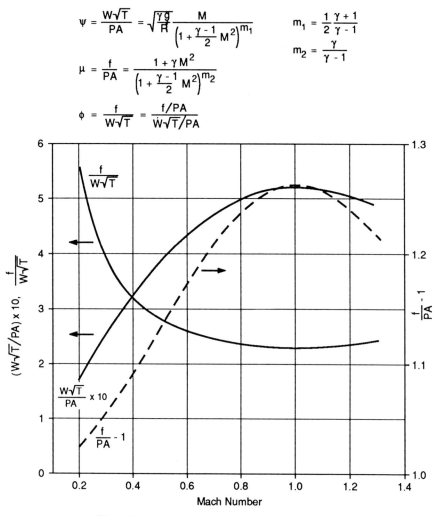

$$\psi = \frac{W\sqrt{T}}{PA} = \sqrt{\frac{\gamma g}{R}} \frac{M}{\left(1 + \frac{\gamma-1}{2}M^2\right)^{m_1}}$$

$$m_1 = \frac{1}{2}\frac{\gamma+1}{\gamma-1}$$

$$m_2 = \frac{\gamma}{\gamma-1}$$

$$\mu = \frac{f}{PA} = \frac{1+\gamma M^2}{\left(1 + \frac{\gamma-1}{2}M^2\right)^{m_2}}$$

$$\phi = \frac{f}{W\sqrt{T}} = \frac{f/PA}{W\sqrt{T}/PA}$$

Fig. 1.31   Flow rate and momentum functions.

as divisors. None of these completely eliminates the influence of Mach number on the loss, but $(P-p)$ seems to be as good as any.

Origin of Losses

We arbitrarily divide losses into four types: ideal shock losses, friction losses, diffusion losses, and those due to the secondary flow caused by pressure gradients normal to the velocity vector. Although any one of these rarely exists alone, each can trigger a loss that might be avoided if we had a better understanding of the phenomena involved.

Ideal shock losses. These losses are unique because they can be predicted by the ideal equations for the conservation of mass, momentum,

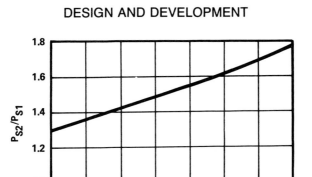

Fig. 1.32   Separation static pressure ratio vs Mach number on a straight surface, single-shock.

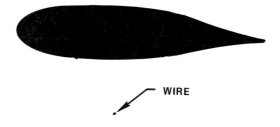

WIRE

Fig. 1.33   Drag experienced by the airfoil and wire is the same.

and energy. Equations (1.26) are also appropriate here when the initial flow is supersonic and the "mixed" flow is subsonic. Even though thermal conductivity and viscosity are involved in the mechanism of the losses, these properties do not determine the magnitude of the ideal losses. Adequate discussions of this subject can be found in any text on gasdynamics.

Experiments using flat surfaces have shown that as long as the relationship between static pressure rise through a shock and the upstream Mach number is below that shown in Fig. 1.32, the observed losses are approximately equal to the ideal losses. A maximum static pressure ratio of 1.4 across the shock is a better guide when the surface is curved. Additional losses arise due to shock/boundary-layer interactions when higher static pressure ratios are encountered. Gradients in flow properties appear, and the procedures of Eqs. (1.26) are necessary to specify the total pressure loss.

Friction losses.   Most other losses have their origin in viscosity. Viscosity exerts a shear stress on a fluid. The magnitude of the stress is

$$\tau = \mu \frac{du}{dn}$$

At any solid boundary of a fluid (e.g., at the inner wall of a pipe), the relative velocity of the fluid must be zero. Viscous forces are brought into play when finite relative motion exists elsewhere in the flow, since du/dn must then be finite over at least some interval of the flow.

The losses directly due to the friction stresses just defined are only a small portion of the total losses found in turbomachinery. Friction forces, however, do initiate sequences of events which result in losses that are several times the magnitude of those due directly to the friction forces themselves. This is the result of the mixing losses discussed in connection with Eqs. (1.26).

Diffusion losses.   These losses accompany any increases in static pressure resulting from decreases in fluid velocity. They are really amplifications of friction losses.

Figure 1.33 vividly illustrates this subject. The shape of the airfoil shown resulted from many years of study and tests. Air flows about it very smoothly. Nearly all of the drag on the airfoil is the result of the friction forces alone, whereas that on the wire is due to mixing.

A visual contrast between flow around a good airfoil and that about a cylinder is shown in Figs. 1.34a and 1.34b. Figure 1.34c is included in the series to show that a good airfoil can be misused. In Figs. 1.34a and 1.34c, the flow path markedly departs from some sections of the surface of the body. This condition is known as flow separation.

Flow separation results from the sequence of events illustrated in Fig. 1.35 (these photographs are from Prandtl and Tietjens). Figure 1.35a shows the distribution of flow immediately after it was almost instantly started from rest. Notice that with incompressible flow, the theoretical maximum velocity occurs at the top and bottom of the cylinder. Its value is twice that of the velocity far upstream and downstream. The ideal minimum velocity is ± 90 deg away from the top and bottom; its theoretical magnitude is zero. Between the top of the cylinder and its downstream surface element, the flow theoretically encounters a static pressure rise equal to

$$\Delta p = \frac{1}{2}\rho(2V_0)^2 = 2\rho V_0^2$$

The shear forces of friction retard the flow at the surface, but they require a finite time interval to develop. The friction forces in Fig. 1.35a were small and velocities near but not adjacent to the surface were almost equal to the ideal. The average velocity available very close to the surface, however, was not quite sufficient to negotiate all the pressure rise required. Consequently, the flow came to rest near the point indicated by A.

Figures 1.35b and 1.35c were taken successively as the flow developed with time. The latter picture clearly shows a "separation eddy." Such a vortex would cause even viscous-free flow to stagnate at point B. The resulting change in the pressure field is communicated upstream, which causes the retarded flow to come to a halt near point C. This situation is a typical point in a chain of events that causes the separation eddy to continuously advance upstream.

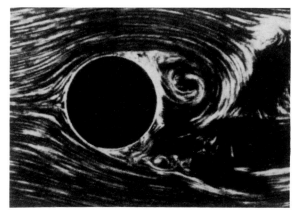

a) Flow about a cylinder (from Prandtl-Tietjens).

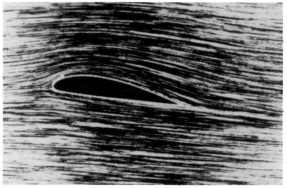

b) Smooth flow about an airfoil (from Prandtl).

c) Separated flow about an airfoil (from Prandtl).

Fig. 1.34 Examples of flow about isolated bodies.

Figure 1.35d was taken a short time later. The disturbed flow in this view had just about reached the top of the cylinder. Right after this point in time, the eddy was washed downstream by the surrounding flow. The process was then more or less repeated over and over again. In general, the opposite sides of a cylinder take turns in generating and discarding eddies. The periodic shedding of vortices subjects the cylinder to an oscillating force. The flow is unsteady for a short distance downstream.

The time-averaged vorticity shed from the cylinder must be zero. This means that far downstream, after all the disturbed flow is mixed together, the mixture can have no angular momentum. If this condition did not prevail, the cylinder would experience an average force perpendicular to the flow.

The flow near the trailing edge of Fig. 1.34b resembles that of Figs. 1.34a and 1.35d, but on a smaller scale. If we applied the relations of Eqs. (1.26) to this distribution, we would have to assign a negligible velocity to a finite element of area. This would result in a total pressure loss over and above the friction loss. The magnitude of this resulting mixing loss depends on the ratio of the area of low flow to the total area available. These losses result from an attempt to decelerate the flow and are, therefore, called diffusion losses.

The separated flow in Fig. 1.34c is one sided, but a finite time is also required to develop it. Immediately after the flow is started, a small eddy appears near the trailing edge. A stagnation point then forms at the leading edge of the eddy. This local high pressure causes the flow upstream of the eddy to leave the airfoil surface and join the existing eddy. Thus, as in Fig. 1.35, the eddy continues to move upstream until it is very close to the leading edge. The separation eddy caused by the retarded surface fluid induces other vortices to form downstream. From time to time, parts of these eddies are washed away. Again, the time average of the shed vorticity must be zero. At any instant, however, the vorticity shed by one vortex may exceed that of the other. The airfoil then experiences both fluctuating lift and drag forces.

The previous discourse illustrates the following points about the amplification of friction losses:

1) When an attempt is made to convert part of the kinetic energy of a flow into static pressure, a portion of the flow is retarded and may separate. The retarded flow augments the losses.

2) Separated flow almost invariably causes the downstream flow to be time-unsteady.

3) Because flow separation requires a finite time to develop, its behavior in an imposed unsteady flow can be different than that in steady flow.

4) Separated flow can produce fluctuating forces on a body.

5) Severely separated flow produces high drag or losses as a result of mixing.

Finally, we should appreciate that when a separation point has become more or less stabilized, the local adverse pressure gradient includes the effect of a not inconsiderable separation eddy immediately downstream. There is indirect evidence, from slotted diffusers and slotted compressor casings, that greater pressure recovery could result from evading the added

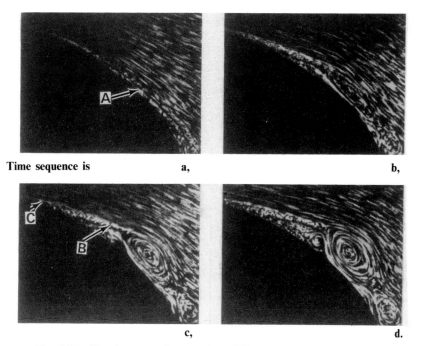

Time sequence is   **a,**   **b,**

                                        c,   **d.**

Fig. 1.35   Development of separation of flow about a circular cylinder.

pressure gradient of this eddy. If an eddy forms initially as a transient instability and if the resulting local pressure rise can be partly dissipated through, for example, a slot, the upstream advance seen in Figs. 1.34 and 1.35 would be arrested. Research in this area has been neglected. We should also note that flow restrictions, such as heat exchangers, placed downstream of wide-angle diffusers increase the pressure recovery available from diffusers. The adverse pressure gradients of the potential flow are not solely responsible for flow separation.

Secondary flow losses.   Pressure gradients normal to the direction of the throughflow are found whenever the flow is curved. Fluid elements retarded by friction are then not in equilibrium with the pressure gradients produced by the less retarded flow and are, consequently, moved along the path of this gradient. Flow about the ends of vanes and blades are included in this category.

Flow that is strongly curved creates both beneficial and penalizing conditions. No rigorous rules are yet known, but the following qualitative information is valid.

One problem arises because the flow direction must be changed in turbomachinery and flow changes direction only when a static pressure gradient is imposed perpendicular to the flow velocity. It can be shown that if this static pressure gradient is in the same direction as the local entropy gradient, the flow is unstable. The degree of instability is proportional to the product of the two gradients.

Unstable flow encourages the interchange of momentum between the mainstream and the retarded flow near the flow boundaries; the value of du/dn becomes high, friction forces are also relatively high, and heat-transfer rates are enhanced. This type of flow can delay flow separation. An overhanging cold front with its accompanying strong surface winds, including tornadoes, is an example of instability.

Stable flow has the opposite characteristics: momentum interchange is discouraged, friction and heat transfer are reduced, and flow separation is less easily averted. An atmospheric inversion with its associated calm smoggy air is an example of stability.

Consider now the flow in the curved duct shown in Fig. 1.36. Let the flow be in the direction indicated. If the mean effective velocity of the main flow is V, a static pressure gradient V^2/R_c (R_c is the radius of curvature of the effective velocity) will be produced that is perpendicular to V. The pressure on the surface indicated by H is greater than that at L.

The shading in the sketches of the inlet and outlet flows passages (X and Y) schematically indicates retarded flow. The retarded flow at the inlet is assumed to have uniform circumferential distribution.

The previous remarks about flow stability suggest that the retarded flow along H will mix with the main flow. Except for a narrow region along H where the entropy increases in the direction of the flow because of friction, all entropy gradients will be perpendicular to the flow in the direction of H to L. Retarded flow thus diffuses from H to L. Meanwhile, the difference in pressure gradients between the main and retarded flows along surfaces A and B causes this retarded flow to be swept toward L. This flow is known as secondary flow. Experimental evidence on two-dimensional bends shows that the swept flow tends to roll up into vortices denoted by E at the corners. The retarded flow at the outlet of Fig. 1.36 is exaggerated to

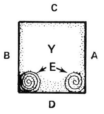

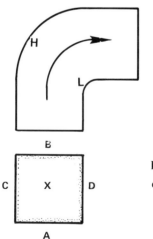

Fig. 1.36 Two-dimensional curved duct illustration of secondary flows.

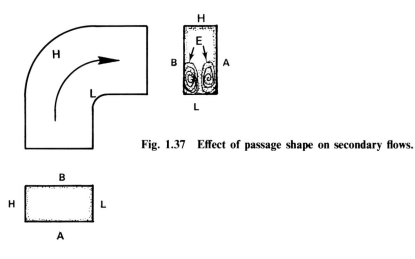

Fig. 1.37   Effect of passage shape on secondary flows.

emphasize the trends. The gradients of entropy and static pressure within these vortices have opposite directions. They are, therefore, very stable and resist any interchange of momentum with the main flow. Thus, they interfere with subsequent attempts to diffuse the flow. The rest of the retarded flow along L is also stable. If the area of the duct is constant, the main flow near L must first accelerate and then decelerate. Inasmuch as the stable boundary layer and vortices cannot be supplied with the necessary momentum to combat the rising static pressure, the flow will separate and generate mixing losses.

Secondary flow occurs whenever there is a pressure gradient normal to the mainstream in either the main or retarded flow. When the bounding surface itself is rotating, secondary flow adjacent to the boundaries will move outward, provided the rotational speed of the retarded flow exceeds that of the mainstream. Otherwise, the motion is inward. The normal pressure gradients in turbomachinery are complex and produce complex secondary flow patterns. Some additional comments are presented in Chapter 6 of Ref. 2.

The shape of the flow passage indicated by X and Y plays an important role in deciding the magnitude of the disturbing influence of secondary flows. Consider the shape shown in Fig. 1.37. This has half the flow area of Fig. 1.36 and a lower hydraulic diameter, $2(H \times B)/(H + B)$. It therefore suffers comparatively higher friction losses and surfaces A and B of Fig. 1.37 provide most of these losses. Thicker regions of retarded flow will accumulate at L because the volume of the retarded flow is comparable to that in Fig. 1.36, while L is only half as long.

Shapes of this type are deemed to have low aspect ratios. An attempt to diffuse this flow will cause the retarded flow area to grow rapidly, because it resists being energized. The potential mixing losses exceed those of Fig. 1.36. Shapes having high aspect ratios ($H > A$) reduce secondary flow losses. They are generally preferred as long as the friction losses are smaller than the mixing losses.

Secondary flows in cascades.   A unique secondary flow appears near the blades and vanes of both compressors and turbines. Boundary layers on the hub and casing do not have the momentum or energy to overcome the pressure field created at the leading edges by the main flow. The boundary layer is consequently brought to rest and pushed away from the surface by the reversed flow coming from the leading edge. This reversed flow originates at various radii along the span and therefore contains a large amount of vorticity. At the same time, the displaced gas in the boundary layer experiences the tangential pressure gradients produced by the airfoils to turn the flow. The retarded flow rolls up because of its vorticity (similar to the trailing vortex of an airplane) and moves to the low-pressure side of the adjacent blades as a secondary flow.

Losses caused by flow about blade ends.   The outer ends of rotor blades are near stationary surfaces while the inner ends of stationary vanes are near rotating surfaces; some clearance between stationary and rotating surfaces is necessary to avoid damage. Flow would ordinarily squirt from high to low pressure through this clearance, and kinetic energy would be converted into heat. Gas flowing backward over rotor tips of compressors re-enters the flow at lower levels of useful energy. Retarded flow tends to thicken and the entropy is increased. Moreover, additional thermal energy is thus added to this flow. Clearance flow also makes its own contribution to secondary flow eddies. The resulting losses have several adverse effects on turbomachinery performance: the attainable pressure ratio is reduced, the entropy rise increased, and the surge-free margin of a compressor eroded. Flow over blade ends, together with secondary flow, has been the subject of much inquiry, but many important quantitative aspects are still not clearly understood. Several articles in Ref. 8 review some current thinking on this subject.

In order to avoid clearance losses, the outer ends of rotor blades are often attached to corotating bands known as shrouds. Stator vanes may be shrouded also. In this case, the bands are attached to the inner ends of the vanes. Circumferential seals adjacent to a shroud then limit, but do not prevent, the flow from the high- to the low-pressure side of a blade or vane. Many varieties of shrouds and seals have been developed.

Shrouds are known to be aerodynamically beneficial in many turbine blade designs where the useful energy of the clearance flow is irrevocably lost and the secondary flow sets up eddies that rotate in about the same direction as the clearance flow. The general aerodynamic value of shrouds in compressors still has to be decided. All experimental data do not agree, possibly because clearance flow vortices oppose those from the secondary flow.

Summary.   Losses arising from nonuniform velocity distributions are the principal ones that concern the aerodynamicist. These adverse distributions are usually brought about by the necessary diffusion during the compression process, by changing the direction of the flow, and by leakage around the ends of blades and vanes.

Losses in nonuniform flow are aggravated by decelerating the flow. Remember also that the retarded flow is propelled by an interchange of the momentum of slowly moving fluid particles with that of faster moving particles. Numerous experimental studies of jets and diffusers have shown that the rate of this interchange is limited and that it depends upon the passage shape. The rate of momentum exchange is obviously hastened in unstable flows and delayed in stable ones.

Probably the greatest sources of losses result from the necessary diffusion of partially stabilized flow. Such flow frequently rolls up into a localized vortex. The entropy gradient in the vortex is in the opposite direction from its pressure gradient. The flow is stable and momentum interchange with the rest of the flow is resisted. Attempting to diffuse the resulting flow aggravates nonuniform velocity distributions. Subsequent mixing amplifies the losses.

Until recently, this situation had to be reluctantly accepted, particularly where compactness was required. Emerging techniques for numerical aerodynamic analysis are demonstrating, however, that designs can be made that will alleviate the accumulation of stable impregnable regions of retarded flows. Special types of tangential, axial, and radial pressure fields have been created by stator vanes. These fields minimize transverse movements of retarded flows. An exciting improvement in a turbine has been obtained. Even greater accomplishments are expected in the near future by alert students of this work. This subject was also noted in Chapter 4 of Ref. 2.

Shrouds are often necessary on compressor stators, however. An accidental rub by a stator heats and expands both the stator and its adjacent rotating hub. As a result, they move toward each other. Rubbing is then intensified, sometimes with disastrous results. The clearance required to eliminate this possibility can be so great that the resulting clearance flows severely reduce performance. In this case, any adverse effects of the shrouds are more than compensated for by eliminating the flow about the blade ends.

Shrouds have sometimes been used on rotor blades of axial compressors. The stationary surface next to the rotor blades, however, can be designed to move away from the blades during rubbing. Closer operating clearances can thus be tolerated. Losses due to blade clearances can probably be reduced by changing the shape of the blade ends when more is known about the flow there. The magnitude of some of these losses, as well as the losses due to leakages through seals, frequently depends upon controlling the relative movements of the rotating and stationary units, thus evading some of the aerodynamic problems. This demands the recognition of problems that are germane to other disciplines.

Blade and vane vibrations can supply another motive for using shrouds. The possible impact of large birds may also require the use of shrouds, particularly on the rotor. "Part span" shrouds may be used. These are placed between the hub and the shrouds of rotors at radii where they can be most effective mechanically.

1.7 INTERRELATIONS AMONG AERODYNAMIC COMPONENTS

The other chapters of these volumes describe how the design of engine components may be implemented once the engine type is decided and the pressure ratios and turbine inlet temperatures have been tentatively selected. Further elaboration is not required here. Some observations of mutual interactions among the components remain to be discussed, as well as some incidental information about individual components that influences their effectiveness.

Most of the problems in this category were anticipated in the discussions of the previous sections. Any component that delivers a nonuniform distribution to components downstream contributes losses and other problems to those components. Conversely, events in a downstream component may affect an upstream one. Because of these interactions, potentially beneficial concepts for one component sometimes have to be deferred to help another.

Several of the authors in this series have emphasized the close cooperation among the many disciplines that is mandatory if a design is to be efficiently executed. Pertinent topics have been discussed in Refs. 1 and 2, and in the remaining chapters of this volume. One object of this section is to identify some of the compromises involved in selecting turbine inlet temperatures, rotor speeds, and rotor diameters. Important interactions among the aerodynamic components are also reviewed in this section.

Temperature and Speed

Most modern engines can profitably use the highest achievable turbine inlet temperatures. An obvious exception is a jet engine for subsonic or low-supersonic speeds. If the bypass ratio of the fans is limited by problems such as ground clearance, Eq. (1.20) shows that aerothermodynamics can again constrain the desired temperature. In the other cases, temperature is limited by material strength, thermal corrosion, and aerodynamic and cooling losses necessary for surviving high temperatures.

Section 4.3 of Ref. 2 offers a good discussion of the design restraints imposed by temperature and stress. The same chapter shows how stress problems are eased by cooling; it also shows how the act of cooling introduces new losses and how the blade shapes required for cooling invite still further losses. Chapter 5 of the same reference provides further material on this subject; Sec. 5.2 in particular gives a good description of the problems of designing for high temperature.

An effect of temperature on material strength is quantified by Fig. 1.38. The abscissa is the Larsen-Miller parameter, which correlates the combined effects of the level and duration of stress and temperature. The ordinate shows the stress allowed for each value of the abscissa.

Two properties are noteworthy. Temperature has an almost exponential effect on useful life—the life is about halved for a 25°F increase in temperature. The second feature concerns the uncertainty of the data. The discussions cited in Ref. 2 provide evidence about the lack of uniformity in

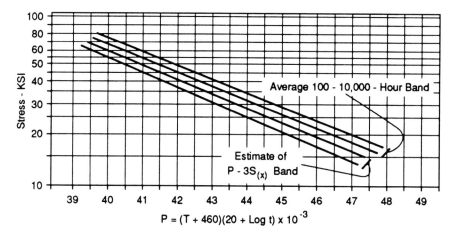

Fig. 1.38   Typical 0.2% Larson-Miller plastic creep curve of cast Inco-713LC.

the temperatures approaching turbines. An educated guess is necessary to avoid either an unreliable or uncompetitive design. The indicated uncertainty of the material data is another challenge. Some designers use the three-sigma data and add an arbitrary value, e.g., 50°F, to the expected temperature. Since life is so sensitive to temperature, experience with similar tasks and with one's associates is a valuable asset for making proper choices.

Figure 1.38 alone would strongly support a choice of low rotor speeds to avoid high stresses. Low linear and angular blade speeds would result from this decision. Euler's turbine equation, however, leads to the observation that power is proportional to the cube of the linear blade speeds and that high speeds are required for lightweight units. These conflicts are resolved by making many design iterations to arrive at an attractive balance. Experience with other engines is again an asset.

Another restraint on speeds is the Mach number relative to the rotors of fan and compressor blades. This problem is discussed in Chapter 3 of Ref. 2. Fan blade stresses are also a determinant of allowable speed, particularly when the added weight of part span shrouds (or dampers or clappers, visible in Fig. 1.27) are used as protection from flutter or to re-enforce the blades in the event of an impact with a large bird during flight.

Diameters

The diameters of the components are determined principally by the gas flow rate. Inasmuch as the radial force on a blade is proportional to the flow area, there is a natural desire to keep this area as low as possible by maintaining the axial component of the Mach numbers at high subsonic values.

This solution opens the door to other problems, however. When a one-dimensional calculation indicates that the axial component of the

Mach number leaving a blade row is about 0.6, the row is probably at limiting loading. What happens is that the effective axial component of the Mach number is about 1—the wakes from the blades and the combined thicknesses of the boundary layers at the tip and casing contract the effective flow area by the 19% needed to account for the difference.

The Mach number of the flow at the compressor inlet affects the relative position of the corrected speed lines; see, e.g., Fig. 1.12a. The gas speed is more or less proportional to the rotating speed, but the rate of change of $(W\sqrt{T}/P)$ with the airspeed decreases to zero as the throughflow Mach number approaches 1. (See Fig. 1.39.) The rate of change of corrected weight flow behaves the same way. This accounts for the crowding of the speed lines in Fig. 1.12a at their higher values. This situation can be troublesome and must be avoided.

There are other reasons for using less than the maximum flow rates. Efficiency suffers when the blade heights become too small and the area of the flow boundaries offering nothing but friction becomes comparable to or greater than those areas responsible for the transfer of useful energy. Again, high velocities approaching combustors must be diffused at the expense of space and, sometimes, available energy. Moreover, a large acceleration of flow ahead of an engine nozzle throat usually aids the thrust coefficient by improving the flow distribution; low upstream velocities are then beneficial. These are a few of the reasons why diameters are also selected only after an extensive inquiry of the design iterations.

Inlets

Only a few items are noted here, inasmuch as the subject of inlets is thoroughly covered in other chapters of this series. Subsonic inlets rarely

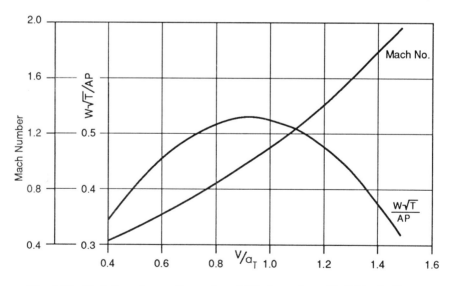

Fig. 1.39   Variation of mass flow rate and Mach number with fluid velocity.

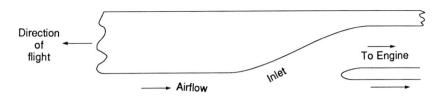

Fig. 1.40   Schematic of flush engine inlet.

impose formidable problems. When airplane speeds are low, however, some flow configurations cause extra internal losses and flow distortions.

A special problem arises in some inlets when an engine is started while the airplane is traveling at high speeds. Flush side inlets such as that in Fig. 1.40 are particularly affected. This sketch shows an inlet used on a vehicle that is launched from an airplane. Flight at high subsonic speed is initially sustained by a rocket. During this period, the turbo-jet engine carried by the vehicle is started; this engine then provides the necessary thrust for cruise to the target. The engine is located near the rear of a cigar-shaped fuselage having a length-to-diameter ratio of about 13. The leading edge of the inlet is about 5 diam downstream of the nose. Although this inlet design is adequate for steady-state operation near the intended cruise speed, it cannot provide an engine with anything like uniform flow while the engine is starting and accelerating. Boundary layer from the fuselage then comprises most of the engine mass flow rate, and the effective total pressure at the engine inlet is low and far from uniform. Many inlets suffer from this difficulty when the ratio of inlet air velocity to flight speed is low.

We have noted that compressors tend to be plagued by rotating stall and surge at low engine speeds. Imposing velocity and entropy gradients as well as time unsteady flow at this time make this situation worse. The best solution requires cooperation between air frame and engine designers early in the program.

Some engines designed for this application have been unable to achieve design speed. Inlets for supersonic flight speeds can be difficult to design. Their leading edges must be relatively sharp. Figure 1.41 illustrates configurations used on four different supersonic aircraft. They represent practical compromises for realizing effective recovery with a minimum of distortions during operation at subsonic and supersonic speeds at level flight or during combat maneuvers. The pressure field of the airplane is used in different ways to assist the pressure recovery.

A flow mismatch with engines often creates unsteady and nonuniform flow distributions that again penalize compressors. Present knowledge really permits us only to install devices in the inlet that serve as bandaids for the problem. What is really needed is variable geometry that enables the engine to efficiently inhale the natural airflow rate through the capture flow area.

a) USAF F-111 in flight.

b) USAF F-5E in flight.

c) USAF F-15 in flight.

d) USAF F-16 just after takeoff.

Fig. 1.41 Configurations for supersonic inlets.

Compressors and Fans

Compressors prefer time-steady and uniform flow in their inlets. If they do not receive flow that reasonably meets these specifications, they respond by surging or with reduced performance, as noted above. At the present time, compressors are prima donnas because they are essentially diffusers and the best diffusion processes occur just before flow separation (or stall) is encountered. When questionable inlet flow is expected, compressors must be given ample stall margins, even at the expense of efficiency. The amount and rate of static pressure rise asked of the critical stages is then reduced. One objective of future research should be aimed at identifying the appropriate variable engine geometry and accompanying controls that are capable of coping with the temperamental nature of a compressor (for example, varying the flow rate when necessary).

Compact transition ducts between compressors (such as that shown in Fig. 1.27) can be a further source of nonuniform flow. Curved surfaces just ahead or just behind the blades or vanes alter the radial distribution of the flow. The local adverse pressure gradients might be higher than otherwise expected, causing the performance of the compressor to deteriorate. If these ducts also have significant localized total pressure losses, the resulting gradients in the flow impair the performance of the following components and increase the design problems. Careful planning is essential in the layout of a flow path.

A compressor can cause problems for both itself and other components. An important one stems from the ability of a compressor to trap an anomaly in the flow. This phenomenon may be visualized with the aid of Fig. 1.42, which sketches a partial two-dimensional development of three stages of an axial flow compressor.

Observe that the close spacing of the blades and vanes limits the horizontal spread of the flow. Suppose we inject a plume of smoke such as that indicated by A into the compressor. The smoke will be confined by a group of the first row of rotor blades and will move toward the right. A little bit of the smoke may escape between the rotor and stator, but most of it will be captured by just a few passages in the next row of stator blades. The continuation of these events through each stage eventually will cause the smoke to emerge from the three stages at position B. It is displaced in the direction of rotation and its circumferential extent is only slightly expanded.

This trapping of the flow is responsible for some important interactions between a compressor and the rest of an engine. If the flow entering a compressor is not uniform, it behaves almost as if it were a number of independent compressors, each having different inlet conditions. Then, each path would have the same map, such as that sketched in Fig. 1.43. The processes actually involved with nonuniform entering flows are complex and some of them are imperfectly understood. The trapped flow concept does help to explain some important observations, however.

Suppose that the flow in region A of Fig. 1.42 has a relatively low total pressure due to a flow obstruction upstream. Suppose further that the compressor would normally operate near point X of Fig. 1.43 if the flow

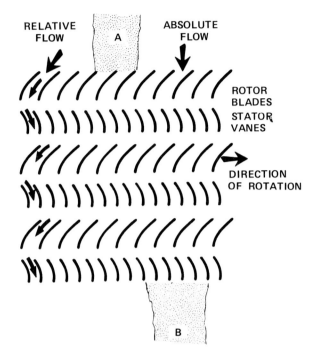

Fig. 1.42   Schematic development of three blade rows of an axial flow compressor.

were uniform. Observe that tangential static pressure gradients immediately downstream of the last blade row would require curvatures of the flow path that cannot be justified, but that static pressure gradients created upstream are easily tolerated. It is reasonable to expect the upstream flow to arrange itself so that uniform static pressure appears at the exit of the last blade row. The compressor thus responds to the flow through region A by shifting its operating point to one having a higher total pressure ratio, say point Y. If region A is large enough and if its total pressure is sufficiently depressed, the compressor will surge. (Point Y would cross the surge line.) If the low total pressure in region A is the result of flow separation in the inlet, the magnitude and extent of the losses probably vary with time. If the transient losses are large enough and persist long enough, compressor surging can be expected.

The trapped flow illustrated in Fig. 1.42 also works backward. A strut or other obstacle placed in a flow creates a high static pressure field near its leading edge. Ordinarily, the pressure of such a field decreases rapidly with distance. If the strut is placed close enough to a row of vanes, however, the local increase in pressure will retard the flow through only one or a few of these vanes. This retardation is propagated upstream with limited circumferential spreading. The strut will thus cause a small portion of the compressor to assume the operating point symbolized by Y in Fig. 1.42, while the rest of the compressor operates at point X. This behavior rarely produces surging problems, but it can elevate local temperatures by as

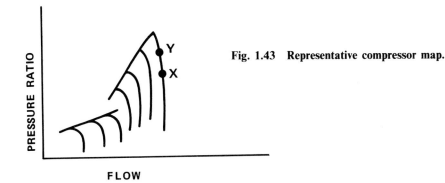

Fig. 1.43   Representative compressor map.

much as 25°F and produce regions of low airflow. The variations will also affect the distribution of turbine inlet temperature. Inasmuch as the rotor blades passing through a trapped-flow region experience a variation in force, blade vibrations can be induced at a considerable distance upstream or downstream of a disturbing device.

In order to control mixing losses at the ends of the blades and vanes of a compressor and preserve a reasonably uniform axial component of velocity, extra energy is often deliberately added to the air near the blade ends to compensate for momentum deficiencies there. This extra energy produces radial temperature and entropy gradients at the compressor exit, but these properties are approximately axisymmetric if the rotor and stator clearances are uniform. However, the temperature gradients could persist through a combustor and into a turbine.

Combustors

The relationship between the combustor and its adjacent components is thoroughly discussed in Chapter 2. The remarks presented here are included to preserve the continuity of this section.

The temperature distributions of the gases within and leaving a combustor depend upon the distributions of the temperature of the approaching airflow as well as the distribution of fuel and airflow into the combustor itself. Local deficiencies in the approaching airflow or excesses in the inlet air temperature cause local hot spots within and downstream of a combustor. Minimizing these hot spots is a goal of engine (not just combustor) design.

One should be aware that compact compressors must be designed so that their discharge velocity is greater than the velocity that combustors can tolerate. Therefore, a diffuser must be placed between the compressor and the combustor even though part of the combustor can be used to aid diffusion. Some of the previous discussions apply to this diffuser. If the compressor dispenses nonuniform flow, the diffuser problem can become very difficult. The unevenness in the flow is amplified by the diffuser and the flow can even become time unsteady. Such a situation must be rectified before an efficient combustor furnishing acceptable discharge conditions can be designed.

Turbines

The temperature of the gas entering a turbine should be as circumferentially uniform as possible. There is a maximum temperature that any portion of a turbine stator can withstand before it loses strength or is corroded by chemical processes. The maximum temperature of the flow from the combustor is thus determined. However, the output from the turbine depends on the average temperature. The need for circumferential uniformity is obvious.

On the other hand, there is always an optimum, but nonuniform, radial temperature distribution that yields the best power with a minimum expenditure of cooling air and still subjects the rotors to the same mechanical duress at all radii. Ample discussions of this subject are included in Chapters 4 and 5 of Ref. 2.

Any flow disturbance that inhibits uniform circumferential temperatures or proper radial temperature distribution imposes a penalty on engine performance. Close cooperation is required, therefore, between the designers of inlets, compressors, and combustors so that any necessary compromises are consistent with the objectives of the engine. This is another example where specialists should be familar with peripheral disciplines.

A large change in flow direction is required in the vanes and blades of many turbines. This is particularly true near a turbine hub. Large secondary flow vortices are often generated there. The extent of these vortices depends upon the quality of the flow entering the turbine. They can be partially controlled by adjusting the radial gradients of the energy abstracted by a turbine stage.

It is fortunate the turbine rotors are usually exposed to lower temperatures than the upstream stators are. The motion of the rotors reduces the relative total temperature of the gas, which in turn controls the temperature of uncooled blades. The ideal difference is

$$T - T' = (V_\theta U - \tfrac{1}{2}U^2)$$

But notice that retarded flow from the stators at the blade ends reduces V_θ, with a consequent increase in T'. Unexpected erosion at the blade ends has been traced to this phenomenon. This is another reason to carefully watch the details of turbine flow.

Secondary flow vortices at the blade ends have exhibited another disagreeable characteristic. If the curvature of the shroud or hub boundaries of the flow passages are convex, the vortices migrate into the area used by the main flow. They then act as a barrier, which causes further increases in loss. Such convex curvatures should thus be avoided.

The gas velocities behind the final turbine in the gas path need to be reduced in many engine designs. The boundary layers near the hub are invariably rotating and the flow is stable. The secondary flow eddies are also stable, so avoiding some flow separation along the hub is difficult.

Since the average entropy and static pressure gradients at the casing are in the same direction, the flow there is unstable. Gas velocities in this region can usually be decreased efficiently. The radial distribution of static pressure at a turbine exit can often be controlled so that much of the required reduction in velocity occurs at the casing, thus sparing the hub.

Emerging techniques for calculating three-dimensional viscous flow will make it possible to diminish many of these unwanted secondary flow effects. More intelligent choices will then be available to a turbine designer.

Afterburners

The interactions with afterburners have already been covered in Ref. 2 and in Sec. 1.4. It has been noted that flow entering this component must be slow enough to enable stable combustion to be produced. This affects the design of the last turbine and the transition duct. When fan flow is included, greater attention must be given this flow because the lower gas temperatures increase ignition difficulties. The temperatures at the exit ought to be uniform.

Nozzles

The discussion associated with Eqs. (1.26) describes the requirements of nozzles. The greatest thrust for the enthalpy and flow rate is measured when all the properties of the flow are uniform. Nozzles receive all the debris from the rest of the engine, so there is no substitute for careful attention to all flow passages. We must also be sure that the static pressure at the nozzle exit is the one assumed. Pressure fields from the airframe, or from the curvature of the flow from or about a nozzle, may alter the effective value of this pressure.

Summary

Besides the obvious effects of a mismatch in the flow capacity and efficiency of engine components, the main sources of unfavorable interactions among components come from nonuniform and unsteady flows. Sometimes a deliberate manipulation of flow will make one component look good at the expense of its neighbors. Such schemes are usually unsatisfactory—the overall effect is all that matters. Each component exists for the engine and not vice versa. *There is always a need to compromise.*

1.8 INTERACTION WITH OTHER SPECIALTIES

It cannot be mentioned too often that there are many disciplines involved in designing, developing, and manufacturing engines. How well one discipline understands the problems and technology of other disciplines plays an important role in determining costs as well as the elapsed time between the initiation of design and the delivery of an approved product.

An aerodynamicist's lack of appreciation of stress and material technology may require him to design and redesign a blade a number of times before the aerodynamic and endurance requirements are satisfied. If manufacturing technology is ignored, an unnecessarily expensive process may have to be devised to make and inspect the blade. It is especially noteworthy that a part which can be thoroughly inspected is often preferred to a potentially superior part which does not lend itself to inspection. Frequently, the potentially superior part turns out to be inferior because it does not conform to design requirements. Simple design changes that improve the ability to reproduce a product, perhaps with a slight penalty in latent performance, should always be in the thoughts of an aerodynamic designer. This emphasizes the requirement for the designer to understand the real needs of the customer.

As mentioned in Sec. 1.2, the purpose of an airplane is to render a service that enough people want and can afford. The costs to the eventual customer include his share of the expenses associated with the initial design and development, manufacturing, maintenance, and availability. (Availability signifies, among other connotations, whether nine or ten engines must be purchased to be sure that at least eight are available when needed.) By understanding the total picture, an honest evaluation can be made about whether the added cost of a supposedly more efficient part is worth its latent contribution to reduced fuel consumption, reduced weight, or increased thrust. This knowledge, of course, cannot be acquired instantly. The successful aerodynamic designer will always be alert to opportunities that will improve his understanding of these interrelationships.

Besides adopting this long-range philosophy, there are many areas where a good understanding of peripheral technology should be sought almost immediately. A few subjects and problem areas have been selected to illustrate the need for the aerodynamicist to be involved in many activities. The important subject of controls is not covered here since some attention was directed to them earlier.

Dimensional Integrity

Notice was taken in previous sections of some effects of the dimensional changes that accompany temperature variations throughout an engine. The existence of variations in temperature and dimensional changes along the length of the engine during steady-state operation are readily appreciated. The effects of transient operation on changing these variations of temperature and dimensions with time should also be recognized.

Circumferential variations in temperature near an outer casing can cause bulges in the casing, with a resulting local increase in clearance. Simple calculations can indicate how easily noticeable increases in such clearances are achieved.

It is practically impossible to keep uniform clearance between stationary and rotating parts at all times. There is, however, always some operating condition and one or more areas of an engine where the clearance is a minimum. The circumstances depend upon the temperatures and vibrations. Airplane maneuvers are also involved. The requirement of mechanical integrity sets the magnitude of the minimum clearances. The clearances at other operating conditions are then automatically defined.

The relative growth of the rotating and stationary parts depends to some extent on the aerothermodynamic design. Heat-transfer rates and blade shapes are involved.

Changes in the flow area of turbine stators and exhaust nozzles with temperatures are also pertinent items. If these areas are even moderately different than the design intent, the engine will be mismatched. Improper pressure ratios will be imposed upon the turbine and the engine exhaust nozzle. The bypass ratio will be affected and the thrust of an otherwise well-designed engine will deteriorate.

Experimental Testing

It is easy to fall into an organizational trap of four isolated groups. One group designs a component, a second group supervises its manufacture and assembly, a third tests it in the laboratory, while a fourth interprets the data. If the performance is lower than anticipated, the resulting discussions, while stimulating, do not hasten the delivery of a product or lower its cost. Was the design bad? Was is poorly made or assembled? Are the test data bad? Are the interpretations inadequate?

The four activities stated above constitute the source of nearly all pertinent aerothermodynamic data. The elements in each process should be carefully monitored and understood by the aerothermodynamic designer. Ideas for changing and improving procedures will always be uncovered. More valuable information has more often been learned by watching or participating in an assembly or a test than by looking at an inspection report or examining performance curves.

Flow Blockage by Struts and Accessory Drive Shafts

The rotating parts of an engine require bearings that must be supported by the external casing. Mechanical support is provided by struts that pass through the flow passages. At first blush, an aerodynamicist prefers the struts to be thin, few in number, and remotely located from any blade or vane rows. One reason for this has been called "flow trapping."

Unfortunately, this aerodynamic arrangement does not work. When an airplane executes some required, but violent, maneuver, the struts must be strong enough to resist the inertial (including the gyroscopic) reactions of the rotor with minimal deflections. Excessive deflections require blade and vane end clearances that consume more efficiency than would the mechanically appropriate struts.

Designers must also anticipate the loss of one or more blades due to unanticipated blade vibrations or to foreign object damage. Rotor vibrations thus created are greater than those normally anticipated. The choice of strong struts is again preferable to excessive blade and vane clearances.

Supplying and scavenging oil from the bearings requires struts having adequate cross-sectional area. These struts also pass through the engine flow passages. The use of thick struts is frequently necessary, but they should not be placed in sensitive aerodynamic areas. A mutual understanding of the problem by specialists in aerodynamics, stress and vibration analysis, lubrication, bearings, and weight control is the key for arriving at the proper tradeoffs between clearances, blockage, and assembly.

It is noted in passing that some struts may be aerodynamically useful.

For instance, they can be shaped and located to remove part of the swirl from the flow leaving a turbine.

Similar problems arise when mounting accessories such as oil pumps, fuel pumps, starters, generators, and other rotating accessories along the external side of the engine casings. Some of these have been shown in engine photographs. A so-called tower shaft must then pass through the airstream to connect the accessories to an engine shaft. When there are high- and low-pressure compressors and turbines, most connections are made to the high-pressure shaft because its speed varies less than that of any other shaft during flight. Moreover, this is the logical unit to be attached to the starter. The tower shaft is then mounted radially near the compressor. (Putting it between turbines invites additional problems due to high local temperatures.)

Because of the need for compactness, there are often strong motives for placing the tower shaft close enough to the downstream stators of the first compressor, thus introducing the flow trapping problems. Determining the best location and the shape of the housing is another problem that requires intelligent compromises by all the specialties concerned with the design.

Foreign Object Damage

There are many causes of foreign object damage (FOD). When some airplanes begin their takeoff run, a veritable tornado often develops between the engine inlet and the runway. Loose objects are picked up and sucked into the inlet where they impact the rapidly moving compressor blades. Rags, tools, and other materials are sometimes carelessly left in the inlet by maintenance attendants. These materials hit the blades when the engine is started. Birds can be sucked into the inlet while the airplane is in flight. Special provisions must be made for large birds or the devastation can be catastrophic.

Engines must pass certain tests prescribed either by law or by contract to prove their ability to withstand particular types of foreign object ingestion. Some of the tests are so severe that aerodynamic performance of the fan or the first-stage compressor may have to be sacrificed. A mutual problem of the aerodynamicist and the stress analyst is to minimize this sacrifice. Unless each understands the other's field to some extent, intelligent dialogue is difficult and the probability of finding an optimum solution becomes remote.

In order to make intelligent decisions from the beginning, the stress, aerodynamic, and performance analysts must anticipate critical problems and their solution before the engine specifications are solidified. A great deal of time and money is otherwise spent overcoming unexpected obstacles.

Effects of Advanced Aerodynamic Technology

Initial efforts with advanced computational techniques have already shown that marked departures from conventional vane shapes will produce improvements in efficiency which have great financial value, particularly when fuel availability is decisive. A logical extrapolation of this result indicates that the vane and blade shapes of future engines will have different shapes than the ones now in use. The first reaction of stress analysts and manufacturers will probably reflect a complete lack of enthusiasm. Acceptance and use of new and improved ideas will come only when the three specialties share sufficient information to permit them to mutually overcome the apparent difficulties. The effectiveness of the aerodynamicist is again enhanced by his growing ability to contribute to solutions of the problems that stand between his good ideas and their ultimate use in engines.

Ceramic Turbine Blades and Vanes

The high turbine inlet temperatures of newer engines are obtained through the use of cooled turbines in which the principal material is nickel. This situation has certain disadvantages. Nickel is not native to the United States and the supply could be curtailed at any time by the whims of another nation. Moreover, the total known supply is not inexhaustable and there are competing demands for it. Nickel is heavy and expensive. The turbine cooling process limits the advantage of elevated temperatures because it lowers turbine efficiency. Recall that a good first approximation to a turbine's value is the product:

$$\text{(Turbine efficiency)} \times \text{(turbine inlet temperature)}$$

For these reasons, a great deal of time and money has been spent on the search for cheap and readily available materials that do not require cooling. Ceramics have been identified as promising candidates. These materials are extremely brittle in comparison with presently used turbine materials, which are already brittle to a troublesome degree. Extreme care must be taken in mechanical design. Even so, reductions in the allowable tensile stresses must be contemplated. These reductions can be accomplished by reducing the centrifugal forces on the blades and by minimizing bending stresses due to centrifugal and aerodynamic forces.

Either scheme offers a threat to turbine efficiency. Larger aerodynamic blade loadings may be necessary if comparatively low blade speeds are required. In addition, the radial distribution of the axial and circumferential components of flow entering and leaving the blades and vanes may differ from that encompassed by our present experience. The need for new blade shapes must, therefore, be anticipated. The development of new turbine aerodynamic technology to keep the efficiency within tolerable levels thus becomes a challenge to the aerodynamicist. Emerging computation techniques may enable this development to be realized at an affordable price. A continuing appraisal of such opportunities is requisite.

Brittle ceramic materials may not be the answer. The law of the "perversity of nature" probably applies, however, and increases in brittleness seem to be an inevitable accompaniment to increases in allowable material temperatures. An aerodynamicist should understand the interaction between his blade designs, the corresponding stresses, and their effect on forthcoming materials. He is then prepared to pursue the courses necessary to extend his degrees of freedom for the design of efficient high-temperature turbines. Remember, however, that a small reduction in the basic aerodynamic efficiency can be an almost trivial ransom for retrieving power that would otherwise be lost to cooling at very high turbine inlet temperatures.

Environmental Problems

The design freedom in propulsion engines has recently been restricted by laws and regulations enacted to protect our environment. One area of protection limits the allowable noise that various types of aircraft can generate. The other concerns certain polluting compounds that are generated in a combustor and then discharged in the exhaust.

Noise is discussed in Chapter 7 of Ref. 2 and emissions in Chapter 1 of Ref. 2. The noise considerations described below illustrate one of the conflicts of interest in this area.

Acousticians and vibration analysts would like to see successive blade rows placed far apart. Time-steady flow in a rotor produces time-unsteady flow in a stator and vice versa. The closer the two elements are to each other, the greater the forces causing vibration and noise become.

The aerodynamicist and the weight analyst, on the other hand, prefer to have them close together. The effect on engine weight is obvious. As far as aerodynamics is concerned, there is experimental evidence that placing rotors and stators close together benefits performance. The optimum solution is found when all four specialties share their expertise to meet all of the objectives of the engine. A new interface is thus introduced by environmental considerations.

As a general note, proposed regulations are published and comments by the affected parties are invited. The participants in these areas should have a thorough understanding of existing regulations and be constantly alert to proposed changes. They must also be aware of the attitudes of key people involved in offering and adopting regulations, so that improvements in technical competence can be directed to those areas where problems are anticipated.

1.9 ADVANCED FLOW CALCULATIONS

The art of flow calculations is advancing at a rapid rate. The revolution in digital computers has inspired an accompanying revolution in the art of numerical analysis. Even a lengthy treatise at this point would do an injustice to this subject. Just enough information will be given to describe what lies behind some of the ongoing activities.

Most of the new procedures embody some form of the Navier-Stokes equations that include the Reynolds turbulence stresses. Some models for estimating the distribution of turbulence are also required. The available models are approximate and subject to opinions and improvements. Nevertheless, some very useful advances have been made with the crude turbulence models now existing. Several methods of calculating a flowfield can be developed from this base. Some procedures use the concept of boundary layer.

Another technique that was recently introduced borrows an idea from electromagnetic theory. The flow vector is divided into two components—one with no curl and the other with no divergence. The latter vector describes all the vorticity in the flow. In some respects, it is really a generalization of the boundary-layer concept. Compatibility conditions between the two vector fields are required. The first field is mixed elliptic-hyperbolic and can be solved by known "relaxation" procedures. The second field is essentially parabolic and is solved by "marching methods." (An upstream vector does not know what a downstream vector is doing.)

Two classes of computational techniques are being pursued. One class begins by assuming the flow to be at rest and imposes the boundary conditions. The time-unsteady equations of motion then estimate the acceleration of the flow until it reaches equilibrium. The other class uses finite-difference or finite-element methods. Transformation of coordinates as a function of the local Mach number and the location of the point of interest in the flowfield may be used. Methods of taking the finite differences or of describing the finite elements also vary with both the local Mach number and the point in the field.

All the programs are very involved and their preparation presently requires an outstanding ability in computer programming, numerical analysis, and fluid mechanics. Because of the enormity of the problem, existing programs must be considered to be only partially developed. Even in this crude state, their use has, for example, indicated ways of designing turbine vanes so that the secondary-flow losses are reduced. The experimental test of the resulting turbine was more than gratifying.

This field is moving forward. Useful new concepts are continually being disclosed. As a result, the ability to accurately analyze complex flows in detail is noticeably improving from year to year. Close attention must be paid to this activity.

The pursuit of advanced three-dimensional analysis will bring about step improvements in turbomachinery performance. Side benefits will be the reduction of expensive testing and the delineation of forcing functions that affect blade vibration.

1.10 BIOGRAPHY OF A TYPICAL ENGINE

This concluding section of the chapter is presented to show that all specialties have a long way to go before we can truly say that engines can be accurately designed. A lot of development work needs to be done in order to produce a successful engine. Most of it has to be performed by "greasy-fingered" engineers, who are often forced to work with trial-and-error methods. They need all the useful help they can get from the various specialties for reducing weight, improving fuel consumption, and increasing both reliability and component life.

With respect to the biography, observe first that research is continually being directed toward improving the capabilities of aircraft and engines. General analyses of these results are periodically made to examine the feasibility of improving existing airplanes, of making an airplane that will either perform a useful service more economically, or of providing one that was hitherto unavailable.

When the feasibility of an improved engine/airframe system is determined, an analysis similar to that presented in Chapter 3 is initiated. Engines and airplanes having the proposed advanced concepts are simulated on large digital computers. Estimates are made of the properties and behavior of every significant part of the engine and aircraft. Vital features of the synthesized airplane are calculated along its proposed flight paths. Numerous modifications of the synthesized engines and aircraft are examined until the calculated optimum reliable airframe/engine combination for the proposed mission is found.

The results are then reviewed to determine the worthiness of the airplane. This review includes an estimate of the cost of designing and developing the engine and airplane, the cost of manufacturing the desired number, and the cost of operating and maintaining them over their expected life. These economic evaluations (which are often identified by the words "cost of ownership," "life-cycle-costs," or "return on investment") are repeated throughout the life of the enterprise.

If a decision is made to proceed with the design and development of an engine, the initial specifications are provided by the preceding studies. Tests are begun on the various components to evaluate their effectiveness: i.e., how does the performance compare to that expected? Several design modifications are frequently required to realize the initial expectations.

After the components are functioning reasonably well, they are assembled into an engine to evaluate the overall performance and reliability. Parts of new engines usually operate in more hostile environments than any of their predecessors and unanticipated interactions among the parts (both aerothermodynamic and mechanical) are occasionally encountered. Many events can and do happen. Some problems can be rectified by minor changes in the engine design. Others may require major redesigns that need to be re-evaluated on component rigs. In some instances, added required costs of time and money can put the project in jeopardy.

When the engine performs its functions on a test stand with sufficient reliability, it is ready for flight tests. New interactions between the airframe and the engine introduce additional problems. This is particularly true when there is no previous experience with the maneuvers and flight regimes demanded of the airplane. Expedited rebirth and development of critical engine components have been necessary even at this late phase of development. Dedicated and competent engineering teamwork usually solves the problems, finally producing an airplane that meets its specifications.

After an engine has been certified by a government bureau, it enters its intended service. When many hours of flight have elapsed, new major problems have suddenly appeared. Although most of these are mechanical (related to fatigue), they probably represent a poor tradeoff, initially made with incomplete data, between aerodynamics and stress. New evaluations and new designs have to be made and developed in a hurry to solve the problems. Again, it is the teamwork of many talents that overcome the difficulties.

Eventually, even better performance is wanted and advanced development begins. The engine is gradually and continually improved by new technology. This phase may last 20 years or more. Weak parts are strengthened to increase the time between overhauls (TBO). The weights of other parts are decreased by improved design or by material substitution. Superior aerothermodynamics are developed that can, for instance, raise the output without objectionable increases, or even reductions, in fuel consumption. By the time the engine is eventually superseded by one from a new generation, its output may have been increased 100% with only small increases in size and weight and little, if any, increase in specific fuel consumption.

The events described in this review are not atypical. Including them is not meant to disparage any work of the past or discourage bold ventures in the future. Rather, it is intended to emphasize the need for recognizing problems as early as possible during design and development. It also points out the value of engine/airframe designers having the wide scientific background necessary for recognizing the source of a variety of interrelated problems and solving them.

Of equal importance, it emphasizes the need of basic research for anticipating and solving problems on inexpensive rigs rather than expensive engines. Most of this research is not glamorous and does not provide headlines. Its ultimate financial value cannot, however, be overemphasized.

APPENDIX: CENTRIFUGAL COMPRESSORS

Centrigugal compressors are being used more and more in some types of propulsion engines. This trend was not foreseen when the original outline for these volumes was conceived, but it is now evident that some remarks on this subject are necessary. This chapter seemed the suitable place for these comments.

Background Information

Centrifugals were the first usable compressors that did not rely on a positive displacement principle. The early ones were extrapolations of centrifugal pumps and the designers were unaware of the implications of compressibility, boundary layers, and three-dimensional flow gradients. Because of centrifugal force, however, usable increases in pressure could not be denied. Their first appearance in aircraft propulsion systems was as superchargers for reciprocating engines. As a result of continued development, it became reasonable to try to build a jet engine around them. This venture was successful; as a result, the first generation of gas turbine propulsion engines used centrifugal compressors. However, subsequent large engines used axial flow compressors. They were much smaller in diameter and a simple design theory derived by perturbating two-dimensional airfoil theory led to higher efficiencies. Axial compressors proved to be less rugged and very sensitive to tolerances when their size was reduced for low flow rates, so centrifugal units continued to be used in small gas turbine engines.

There is no sharp line that distinguishes centrifugal and axial flow compressors. One rarely finds a pure example of one or the other. The same equations of motion apply to both, but many of the physical laws are most usefully expressed in different forms. This is illustrated in the following discussions.

Blade Loading and Static Pressures Rise in Axials

As indicated by Fig. A1.1 the flow passages between the blades of an axial unit are really diffusers. The increase in available flow area from A_1 to A_2 is responsible for any rise in static pressure. The curvature of the streamlines accompanying changes in the flow direction instigate the pressure and velocity gradients normal to the flow. The tangential forces on the blades are thus established. The velocity gradient normal to any streamline is

$$\frac{dV}{dn} = \frac{V}{R_c}$$

Imagine a mean streamline with a radius of curvature of $R_{c,M}$ and a mean fluid velocity of V_M. The difference between the velocities on the two sides of the passage is proportional to $\Delta n(V/R_c)_M$. The magnitude of the blade loading depends on the differences of the squares of the velocities on either side which is

$$\frac{1}{2}(V_1^2 - V_2^2) = V_M \, \Delta V \approx V_M^2 \frac{\Delta n}{R_{c,M}}$$

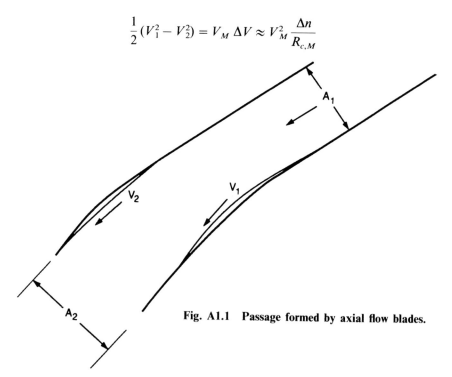

Fig. A1.1   Passage formed by axial flow blades.

The important features of the axial compressors just noted are:

1) The magnitude of ΔV is proportional to the velocity itself.

2) Any event, such as a thickening boundary layer, that reduces streamline curvature also reduces ΔV.

3) The static pressure rise across the vanes depends upon the ratio A_2/A_1 and upon the square of the relative velocity; boundary-layer growth is detrimental to this result also.

Blade Loading on Centrifugal Units

Figure A1.2 shows the passage between two adjacent radial blades of a centrifugal impeller. The blades rotate at an angular velocity of ω_0. The tangential component of velocity relative to the blades is practically zero, although the flow between the blades is actually whirling with the opposite angular velocity as the blades. The field of centrifugal force $\omega_0^2 r$ produces a static pressure rise in the direction of flow. Note also that the flow area increases in the direction of the flow because this flow is assumed to be two-dimensional, so diffusion is also involved. This causes an added increase in static pressure rise, which is small in comparison to that due to centrifugal force.

A tangential pressure gradient proportional to the Coriolis force ($2\omega_0 V_r$) is responsible for the tangential blade loading. The relative flow has the vorticity of $-2\omega_0$ when the entering flow has an absolute vorticity of zero. (Be aware that tangential gradients in entropy can also produce vorticity.) We may postulate that this relative vorticity causes the velocities adjacent to the two sides of the blades to differ by a quantity that is proportional to both ω and the width of the passage. Consider now the element ABCD of Fig. A1.2a. We have presumed the relative velocities to be radial, so the circulation about ABCD is

$$\Gamma = -2(\Delta V)(\Delta r)$$

and the relative vorticity is

$$-\frac{2(\Delta V)(\Delta r)}{\ell \, \Delta r} = -2\omega_0$$

Hence, $\Delta V = \ell \omega_0$.

Notice that the magnitude of ΔV is independent of the relative velocity. This is completely different from our observation for axial equipment. If V_M happens to equal ΔV, the flow is brought to rest. Lower values of V_M or higher values of ℓ cause a reversal of the flow along the leading side of the blade. Finite-difference analyses show that an eddy similar to that indicated in Fig. A1.2b would form an ideal flow. In real flows, this eddy would be washed away and a stable flow with entropy gradients, such as that shown in Fig. A1.2c, would appear.

This conjecture is supported by measurements made on a large rotating impeller (Ref. 9) and shown in Fig. A1.3a. Lines of constant local adiabatic

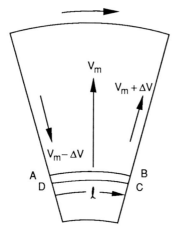

a) Velocity induced by relative vorticity.

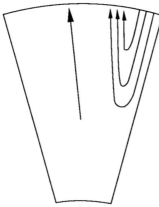

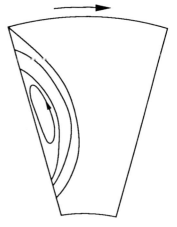

b) Ideal eddy of low throughflow rates.

c) Separation of real flows at low flow rates.

Fig. A1.2 Effects of relative circulation in radial flow impeller.

efficiency are plotted and we observe that low efficiencies denote regions where the velocities are low or even reversed. Note also that entropy gradients are almost perpendicular to the streamlines. These gradients relieve at least part of the relative vorticity requirement.

Figures A1.3b and A1.3c illustrate two experiments to increase V_M in the outer half of the passages by altering the flow areas. Some success is achieved, but flow separation seems to appear at the trailing surface near the tips due to the adverse velocity gradients that arise as the blades become unloaded.

The difference in the pressures on the two sides of a passage is proportional to the difference in the squares of $(V_M \pm \Delta V)$ and is

$$V_B^2 - V_A^2 = 4V_M \, \Delta V = 4\omega_0(V_M \ell b)/b$$

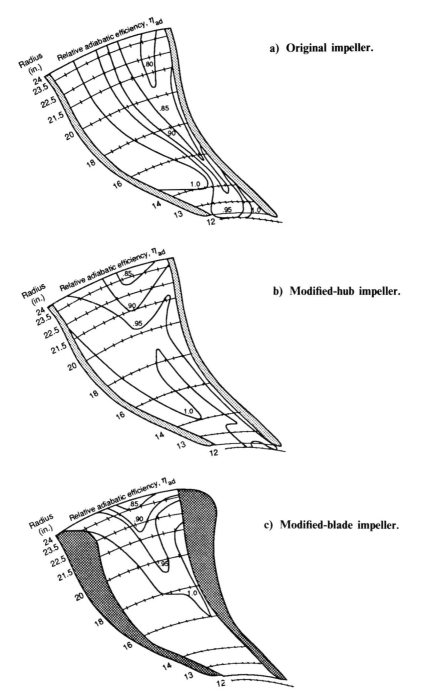

a) Original impeller.

b) Modified-hub impeller.

c) Modified-blade impeller.

Fig. A1.3 Relative adiabatic efficiency distribution throughout impeller rotating passage at corrected weight flow at 26 lb/s.

The quantity in the parentheses represents the volume flow rate in two-dimensional flow and the pressure difference is proportional only to the angular velocity and the volume flow rate. The configurations of Figs. A1.3 experienced the static pressure distributions shown in Fig. A1.4. Although the magnitude of the pressures varies from one unit to another, the pressure differences on the two sides are comparable at a given flow rate. This is another departure from axial flow experience We also note that this result implies a correlation between b and the effective value of V_M.

The change in the magnitude of pressures for the three configurations reflects the difference in the radial velocities caused the the changes in flow area. Figure A1.5 is the same as Fig. A1.4, except for the added calculated lines of constant relative Mach numbers shown. The effectiveness of changing this Mach number are readily perceived. These lines show the static pressure rise resulting from centrifugal force; it is obvious that extreme disturbances in velocity are necessary to negate this effect.

Factors Affecting Axial Flow Distributions in Centrifugal Rotors

The large changes in fluid angular velocity found in centrifugals produce gradients in axial velocity that can be bothersome. The cause is a requirement of radial equilibrium, discussed in Chapter 3 of Ref. 2. Space is devoted to the subject here to stress the implications that particularly afflict centrifugal compressors. Simple assumptions are made in order to focus directly on the problem.

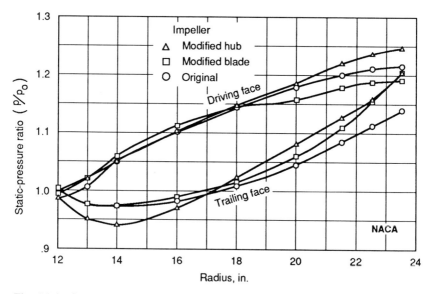

Fig. A1.4   Static pressure distribution near blade surfaces for three impeller configurations at corrected weight flow of 26 lb/s.

Assume we have a set of rotating blades that increase the angular velocity of the fluid passing through them from zero to that of the blades. Assume further that the throughflow has an axial direction, the radial components being negligible. An applicable equation for equilibrium is

$$-\frac{1}{\rho}\frac{\partial p}{\partial r} = -\frac{V_\theta^2}{r} = V_z\frac{\partial V_z}{\partial r} + V_\theta\frac{\partial V_\theta}{\partial r} - \frac{\partial}{\partial r}(\omega_0 r V_\theta)$$

Suppose that the tangential component of velocity is expressed by

$$V_\theta = \Gamma/r + \omega_a r, \qquad \Gamma = \text{const}$$

We can then derive the equation

$$V_z\frac{\partial V_z}{\partial r} = 2\omega_a(\omega_0 - \omega_a)r - 2\omega_a\frac{\Gamma}{r}$$

We note that V_z is unchanged when $V_\theta = \Gamma/r$, which is a design objective for many axial turbomachines. This equation becomes

$$\frac{1}{2}(V_{z,2}^2 - V_{z,1}^2) = \omega_a(\omega_0 - \omega_a)(r_2^2 - r_1^2) + 2\omega_a\,\pi\ell n\left(\frac{r_1}{r_2}\right)$$

Subscripts 1 and 2 refer to the hub and casing, respectively.

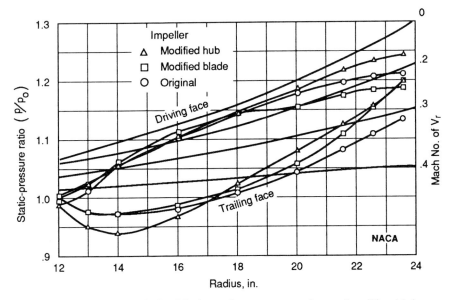

Fig. A1.5 Constant relative Mach number curves superimposed on Fig. A1.4.

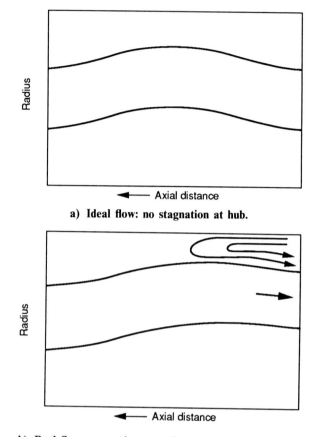

a) **Ideal flow: no stagnation at hub.**

b) **Real flow: separation at casing precludes stagnation at hub.**

Fig. A1.6   Streamlines when tangential velocity for solid-body rotation is increased from zero to wheel speed.

The following remarks apply when $\Gamma = 0$. Observe that $V_{z,2} = V_{z,1}$ when $\omega_a = 0$ or ω_0. The difference in velocities is a maximum when $\omega_a = 0.5\,\omega_0$. Figure A1.6 suggests a shape required for the streamlines in order to effect changes in the axial velocity. The necessary radial flow is not negligible and it does reduce the calculated difference in the axial velocity. The existence of significant radial gradients in the axial component of the velocity that are more or less independent of the magnitude of the mean velocity is still valid, however. The situation is not unlike that shown in Fig. A1.2. When the mean velocity becomes too low or the difference in radii too great, the equilibrium equation suggests an imaginary velocity at the hub. Detailed calculations with exact inviscid equations indicate that a stationary eddy forms there. Such an eddy would be unstable, so the real flow behaves as shown in Fig. A1.5b. Flow separation at the casing removes the necessity for the eddy at the hub.

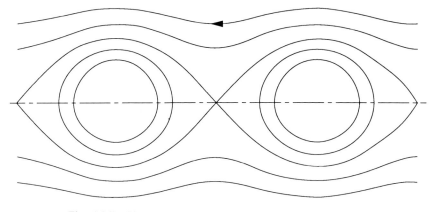

Fig. A1.7 Flow about equispaced blades of an axial unit.

Special Flowfields in Centrifugals

The effect of an element of surface on a flow can be represented by a vortex of suitable strength. For the blades of an axial unit, it is correct to use an infinite row of equally spaced vortices. Figure A1.7 shows the flowfield of this array. An almost uniform stream to the left is induced at a distance above the vortices; an equal but opposite flow appears below. Radial flow equipment requires the vortices to be equally spaced on a

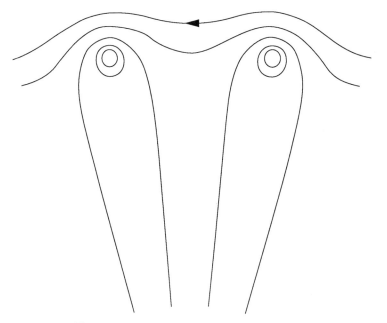

Fig. A1.8 Radial flow vortices around a circle.

circle. Figure A1.8 shows the field generated by 19 such vortices. In both figures, the field in the immediate neighborhood of a vortex resembles that of an isolated vortex. The external flow of Fig. A1.8 is not too dissimilar from that on one side of Fig. A1.7. The internal flow is entirely different. One is thus suspicous about the use of simple procedures for applying two-dimensional cascade data to situations having strong radial velocity components.

In summary, the following points can be made:

1) Part of the blade loading in centrifugal compressors is due to factors that are independent of gas velocity. Gas may appear to stagnate on one surface of a blade, but a disturbed flow, with entropy gradients, appears on the opposite surface.

2) Angular accelerations, with radial gradients in axial velocity may induce flow to stagnate at the hub; the resulting increases in entropy appear at the casing.

3) Most of the static pressure rise in the impeller is a result of centrifugal force rather than diffusion.

4) Blade surface elements of centrifugals influence the flow differently from similar elements in axials.

Fig. A1.9   Centrifugal compressor and diffuser.

Outline of Design Procedures

Centrifugal compressors are potentially capable of efficiently generating a high pressure ratio in a short single stage. Their diameters must exceed those of equivalent axial compressors, but they can be smaller than those of other engine components—a fan, for example. Figure A1.9 shows a centrifugal compressor that has been in production for a number of years at the Garrett Corporation. Its principal parts are identified in the photograph. The pressure ratios and efficiencies produced by this unit are shown in Fig. A1.10. The blades in this impeller are curved at the outlet. Because of the direction of curvature, this is known as a "backward-curved impeller." Many other impellers have straight radial blades at the outlet.

One of the principal virtues of centrifugal compressors has been their simplicity. One continuous set of blades often comprises the rotor. Another continuous set frequently forms the stator or diffuser. The number of blades of the rotor and stator are limited by the area required at their inlets. (The allowable height of the inlet blades of the rotor is determined by Mach number limits and stress.) If too many blades are used at the inlets, the compressor will not move the required amount of air.

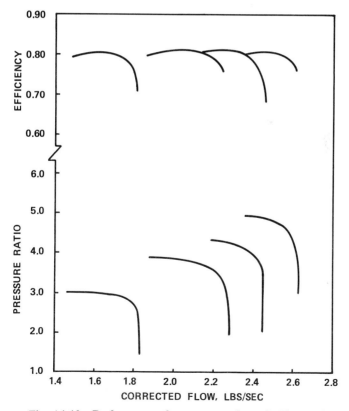

Fig. A1.10   Performance of compressor shown in Fig. A1.9.

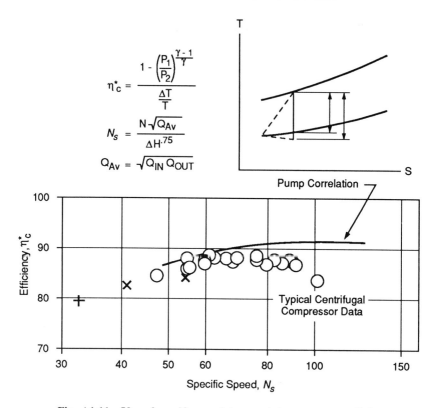

Fig. A1.11   Use of specific speed to correlate compressor efficiency.

Impellers.   Selecting a speed and size conducive to good efficiency is the first problem. Remarks on the subject of speed selection were made earlier in this chapter, but we have a benefit from the legacy of pump design to assist us with centrifugal compressors. An excellent correlation of pump output, flow rate, and speed was achieved with a quantity called specific speed and defined by

$$N_s = N \frac{\sqrt{Q}}{(\Delta H)^{3/4}}$$

This ratio is dimensionless when ΔH represents energy per unit mass.

We have a problem in applying this to compressors because volume flow rate decreases as gas flows through a compressor. The definition of efficiency is also unclear because η_{ad} and η_c^* are indistinguishable in pumps, but differ for gases. Trial and error has yielded the correlation shown in Fig. A1.11. The agreement with accepted pump correlation is instructive. Note that the pump results included mixed-flow units which are very efficient, especially at the higher specific speeds. The use of this curve or similar data enables a rotating speed to be selected with the expectation

that good efficiencies can be obtained with proper design. Stress requirements are assumed to be satisfied.

The diameter of the impeller depends on the backward curvature, if any.
Our present knowledge suggests that the maximum obtainable efficiency
does not depend on this curvature. The maximum efficiency with straight
radial blades, however, is practically at the surge point; the point of
maximum efficiency shifts to higher flow rates as the backward curvature
increases. The backward curvature is selected from this consideration. A
deviation angle (or slip factor) is estimated and the diameter to produce the
required energy is then calculated.

The velocity distribution through a proposed design should be estimated,
at least by a quasi-three-dimensional design procedure for ideal flows, to
avoid the unhappy consequences of stagnated flows. A three-dimensional
finite-difference or finite-element procedure is preferred because existing
published correlations of deviation and optimum incidence angles are not
always satisfactory. Supersonic relative inlet flows must be estimated, so the
calculation procedure should be valid for supersonic velocities. (Excellent
experimental results have been obtained with relative inlet Mach numbers
exceeding 1.3.)

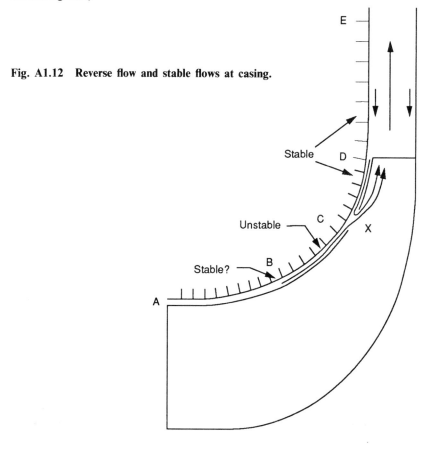

Fig. A1.12   Reverse flow and stable flows at casing.

Defining a design for turning the flow from an axial to a radial direction and providing the angular acceleration is often elusive. The problems illustrated by Fig. A1.2 can combine with those shown in Fig. A1.6. Moreover, the gradients of the latter figure are augmented by changing the velocities from an axial to a radial direction. Approximate analyses can screen a number of proposed arrangements before they are submitted to more detailed calculations.

It is also worthwhile to consider the stability of the interface between the main flow and the boundary layer at the casing. The discussion of secondary flows noted that stable boundary layers are the result of entropy gradients which are opposite to static pressure gradients and that stable boundary layers are displaced by other components of the pressure gradient. Either flow separation or secondary flows can appear. The static pressure gradient at the casing can be directed inward in the area roughly denoted by B in Fig. A1.12. This is a result that should be avoided.

The interface is usually unstable in the region indicated by C because a significant part of the centrifugal gradient is normal to the casing surface. This normal gradient weakens as the flow direction becomes more radial and a stable interface appears at D and continues throughout the rest of the compressor. Oil traces on the casings of many impellers show the flow spiraling inward from the diffuser to a circumference indicated by X in Fig. A1.12. An encounter with flow spiraling outward is visible and a circular ring of oil is left on the casing. The projection of the flow velocities apparently follow the sketched path shown in the figure.

Observe that boundary-layer movement at the casing is affected by the clearance between it and the rotor. Excessive clearance reduces the angular velocities along the shroud and invites radially inward flows. We should also realize that rotors suffer a mechanical deflection at high speeds and that pressure and temperature gradients cause complex displacements of the casing. Monitoring rotor clearance at several places along the shroud during development is strongly recommended, lest rubbing at one place is avoided at the expense of excessive clearance at another.

The inward flow of the casing boundary layer beyond D is a result of the centrifugal force of the swirling flow which diminishes as the casing is approached. This has to be accepted if this general compressor configuration is wanted. The use of mixed-flow impellers (Fig. A1.13) provides relief. The centrifugal gradient is normal to the wall and thick boundary layers are suppressed, provided that stable secondary flow vortices have not been produced previously. Two wheels joined together are shown in the sketch; this reduces weight and hub stresses.

It has been noted that the flow rate determines the minimum spacing of the blades at the inlet. The spacing naturally increases with radius and may become so wide near the tip that adverse velocity gradients are unavoidable. Splitter vanes such as those shown in Fig. A1.14 are then needed. These vanes should not divide the flow passage into two equal parts. Instead, they should be positioned to divide the mass flow rate into two equal parts and to provide equal flow area at the perimeter. Otherwise, the flow will adjust itself so that the static pressure at the perimeter is uniform.

This could require bad incidence angles with losses at the leading edge of the splitters. In some units having very high pressure ratios, another set of splitters could be useful.

Diffusers.   Diffusers reduce the high gas velocities downstream of an impeller to a level that can be managed by the following component. The simplest diffuser is a vaneless diffuser consisting of two nearly parallel plates. Any gas then discharged from the impeller flows radially outward between the plates. Conservation of angular momentum automatically lowers the angular velocity of the gas as the radius increases; the distance between the plates controls the radial component of velocity. These diffusers alone are never used in aircraft engines—the diameters and weight are too large and the losses are high because of the large areas exposed to surface friction.

Vaneless diffusers are useful during initial impeller tests because the range between maximum and surge flow can be made to depend on only

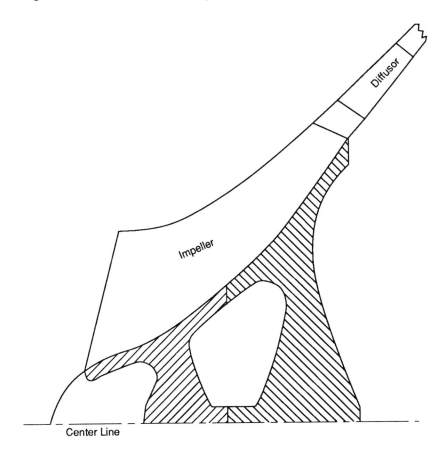

Fig. A1.13   Mixed flow compressor.

Fig. A1.14   Impeller with splitter vanes.

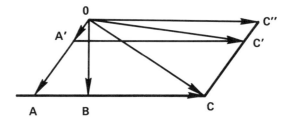

Fig. A1.15   Velocity diagram at rotor exit.

the impeller. The potential of the impeller alone can be developed. Tests with vaneless diffusers also provide valuable insights about flow anomalies created by an impeller. Early tests showed that entropy rose rapidly just downstream of an impeller. Thereafter, the rate of increase could be reduced to the level expected from friction. The initial losses were attributed to mixing, since the laws of angular momentum and flow continuity cause fluids having different velocity levels to be on a collision course and the forced mixing rapidly exchanges momenta. Similar tests showed that when the space between the front and rear diffuser plates converged immediately after an impeller to reduce the flow area by 20% or more, the overall losses were lowered. Too much reduction caused a nonproductive increase in friction losses. It was believed that the initial reduction in spacing naturally reduced momentum losses and also caused the boundary layer on the plates to be energized through an unstable interface. This increased the efficiency of the ensuing diffusion. Other phenomena unique

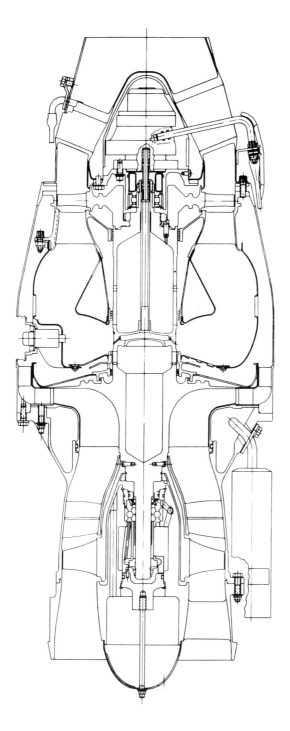

Fig. A1.16 Engine for a guided missile.

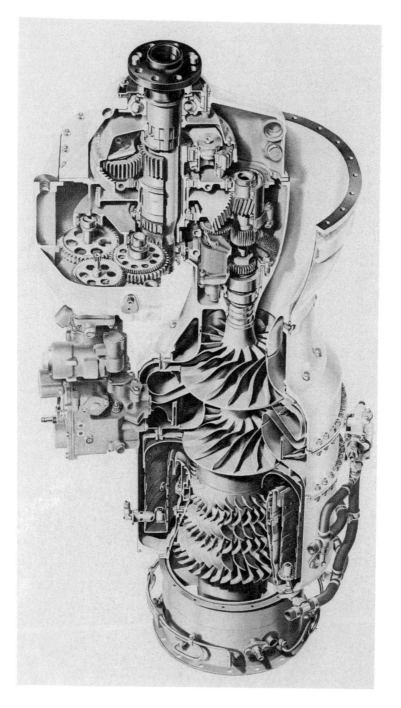

Fig. A1.17 Turboprop engine.

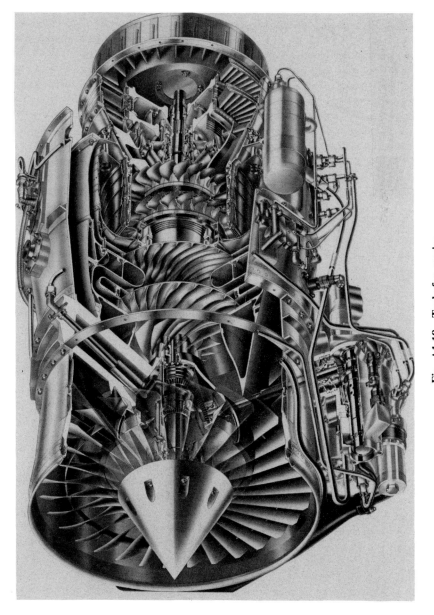

Fig. A1.18 Turbofan engine.

to the design under test are sometimes disclosed during tests with a vaneless diffuser.

Vanes must be used in practical diffusers to recover pressure within a reasonable size. The diffusers illustrated in Fig. A1.9 are known as the vane island type; other types are identified as pipe and cascade diffusers. Pipe diffusers are really conical passages provided at the periphery. Cascade diffusers are formed by several rings of diffusing elements—each ring consisting of a number of equally spaced airfoils.

All of these types offer comparable performance, although the cascade type seems to yield somewhat higher efficiency when each ring is experimentally aligned with the local flow and the leading edge of the first row of blades is twisted to accommodate the gradient in the incoming flow angles. These diffusers are the most expensive to make and are the most easily eroded by high-velocity particles passing from the impeller. The vane island diffuser uses the fewest number of airfoils. It also creates pressure fields that extend almost to the impeller inlet, which can be measured on the casing. Whether the gradients do good or harm is not known.

The familiar limits on diffusion rates seem to apply to these diffusers, although centrifugal pressure gradients do help. The proper orientation of the leading edges of diffusers is, at present, the biggest aerodynamic problem. Proper orientation is accomplished by tests rather than design because diffuser inlet flow is time unsteady and skewed. Consider Fig. A1.15, which indicates the velocity vectors near an impeller tip. Line O-A designates the ideal relative velocity vector leaving the impeller and line O-C the absolute velocity. Retarded flow in the impeller is shown by lines O-A' and O-C', while nearly stagnated flow on the impeller appears as line O-C". The main flow at the diffuser inlet oscillates between lines O-C and O-C" as one blade after another passes. Near the casing, the component A'-C' and the radial component are diminished by friction. Trial-and-error methods using visual aids seem to be as effective a development procedure as any.

Examples of Application to Propulsion Engines

Figure A1.16 shows the application of a centrifugal to a jet engine for a guided missile. The entire engine was designed to meet a difficult requirement for low costs, so the whole assembly is simple. The centrifugal compressor provides most of the compression and its centrifugal pressure gradient enabled the engine to start, accelerate, and run in spite of severe inlet distortions during starting.

Figure A1.17 is a turboprop engine. Two compressors are used to attain a pressure ratio of over 12:1. The passage between the two compressors is called the crossover ducting and it offers interesting design problems. We should remove as much as possible of the swirl from the first unit by the time air reaches the top of the crossover system. Otherwise, the angular momentum would be preserved and the angular velocity entering the second rotor would diminish the flow rate and pressure ratio of that unit. Observe that both impellers were backward curved. Good efficiencies were realized with a comfortable surge margin. Observe that the inlet duct is curved near the impeller inlet. The impeller encountered undesired velocity gradients, although there were practically no entropy gradients. The first compressor was fortunately forgiving and no sensible effect on performance was measured.

It is instructive to compare the size of the turbine engine with that of the gears and accessories. This emphasizes a usefulness of low-speed turbines for driving propellers.

A turbofan, Fig. A1.18, is the third example. This photograph shows that centrifugal units can be produced that do not dictate the diameter of turbofan engines. In comparison to an axial compressor, the compressors are very compact. The overall efficiencies of the centrifugal may be somewhat lower, but this probably reflects the lack of attention that has been given centrifugals. Better performance is expected in the future.

References

[1]Oates, G. C., *Aerothermodynamics of Gas Turbine and Rocket Propulsion* (revised and enlarged edition), AIAA Education Series, Washington, DC, 1988.

[2]Oates, G. C. (Ed.), *Aerothermodyamics of Aircraft Engine Components*, AIAA Education Series, New York, 1985.

[3]Nicolai, L. M., *Fundamentals of Aircraft Design*, School of Engineering, Univ. of Dayton, Dayton, OH, 1975.

[4]Dix, D. M. and Gissendanner, D. A., "Derivative Engines versus New Engines: What Determines the Choice?," Trans. of the American Society of Mechanical Engineers, New York, Oct. 1985.

[5]Keenan, J. H., Chao, J., and Kay, J., *Thermodynamic Properties of Air, Products of Combustion, and Component Gases* (2nd Ed.), Wiley, New York.

[6]Irving, T. F., Jr. and Liley, P. E., "Steam and Gas Tables with Computer Equations," Academic, Orlando, FL, 1984.

[7]Sonntag, R. E. and Van Wylen, G., *Introduction to Thermodynamics, Classical and Statistical* (2nd Ed.), Wiley, New York.

[8]*Journal of Engineering for Gas Turbines and Power*, Transactions of the American Society of Mechanical Engineers, Jan. 1986.

[9]Michel, D. J., Mizisin, J., and Prian, V. D., "Effect of Changing Passage Configuration on Internal-Flow Characteristics of a 48-Inch Centrifugal Impeller. I—Change in Blade Shape," NACA TN-2706, 1952.

CHAPTER 2. TURBOPROPULSION COMBUSTION TECHNOLOGY

Robert E. Henderson and William S. Blazowski*
Wright Research and Development Center, Aero Propulsion and Power Laboratory, Wright-Patterson Air Force Base, Ohio

*Currently with Exxon Research and Engineering Company, Florham Park, New Jersey.

2
TURBOPROPULSION COMBUSTION
TECHNOLOGY

2.1 INTRODUCTION

The evolution of aircraft gas turbine combustor technology over the past 50 years has been extremely impressive. In 1939, the combustion system was the primary limitation in development of the first aircraft gas turbine (see, e.g., Ref. 1). In fact, the first German jet engine, which was bench-tested by von Ohain in 1937, was fueled with hydrogen because of difficulties in designing a suitable hydrocarbon-fueled combustor. Now, however, the complexity and hardware costs associated with current rotating engine components (compressor and turbine) far exceed those of the combustion system. Recent developments, however, have once again caused significant shifts in development emphasis toward combustion technology. New concepts and technology improvements will be continually needed to satisfy legislated exhaust pollutant regulations. Moreover, future emphasis on engines that can utilize fuels with a broader range of characteristics are expected to require additional combustor technology development.

Beyond these externally imposed requirements are the combustion system performance improvements necessary to keep pace with new engine developments. Further reductions in combustor physical size and weight are expected to continue as firm requirements. Performance improvements, especially in respect to engine thrust/weight ratio and specific fuel consumption, will require higher combustor temperature rise, greater average turbine inlet temperatures, and closer adherence to the design temperature profile at the turbine inlet. High-performance designs must also permit greater Mach number operation within and around the combustor to reduce pressure drop and minimize the physical size of compressor exit diffuser hardware. Costs (both initial and operating) must be minimized, as recent experiences with high-temperature engines have confirmed the necessity to consider reliability and maintenance aspects of life cycle cost, as well as performance and fuel consumption.

The purpose of this chapter is to introduce the reader to the hardware aspects of aircraft gas turbine main burners. Fundamental aspects have been addressed in Chapter 1 and afterburners in Chapter 2, both of Ref. 2. A number of reference texts[3-8] address various aspects of turbopropulsion combustion in a detailed manner. In particular, Ref. 8 cites more than 700 reports and technical articles on the topic of turbopropulsion combustion. This chapter will cover the following four topics: 1) description of various hardware types and definition of all terms of importance, 2) review of parameters pertinent to performance, 3) discussion of tools available to the combustor designer, and 4) review of the future requirements of exhaust emission reduction, achievement of greater fuel flexibility, and advancement of burner performance.

2.2 COMBUSTION SYSTEM DESCRIPTION/DEFINITIONS

In order to appreciate and comprehend contemporary turbopropulsion combustor design philosophy, a number of general design and performance terms must be understood. The purpose of this section is to acquaint the reader with the commonly used combustion nomenclature utilized throughout this chapter. A brief description and/or definition of combustion system types, sizes, configurations, and flow path terminology is given in the following subsections.

Types

Turbine engine combustors have undergone continuing development over the past 50 years, resulting in the evolution of a variety of basic combustor configurations. Contemporary combustion systems may be broadly classified into one of the three types schematically illustrated in Fig. 2.1.

Can. A can combustion system consists of one or more cylindrical combustors, each contained in a combustor case. In the small T-63 turboshaft engine of Fig. 2.2, a single combustor can is used, whereas larger propulsion systems use a multican assembly in an arrangement designed to provide a continuous annular gas flow to the turbine section. The combustion system of the J33 engine illustrated in Fig. 2.3 is representative of such multican systems.

Cannular. This combustion system consists of a series of cylindrical combustors arranged within a common annulus—hence, the cannular

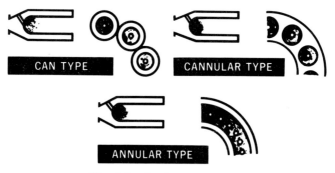

Fig. 2.1 Combustor types.

name. This combustor type is the most common in the current aircraft turbine engine population, but it is rapidly being replaced with the annular type as more modern engines comprise larger portions of the fleet. The J79 turbojet engine main combustor, illustrated in Fig. 2.4, exemplifies cannular systems.

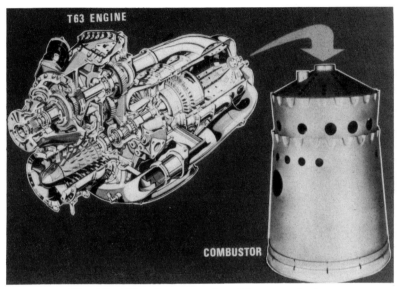

Fig. 2.2   T63 turboprop engine with can combustor.

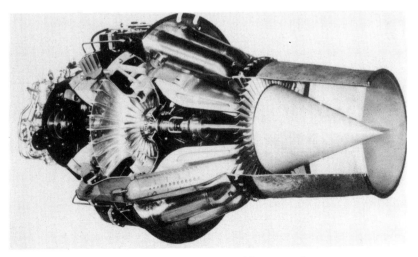

Fig. 2.3   J33 turbojet engine with can combustors.

Fig. 2.4   J79 cannular combustor.

Annular.   Most modern combustion systems employ the annular design
in which a single combustor having an annular cross section supplies gas to
the turbine. An example of this combustor type, the TF39, is illustrated in
Fig. 2.5. The improved combustion zone uniformity, design simplicity,
reduced liner surface area, and shorter system length provided by the
common combustion annulus has made the annular combustor the leading
contender for all future propulsion systems.

Size

Contemporary combustion systems may come in a variety of sizes
ranging from the small 2.3 kg/s (5 lbm/s) annular burner of the WR-19
engine (Fig. 2.6) to the large 110 kg/s (242 lbm/s) annular combustor of the
JT9D engine (Fig. 2.7). The WR-19 combustor is approximately 25.4 cm
(10 in.) in diameter and intended principally for missile and remotely
piloted vehicle engine applications; the JT9D combustor is approximately
91 cm (36 in.) in diameter and is used in engines to power the wide-body
747 and DC-10 class aircraft.

Combustor Configurations

Combustion system configuration may also be classified according to
airflow direction through the chamber.

Fig. 2.5   TF39 annular combustor.

Axial throughflow. The most common configuration is the axial throughflow design where combustion air flows in a direction approximately parallel to the axis of the engine. The JT9D annular burner illustrated in Fig. 2.7 is typical of an axial throughflow configuration.

Reverse flow, folded. Engines with centrifugal compressors often employ compact, reverse flow combustors. In these combustion systems, air is passed along the outside of the burner and then turned to flow through the combustion chamber. The combustion gases are then turned once again to pass through the turbine. Hence, the air is required to make two 180 deg reversals in moving from the compressor to the turbine. The reverse flow configuration is often employed to minimize engine length, especially in small turboshaft and turbofan engines where propulsion system length is an important design factor. Figure 2.8 illustrates the TFE-731 combustor, a typical reverse flow configuration.

Radial inflow or radial outflow. The radial inflow and radial outflow combustor configurations are also well suited to centrifugal compressor propulsion systems. The radial inflow combustor has an outward-oriented dome or headplate with combustion gas flow directed toward the engine centerline, while the radial outflow configuration has an inward-oriented dome with the primary flow direction being away from the engine centerline. For example, the WR-19 combustor illustrated in Fig. 2.6 is typical of compact radial outflow designs.

Flow Path Terminology

This subsection will identify and briefly describe basic airflow distribution terminology for a conventional combustor. Distribution of air in,

Fig. 2.6   WR19 turbofan engine.

Fig. 2.7   JT9D annular combustor.

around, and through the combustor results in the four basic airflow regions illustrated in Fig. 2.9. Effective control of this air distribution is vital to the attainment of complete combustion, stable operation, correct combustor exit temperature profile, and acceptable liner temperatures for long life.

Primary air.   This is the combustion air introduced through the dome or headplate of the combustor and through the first row of liner air holes. This air mixes with incoming fuel, producing the approximately stoichiometric mixture necessary for optimum flame stabilization and operation (see discussion on combustion stabilization in Sec. 1.4 of Ref. 2).

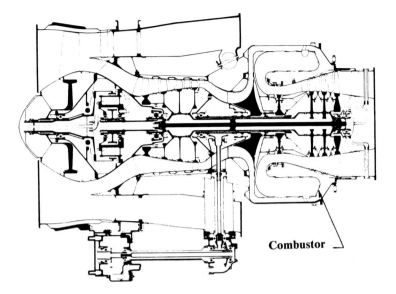

Fig. 2.8   TFE731 turbofan engine with reverse flow combustor.

Intermediate air. To complete the reaction process and consume the high levels of primary zone CO, H_2, and unburned fuel, intermediate air is introduced through a second row of liner holes. The reduced temperature and excess air cause CO and H_2 concentrations to decrease (see the chemical kinetic and equilibrium relationships presented in Sec. 1.1 of Ref. 2).

Dilution air. In contemporary systems, a large quantity of dilution air is introduced at the rear of the combustor to cool the high-temperature gases to levels consistent with turbine design limitations. The air is carefully used to tailor exit temperature radial profile to ensure acceptable turbine durability and performance. This requires minimum temperatures at the turbine root (where stresses are highest) and at the turbine tip (to protect seal materials). However, modern and future combustor exit temperature requirements are necessitating increased combustion air in the primary and intermediate zones; thus, dilution zone airflow is necessarily reduced or eliminated to permit these increases.

Cooling air. Cooling air must be used to protect the combustor liner and dome from the high radiative and convective heat loads produced within the combustor. This air is normally introduced through the liner such that a protective blanket or film of air is formed between the combustion gases and the liner hardware. Consequently, this airflow should not directly affect the combustion process. A detailed discussion of the various design techniques employed to cool the combustor liner is given in Sec. 2.3.

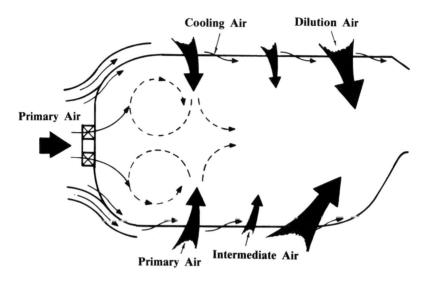

Fig. 2.9   Main combustor airflow distribution.

2.3   COMPONENT CONSIDERATIONS

Combustion System Demands

In recent years, significant engine performance gains have required advancements in turbopropulsion combustion. Advanced strategic and tactical aircraft propulsion systems utilize main burners with the operational flexibility to accept broad variations in compressor discharge pressure, temperature, and airflow while providing an acceptable exit temperature profile with minimum pressure loss and near-perfect combustion efficiency. Furthermore, heat release rates and combustor temperature rise capabilities have significantly increased and will continue to progress toward stoichiometric exit temperature conditions.

A broad list of combustion system performance and design objectives is required of all new combustors as they enter development. Although this list can be quite lengthy, the more important requirements, some of which were alluded to above, are given in Table 2.1. A number of these demands will be discussed in more detail in the following subsections.

Combustion efficiency.   Because propulsion system fuel consumption has a direct effect on aircraft system range, payload, and operating cost, it is imperative that design point combustion efficiency be as close to 100% as possible. Combustion efficiency at the high-power/high-fuel-consumption conditions of takeoff and cruise is always near 100% (usually greater than 99.5%). However, off-design efficiency, particularly at idle, can be in the low nineties. With the advent of chemical emission controls and limitations,

Table 2.1   Combustion system design and performance objectives

Performance objectives:
   High combustion efficiency (100%) at all operating conditions
   Low overall system total pressure loss
   Stable combustion at all operating conditions
   Reliable ground-level ignition and altitude relight capability

Design objectives:
   Minimum size, weight, and cost
   Combustor exit temperature profile consistent with turbine design
      requirements
   Low stressed structures
   Effective hot parts cooling for long life
   Good maintainability and reliability
   Minimum exhaust emissions consistent with current specified
      limitations and regulations

this parameter becomes of particular significance during low-power operation. For example, combustion efficiencies at off-design conditions, such as idle, must now exceed 98.5% to satisfy limitations on exhaust carbon monoxide and unburned hydrocarbons.

As discussed in Chapter 1 of Ref. 2, combustion efficiency can be defined in a number of equation forms as

$$\eta_c = \frac{\text{enthalpy rise (actual)}}{\text{enthalpy rise (ideal)}} = \frac{(h_4 - h_3)_a}{(h_4 - h_3)_i} \qquad (2.1)$$

$$= \frac{\text{temperature rise (actual)}}{\text{temperature rise (ideal)}} = \frac{(T_4 - T_3)_a}{(T_4 - T_3)_i} \qquad (2.2)$$

where subscripts 3 and 4 represent the combustor entrance and exit conditions, respectively.

Combustion efficiency η_c can also be determined from the concentration levels of the various exhaust products. A description of the combustion efficiency calculation based on exhaust product chemistry is given in Sec. 1.4 of Ref. 2.

Combustion efficiency can be empirically correlated with several aerothermodynamic parameters such as system pressure, temperature, reference velocity V_R,* and temperature rise. Two examples of such correlations will be discussed in Sec. 2.4.

---

*That velocity at the reference plane, or plane of maximum cross section, within the combustor under flow conditions corresponding to T_3. Values of reference velocity are 15–30 m/s for contemporary combustor designs.

Overall pressure loss. The combustion system total pressure loss from the compressor discharge to the turbine inlet is normally expressed as a percent of compressor discharge pressure. Losses of 5–8% are typically encountered in contemporary systems. Combustion system pressure loss is recognized as necessary to achieve certain design objectives (pattern factor, effective cooling, etc.) and can also provide a stabilizing effect on combustor aerodynamics. However, pressure loss also affects engine thrust and specific fuel consumption. A 1% increase in pressure loss will result in approximately a 1% decrease in thrust and a 0.5–0.75% increase in specific fuel consumption. Consequently, design goals for pressure loss represent a compromise among the above factors.

Overall pressure loss is the sum of inlet diffuser loss, combustor dome and liner loss, and momentum loss resulting from combustor flow acceleration attendant with increased gas temperature (see Sec. 1.3 of Ref. 2). Since many aspects of combustor performance are dependent on the airflow turbulence generated within the combustor (which in turn depends on the liner pressure drop), rapid and complete burning of the fuel and air is strongly influenced by the extent of the pressure drop experienced as air is introduced into the combustion zone.

Combustion system pressure drop can be expressed in terms of three different loss parameters: fractional pressure loss, inlet velocity head loss, and reference velocity head loss.

Fractional loss. Overall combustor/diffuser pressure loss is most commonly expressed as the fractional loss defined as

$$\frac{\Delta P}{P_3} = \frac{P_3 - P_4}{P_3} \qquad (2.3)$$

where

$\Delta P = P_3 - P_4 = $ pressure drop
$P_3 = $ compressor discharge total pressure
$P_4 = $ turbine nozzle inlet or combustor exit total pressure

This loss generally increases with the square of the diffuser inlet Mach number.

Inlet velocity head loss. This loss coefficient is given in terms of inlet velocity head. It expresses losses in a manner that accounts for the additional difficulties in designing for minimum pressure loss as inlet velocities are increased. Inlet velocity head loss is defined as

$$\frac{\Delta P}{q_3} = \frac{P_3 - P_4}{q_3} \qquad (2.4)$$

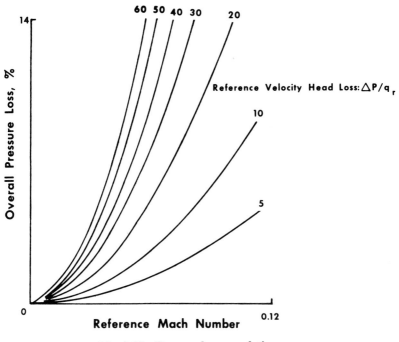

Fig. 2.10   Pressure loss correlation.

where

P_3 = compressor discharge total pressure
P_4 = combustor exit total pressure
q_3 = dynamic pressure at the compressor discharge

Reference velocity head loss.   This loss coefficient is expressed in terms of reference velocity. It represents a measure of the pressure loss normalized by a term that accounts for difficulties associated with high combustor volumetric flow rates. Reference velocity head loss is defined as

$$\frac{\Delta P}{q_r} = \frac{P_3 - P_4}{q_r} \tag{2.5}$$

where

P_3 = compressor discharge total pressure
P_4 = combustor exit total pressure
q_r = dynamic pressure corresponding to V_R

The relationship between fractional pressure loss (or overall pressure loss) and reference velocity head loss is shown in Fig. 2.10 as a function of reference Mach number.

Exit temperature profile.   A third performance parameter relates to the temperature uniformity of the combustion gases as they enter the turbine. In order to ensure that the proper temperature profile has been established at the combustor exit, combustion gas temperatures are often measured by means of high-temperature thermocouples or gas sampling techniques employed at the combustor exit plane. A detailed description of the thermal field entering the turbine both radially and circumferentially can be determined from the data. A simplified expression, called the pattern factor or peak temperature factor, may be calculated from this exit temperature data. Pattern factor is defined as

$$\text{Pattern factor} = \frac{T_{\max} - T_{\text{avg}}}{T_{\text{avg}} - T_{\text{in}}} \qquad (2.6)$$

where

$T_{\max}$ = maximum measured exit temperature (local)
T_{avg} = average of all temperatures at exit plane
T_{in} = compressor discharge average temperature

Contemporary combustors exhibit pattern factors having ranges of 0.25–0.45. Pattern factor goals are based primarily on the design requirements of the turbine first-stage vane, which requires low gas temperatures at both the hub and tip of the turbine—areas where high stresses and protective seals require cooler gas temperatures. Consequently, a pattern factor of 0.0 *is not required.* Durability considerations require high-temperature-rise combustors to provide combustor exit temperature profiles corresponding to pattern factors in the 0.15–0.25 range. One will note that, although pattern factor is an important combustor design parameter, it describes the possible thermal impact on the turbine and is an important factor in matching the combustor and turbine components.

Although pattern factor defines the peak turbine vane inlet gas temperature, the shape of the combustor exit temperature radial profile is the critical factor controlling turbine blade life. Figure 2.11 illustrates typical radial profile characteristics and their attendant relationship with the pattern factor. By proper control of dilution air, the combustor exit temperature field is tailored to give the design pattern factor and radial profile consistent with turbine requirements.

Combustion stability.   Combustion stability is defined as the ability of the combustion process to sustain itself in a continuous manner. Stable, efficient combustion can be upset by the fuel-air mixture becoming so lean that temperature and reaction rates drop below the level necessary to heat and vaporize the incoming fuel and air effectively. Such a situation causes blowout of the combustion process. An illustration of stability sensitivity to mass flow, velocity, and pressure characteristics as a function of equivalence ratio is given in Fig. 2.12. These trends can be correlated with the perfectly stirred reactor theory described in Sec. 1.3 of Ref. 2.

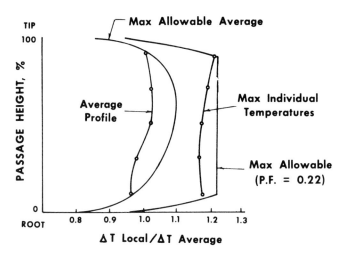

Fig. 2.11   Radial temperature profile at combustor exit.

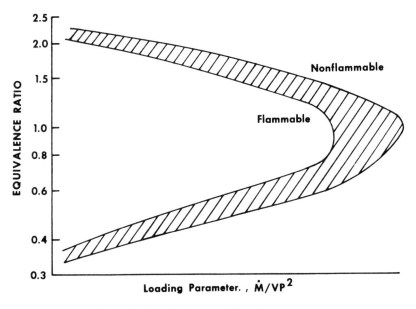

Fig. 2.12   Combustion stability characteristics.

Ignition.   In a turbine engine combustor, ignition of a fuel-air mixture requires inlet air and fuel conditions within flammability limits, sufficient residence time of the potentially burnable mixture, and the location of an effective ignition source in the vicinity of the burnable mixture. Each of these factors has been discussed from a fundamental standpoint in Sec. 1.4

of Ref. 2. Reliable ignition in the combustion system is required during ground-level startup and for relighting during altitude windmilling. The broad range of combustor inlet temperature and pressure conditions encompassed by a typical ignition/relight envelope is illustrated in Fig. 2.13. It is well known that ignition performance is improved by increases in combustor pressure, temperature, fuel-air ratio, and ignition source energy. In general, ignition is impaired by increases in reference velocity, poor fuel atomization, and low fuel volatility. A more extensive discussion of the ignition source itself is deferred to Sec. 2.3.

Size, weight, and cost. The main combustor of a turbine engine, like all other main components, must be designed within constraints of size, weight, and cost. The combustor diameter is usually dictated by the engine casing envelope provided between the compressor and turbine and is never allowed to exceed the limiting diameter defined for the engine. Minimization of combustor length allows reduction of engine bearing requirements and permits substantial reductions in weight and cost. Advancements in design technology have permitted major reductions in combustor length.

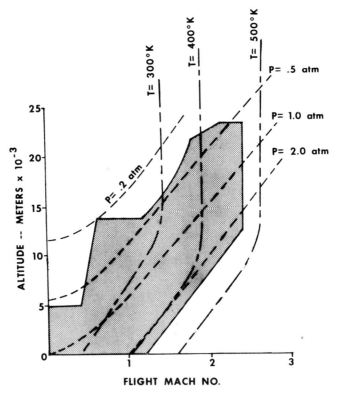

Fig. 2.13 Ignition/relight envelope.

Table 2.2   Contemporary combustor size, weight, and cost

Parameter	TF39	TF41	J79	JT9D	T63
Type	Annular	Cannular	Cannular	Annular	Can
Mass flow (design point)					
Airflow, lb/s	178	135	162	242	3.3
kg/s	81	61	74	110	1.5
Fuel flow, lb/h	12,850	9,965	8,350	16,100	235
kg/h	5,829	4,520	3,788	7,303	107
Size					
Length, in.	20.7	16.6	19.0	17.3	9.5
cm	52.6	42.2	48.3	43.9	24.1
Diameter, in.	33.3	5.3/24.1[a]	6.5/32.0[a]	38.0	5.4
cm	84.6	13.5/61.2	16.5/81.3	96.5	13.7
Weight, lb	202	64	92	217	2.2
kg	92	29	42	98	1.0
Cost, $	42,000	17,000	11,300	80,000	710

[a]Can diameter/annulus diameter.

With the advent of the annular combustor design, length has been reduced by at least 50% when compared to contemporary cannular systems.

While reductions in both size and weight have been realized in recent combustor developments, the requirement for higher operating temperatures has demanded the use of stronger, higher-temperature, and more costly combustor materials (to be discussed in Sec. 2.3). Nevertheless, the cost of contemporary combustion systems including ignition and fuel injection assemblies remains at approximately 2–4% of the total engine cost. A tabulation of the approximate size, weight, cost, and capacity of some contemporary combustion systems is given in Table 2.2. Naturally, the final cost of any component is significantly affected by the level of production.

Durability, maintainability, and reliability.   A principal combustor design objective is to provide a system with sufficient durability to permit continuous operation until a scheduled major engine overhaul, at which time it becomes cost effective to make necessary repairs and/or replacements. In the case of the main burner, durability is predominantly related to the structural and thermal integrity of the dome and liner. The combustor must exhibit good oxidation resistance and low stress levels at all operating conditions if durability is to be achieved.

A maintainable component is one that is easily accessible and repairable and/or replaceable with a minimum of time, cost, and labor. While most combustor liners can be weld repaired if damaged or burned, turbine removal is required for replacement of combustors in many cases. Consequently, a burner life consistent with the planned engine overhaul schedule is a primary objective. Combustor cases and diffuser sections require minimal maintenance, and fuel nozzles and ignitors can generally be replaced and/or cleaned with minimal effort.

Reliability can be defined as the probability that a system or subsystem will perform satisfactorily between scheduled maintenance and overhaul periods. Component reliability is highly dependent on aircraft mission, geographical location, and pilot operation since these factors strongly affect the actual combustor temperature/pressure environment and cyclic history of the components. In that the combustor has virtually no moving parts, its reliability is strongly related to fuel nozzle and igniter performance. While fouling and carboning of these subcomponents are common causes for engine rejection, these problems are relatively easy to correct through normal inspect and replace field maintenance procedures.

Exhaust emissions. With the advent of environmental regulations for aircraft propulsion systems, the levels of carbon monoxide, unburned hydrocarbons, oxides of nitrogen, and smoke in the engine exhaust become important. Naturally, the environmental constraints directly impact the combustion system—the principal source of nearly all pollutants emitted by the engine. Major changes in combustor design philosophy have evolved in recent years to provide cleaner operation at all conditions without serious compromise to engine performance. A detailed discussion of the exhaust emissions area is offered in Sec 2.5.

Design Factors

The turbine engine combustion system consists of three principal elements—the inlet diffuser, dome and snout, and inner and outer liners. In addition, two important subcomponents are necessary—the fuel injector and igniter. These elements are illustrated in Fig. 2.14. This section will describe each of these items and will conclude with a materials summary.

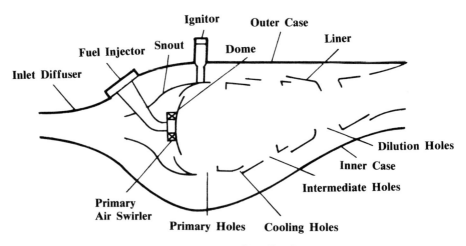

Fig. 2.14   Component identification.

Inlet diffuser. The purpose of the combustor inlet diffuser is to reduce the mean velocity of the air exiting the compressor and deliver it to the combustion chamber as a stable, uniform flowfield recovering as much of the dynamic pressure as possible. Compressor discharge air velocities range 90–180 m/s; consequently, before this high-velocity air is allowed to enter the combustor, it must be diffused to levels consistent with the stable, low pressure loss, high-efficiency requirements of contemporary combustors. Additionally, this resulting flowfield must be introduced in a relatively nondistorted manner to ensure uniform flow distribution to the combustion chamber. The diffuser must accomplish this by effectively controlling boundary-layer growth and avoiding flow separation along the diffuser walls while minimizing length and overall size. A balance must be found between (1) designs with increased size and complexity and their attendant performance penalties, and (2) short-length, rapid divergence designs, which have inherent flow nonuniformity and separation problems. Hence, the inlet diffuser represents a design and performance compromise relative to required compactness, low pressure loss, and good flow uniformity.

A number of performance parameters are commonly used to describe a diffuser and its operation.

(1) *Pressure recovery coefficient C_p.* This is a measure of the pressure recovery efficiency of the diffuser reflecting its ability to recover dynamic pressure. The coefficient is defined as the ratio of static pressure rise to inlet dynamic head,

$$C_p = \frac{P_{S2} - P_{S1}}{\rho V_1^2 / 2g} \qquad (2.7)$$

where P_{S1} and P_{S2} are the inlet and exit static pressures, respectively, and $\rho V_1^2 / 2g$ the inlet dynamic pressure.

For the ideal flow situation, i.e., full dynamic pressure recovery, C_p can be expressed in terms of area ratio:

$$C_p = 1 - (A_1/A_2)^2 \qquad (2.8)$$

where A_1 and A_2 are the inlet and exit cross-sectional areas, respectively.

(2) *Pressure recovery effectiveness.* This parameter describes the ability of a diffuser design to achieve ideal recovery characteristics. Hence, it is the ratio of the actual to the ideal pressure recovery coefficient,

$$\eta = \frac{C_p}{C_{p\,\text{Ideal}}} = \frac{P_{S2} - P_{S1}}{(\rho V_1^2 / 2g)\,[1 - (A_1/A_2)^2]} \qquad (2.9)$$

(3) *Kinetic energy distortion factor (α).* This factor is a measure of the radial nonuniformity of the axial flow velocity profile. The distortion factor

is defined as

$$\alpha = \frac{\int_a (u^2/2)u\rho \; dA}{(V^2/2)V\rho A} \tag{2.10}$$

where V is the mean flow velocity, u the local axial velocity, ρ the density, and A the cross-sectional area of duct.

A factor of 1.0 is equivalent to a flat one-dimensional velocity profile (i.e., plug flow); turbulent pipe flow has a factor of approximately 1.1.

In addition, a number of design parameters are often utilized to predict diffuser performance:

(1) Area ratio A_R: This is the ratio of the exit to inlet areas of the diffuser and defines the degree of area change for a particular design.

(2) Length-to-height ratio L/H: This is the ratio of diffuser length (entrance to exit) to the entrance or throat height and serves as a sizing parameter.

(3) Divergence half-angle θ: This is equivalent to one-half the equivalent cone angle of the diffuser and describes the geometric divergence character-istics of the diffuser walls.

Figure 2.15 relates area ratio, length-to-height ratio, and divergence half-angle to pressure recovery effectiveness.

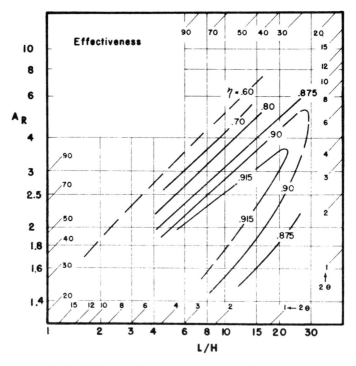

Fig. 2.15   Diffuser design and performance characteristics.

Early inlet diffuser designs were of the smooth curved wall or contoured wall designs. Because of the wide variations in the characteristics of the flowfield exiting the compressor, however, the curved wall diffuser cannot always provide uniform, nonseparated flow at all operating conditions. This can become a critical problem in the short-length diffusers required for many current systems. Consequently, a trend toward dump or combination curved wall and dump diffuser designs is occurring. Although this design results in somewhat higher pressure losses, it provides a known and constant point of flow separation, the dump plane, which prevents stalled operation at all diffuser entrance conditions. Figure 2.16 illustrates these contemporary designs.

The design procedures commonly employed to develop a specific diffuser configuration involve the use of a combination of experimentally generated performance maps, empirical equations, and analytical models. Most available performance maps were generated for two-dimensional straight wall and conical diffusers, the most notable source being the work of Kline and his associates.[9] Until recent years, empirical results such as those illustrated in Fig. 2.15 have been used in the development of annular diffusers. However, with the advent of improved numerical methods and high-speed computers, a number of improved two- and three-dimensional analytical models are now being developed that more accurately describe the flowfield characteristics of the annular diffuser design. Such programs will provide improved analysis of pressure loss characteristics, inlet velocity profile effects, and the influence of turbulence level on diffuser performance.

The need for high performance in short, compact diffusers takes on increasing importance as future engine operating conditions become more and more severe. An advanced compact diffuser design that provides improved boundary-layer control and greater pressure recovery is described in Sec. 2.5.

Curved Wall Diffuser

Fig. 2.16   Contemporary diffuser designs.

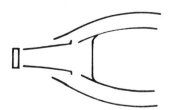

Dump Diffuser

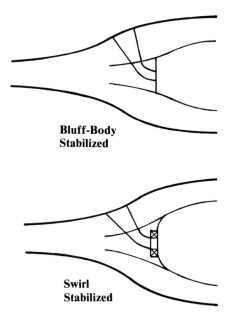

**Bluff-Body
Stabilized**

Fig. 2.17   Combustor dome types.

**Swirl
Stabilized**

Dome and snout.   At the front of the combustion chamber is the snout and dome where air and fuel are initially introduced. The snout is actually a forward extension to the dome dividing the incoming air into two streams—one directly entering the primary zone of the combustor through air passages in the dome and the other entering the annulus around the combustor. The snout also improves diffusion by streamlining the combustor dome, permitting a larger diffuser divergence angle, and providing reduced overall diffuser length. The dome plate contains provisions for receiving the fuel injector and maintains its alignment during operation. Combustor domes are of two basic types—bluff body and swirl stabilized, as illustrated in Fig. 2.17. Early combustors like the J79 (Fig. 2.4) generally fall into the bluff-body class, in which the high-blockage dome plate establishes a strong wake region, providing primary zone recirculation. In effect, the bluff-body dome interacts with the first row of primary zone air holes to establish this strong recirculation region. Most contemporary combustors, however, utilize the swirl-stabilized dome. With this design, the fuel injector is surrounded by a primary air swirler. The air swirler sets up a strong swirling flowfield around the fuel nozzle, generating a centralized low-pressure zone that draws or recirculates hot combustion products into the dome region. As a result, an area of high turbulence and flow shear is established in the vicinity of the fuel nozzle, finely atomizing the fuel spray and promoting rapid fuel-air mixing.

Liner.   The liner provides containment of the combustion process and allows introduction of intermediate and dilution airflow. Contemporary liners are typically of sheet-metal braze and welded construction. The liner is mounted to the combustor dome and generally suspended by a support and seal system at the turbine nozzle entrance plane. Its surface is often a system of holes of varying sizes that direct primary, intermediate, dilution, and liner cooling air into the combustion chamber. While combustion gas temperatures may be in excess of 2500 K, the liner is protected by a continuous flow of cool air (at approximately compressor exit temperature levels) and maintained at temperatures less than 1200 K.

The liner must be designed with high structural integrity to support forces resulting from pressure drop and must have high thermal resistance capable of continuous and cyclic high-temperature operation. This is accomplished through utilization of high-strength, high-temperature oxidation-resistant materials and effective use of cooling air. Depending upon the temperature rise requirements of the combustor, 20–50% of the inlet airflow may be utilized in liner cooling. A number of cooling techniques are illustrated in Fig. 2.18.

(1) *Louver cooling.*   Many of the early jet engine combustors used a louver cooling technique in which the liner was fabricated into a number of cylindrical panels. When assembled, the liner contained a series of annular air passages at the panel intersection points, the gap heights of which were maintained by simple wiggle-strip louvers. This permitted a film of air to be injected along the hot side of each panel wall, providing a protective thermal barrier. Subsequent injection downstream through remaining panels permitted replenishment of this cooling air boundary layer. Unfortunately, the louver cooling technique did not provide accurate metering of the cooling air, resulting in considerable cooling flow nonuniformity with

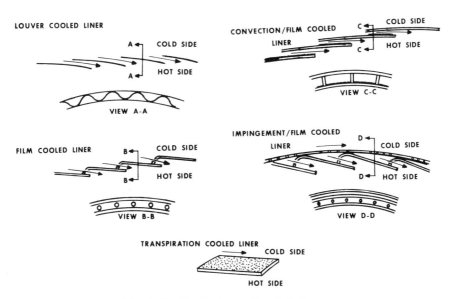

Fig. 2.18   Combustor cooling techniques.

attendant variations in combustor exit profiles and severe metal temperature gradients along the liner.

(2) *Film cooling.* This technique is an extension of the louver cooling technique, but with machined injection holes instead of louvers. Consequently, airflow metering is more accurate and uniform throughout the combustion chamber. Most current combustors use this cooling technique. However, the increased operating gas temperatures of future combustors will result in less air for cooling and more advanced cooling techniques/materials will be required.

(3) *Convection/film cooling.* This relatively new technique permits much reduced cooling airflow (15–25%) while providing high cooling effectiveness and uniform metal temperatures. It is particularly suited to high-temperature-rise combustion systems where cooling air is at a premium. The convection/film-cooled liner takes advantage of simple but controlled convection cooling enhanced by roughened walls, while providing the protective boundary layer of cool air at each cooling panel discharge plane. Although somewhat similar in appearance to the louver-cooled liner, the confection/film coolant passage is several times greater; more accurate coolant metering is provided and a more stable coolant film is established at the panel exit. Principal disadvantages of this design are somewhat heavier construction, increased manufacturing complexity, and repairability difficulties.

(4) *Impingement/film cooling.* This cooling technique is also well suited to high-temperature-rise combustors. When combined with the additional film cooling feature, impingement cooling provides for excellent thermal protection of a high-temperature liner. Its disadvantages, however, are similar to those of the film/convection liner—heavier construction, manufacturing complexity, and repairability difficulties.

(5) *Transpiration cooling.* This is the most advanced cooling scheme available and is particularly well suited to future high-temperature applications. Cooling air flows through a porous liner material, uniformly removing heat from the liners while providing an excellent thermal barrier to high combustion gas temperatures. Both porous (Regimesh and Porolloy) and fabricated porous transpiring materials (Lamilloy†) have been examined experimentally. Fabricated porous materials tend to alleviate plugging and contamination problems, inherent disadvantages of the more conventional porous materials.

Figure 2.19 shows the axial thermal gradient characteristics of each of the liner designs discussed above as a function of relative liner length. As can be seen, transpiration cooling offers better temperature control and uniformity than any other cooling technique.

---

†Developed by Detroit Diesel Allison, General Motors Corp., Patent 3,584,972, "Laminated Porous Material," June 15, 1971.

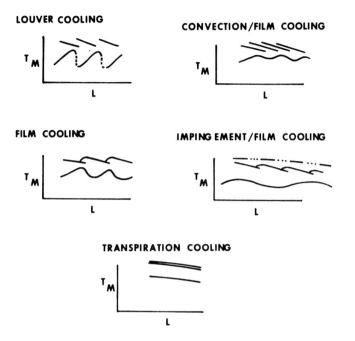

Fig. 2.19   Liner cooling characteristics.

Fuel injection.   Basically four methods of fuel introduction are currently used or proposed for future use. These techniques—pressure atomizing, air blast, vaporizing, and premix/prevaporizing—are discussed below in increasing order of complexity. Each is illustrated in Fig. 2.20.

(1) *Pressure atomizing.*   Most contemporary combustion systems use pressure-atomizing fuel injectors. They are relatively simple in construction, provide a broad flow range, and can provide excellent fuel atomization when fuel system pressures are high. A typical pressure-atomizing fuel injector is illustrated in Fig. 2.21. At least five different design concepts or variations are included in this category: simplex, duplex, dual orifice, variable area, and slinger. These devices typically utilize high fuel pressure (about 500 psi above combustor pressure) to achieve fine fuel atomization. The slinger design, although a pressure-atomizing type, is very different from the conventional fuel nozzle in that the fuel is injected through small holes in the rotating turbine shaft. The high centrifugal forces imparted to the fuel provide atomization. Slinger systems are used in several small engine combustors—the WR19 of Fig. 2.6 is one such system. The principal disadvantages of the pressure-atomizing systems are the propensity for fuel system leaks due to the inherently high fuel pressures required, potential plugging of the small fuel orifices by contaminants entrained in the fuel, and increased difficulty in achieving low smoke levels when fuel system pressures are low.

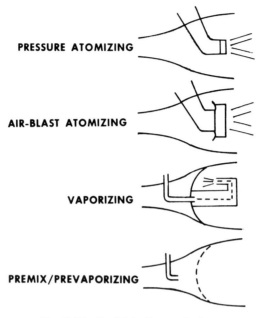

PRESSURE ATOMIZING

AIR-BLAST ATOMIZING

VAPORIZING

PREMIX/PREVAPORIZING

Fig. 2.20   Fuel injection methods.

(2) *Air blast.*   A number of modern combustor designs achieve fuel atomization and mixing through use of primary zone air momentum. Strong swirling motion, often accompanied by a second counterswirl, causes high gasdynamic shear forces to atomize liquid fuel and promote mixing. Low fuel injection pressures (50–200 psi above combustor pressure) are utilized in these schemes. Rizkalla and Lefebvre[10,11] describe air-blast atomizer spray characteristics relative to air and liquid property influences. In addition, the development of a specific air-blast atomizer for gas turbine application is discussed in Ref. 2.

(3) *Vaporizing.*   A number of vaporizing fuel injection systems have been developed; perhaps the most common is the "candy-cane" vaporizer. In this design, fuel and air are introduced into a cane-shaped tube immersed in the combustion zone. During operation, the heat transferred from the combustion region partially vaporizes the incoming fuel, while the liquid/vapor fuel within the tube provides thermal protection for the tube. It is generally agreed, however, that fuel vaporization is very incomplete in this type vaporizer, which is considered by many to be merely an extension of the air-blast principles described above. This design is simple in construction, is inexpensive, and can operate with low fuel injection pressures. The resultant fuel-air mixture burns with low flame radiation, reducing liner heat loads. This design, however, has certain serious shortcomings: poor ignition and lean blowout characteristics, vaporizer tube durability problems during low fuel flows, and slow system response time.

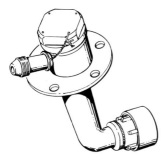

Fig. 2.21   Pressure-atomizing fuel injector.

(4) *Premix/prevaporizing.* The advent of gas turbine emission regulations has resulted in increased interest in premix/prevaporizing fuel injection. In this technique, fuel is introduced and premixed with the incoming air before its introduction to the combustion zone. The design intent is to provide a uniform, low-equivalence-ratio, fully mixed field of vaporized fuel in the combustion region. The result is low smoke and chemical emissions, low flame radiation, improved fuel-air uniformity in the combustion region, and virtual elimination of hot-spot burning. With this system, potential problem areas include incomplete fuel vaporization, danger of flashback through or autoignition of the fuel-air premixture upstream of the combustor dome plate with resulting damage to the combustor hardware, poor lean blowout characteristics, and difficulty with ignition and altitude relight. Staged combustion, utilizing a pilot zone with a relatively conventional stoichiometric design, is often proposed as a method of overcoming stability and ignition difficulties.

Ignition. Ignition of the cold flowing fuel-air mixture can be a major combustor design problem. Nearly all conventional combustors are ignited by a simple spark-type igniter similar to the automotive spark plug. Turbine engine ignition energies are typically 4–12 J, with several thousand volts at the plug tip. Figure 2.22 illustrates a typical spark-type igniter.

Each combustion system is generally fitted with two spark igniters to provide system redundancy. Potential ignition system problems include spark plug fouling with carbon or fuel, plug tip burnoff, electrode erosion with time, and corona-discharge losses along the ignition system transmission lines under high-altitude, low-pressure conditions.

During engine startup, the flame must propagate from can to can in a cannular combustion system via cross-fire or interconnector tubes located near the dome of each can. The large cross-fire ports of the J79 are readily visible in Fig. 2.4. The cross-fire region must be designed to ensure rapid and complete flame propagation around the combustion system. Without proper cross-firing, a "hung start" can occur wherein only one or two combustors are ignited. This condition of poor flame propagation can also occur in annular combustors. In either case, severe local gas temperatures (high pattern factor) are generated that can thermally distress the turbine.

Although the spark igniter is the most common ignition source in use

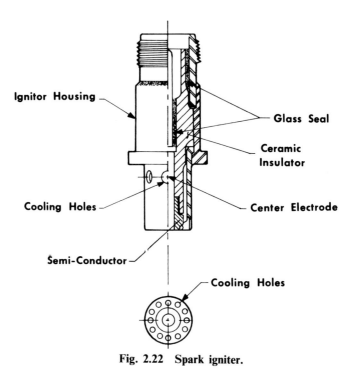

Fig. 2.22   Spark igniter.

today, a number of other ignition or ignition-assist techniques have been employed: the torch igniter, oxygen injection, and use of pyrophoric fuels. The torch igniter, a device combining the functions of fuel injection and spark ignition, is extremely reliable and permits a wide ignition envelope. However, it is more complex and costly and increases maintainability problems relative to conventional ignition systems. Oxygen injection assists ignition by lowering minimum ignition energy requirements. It is especially applicable to facilitating altitude relight. The use of pyrophoric fuels provides perhaps the most positive ignition source available. Pyrophorics react spontaneously in the presence of oxygen and provide excellent altitude relight capability. Unfortunately, pyrophoric fuels are extremely toxic and create special handling and logistics problems. Consequently, these factors have limited its use as a viable ignition technique.

Materials.   The selection of proper materials is a critical element in combustion system design for component structural integrity. Materials possessing high stress tolerance, good oxidation and corrosion resistance, and the ability to withstand the broad cyclic aerothermal loads imposed by the engine during its operation are required. Several high-strength alloy materials are used in combustors today, the selection of which is generally based on the projected operating environment of the propulsion system. This section highlights a few of the more common combustor materials in use today.

(1) *Hastelloy X.*   Hastelloy X is a nickel-base alloy strengthened in solid solution by chromium and molybdenum. It is the most common combustor liner material in use today. Its formability is good, its machinability is difficult but not impractical, and its weldability and brazing characteristics are good. Hastelloy X exhibits good strength and oxidation properties in the 1040–1140 K metal temperature range. Most combustors with Hastelloy X liners are designed to operate at metal temperatures of 1090–1120 K.

(2) *Haynes 188.*   Haynes alloy 188 is a wrought solid, solution strengthened, cobalt-base alloy applicable to static parts operating at temperatures up to 1370 K. It can be readily formed and welded; its oxidation resistance is good, although protective coatings are required for applications above 1250 K; and, like Hastelloy X, its machinability is difficult but not impractical. It is finding increased applicability in the newer combustion systems where liner metal temperatures of 1140–1230 K are necessary.

(3) *TD nickel.*   This superalloy is a non-heat-treatable, high nickel alloy, strengthened by dispersion of fine ThO_2 particles in a nickel matrix. This alloy maintains useful strengths at temperatures up to 1420 K. Its oxidation-erosion resistance is inferior to that of Hastelloy X or Haynes 188 and it requires protective coatings for applications above 1220 K. Its machinability and formability are good. Fusion and resistance welding of this material can be difficult; however, its brazability is considered good. TD nickel offers considerable promise in future high-temperature liner applications where metal temperatures greater than 1250 K may be common. Current material costs and the need for protective coatings, however, have generally precluded serious considreation of TD nickel in contemporary combustion systems. Further, advanced liner cooling techniques have succeeded in maintaining metal temperatures at levels consistent with the Hastelloy X/Haynes 188 material capabilities.

Significant advancements in superalloy technology are required to meet future high-temperature rise combustor requirements. New materials well beyond the capability of TD nickel will be necessary. Improved coatings may provide part of the solution if developed with long life and improved high- and low-cycle fatigue capabilities. Carbon-carbon, ceramics, and advanced thermal barrier and coating materials may also find a role in future combustor design.

2.4 DESIGN TOOLS

The complexity of the aerothermodynamic and chemical processes occurring simultaneously in the combustor prevents a purely analytical approach to component design and performance prediction. Insufficient capability to accomplish measurements of importance within the combustor has precluded all but the most basic understanding of practical gas turbine combustion processes. As a result, one has had little choice but to formulate new designs largely on the basis of personal or organizational experience. Continuation of this approach to combustor design for the sophisticated high-temperature systems under development today and in the future would be extremely costly and time consuming. The turbine engine industry can no longer afford to conduct component development activities on a generally empirical basis. Hence, significant research and development programs are now being directed toward improved analytical design procedures reinforced by more powerful measurement diagnostics.

Combustor Modeling

The principal objective of the combustion system model is to analytically describe and predict the performance characteristics of a specific system design based on definable aerodynamic, chemical, and thermodynamic parameters. Many modeling approaches describing the flowfield and characteristics of a particular combustion system have evolved over the past 30 years. Early models were almost entirely empirical, while the models currently under development are based more on fundamental principles. Improved computer availability and capability, as well as more efficient numerical techniques, have had a significant impact on combustion modeling by permitting the more complex, theoretically based approaches to be considered.

Empirical models.   The empirical model utilizes a large body of experimental data to develop a correlation often using multiple-regression analysis techniques. Such an approach involves a selection of the appropriate design and aerothermodynamic parameters that have been found empirically to influence the performance (e.g., combustion efficiency) of a particular combustor design. Each of the nondimensionalized parameters or ratios are acted upon by appropriate "influence" coefficients or exponents, the value of which reflects the degree of importance of a particular parameter. Since these influence terms are usually derived from test data obtained from combustors generally representing the same basic design family, a major change in design philosophy can require the definition of a new set of influence factors. Consequently, this modeling approach works well on specific combustor designs for which there is a broad base of technical data. Unfortunately, it cannot be used arbitrarily as a general design tool. One example of an empirical correlation model is illustrated in Table 2.3. In this model, combustion system efficiency is defined as a function of the more important combustor design and performance parameters. As can be seen, the model is written in general form and a listing of appropriate coefficients and exponents is provided to permit the computation of combustion efficiency. Again, these influence terms are empirically based and were derived from a bank of combustor data representative of a particular class of combustors.

A second empirical correlation employed by some combustor designers today defines a reaction rate parameter θ based on the "burning velocity" theory of Lefebvre.[3] The θ parameter is given in Fig. 2.23. The resulting correlation establishes the relationship between combustor efficiency, operating condition, and geometric size. One can see that efficiency is not only a function of airflow, inlet pressure, and inlet temperature, but also strongly

Table 2.3   Empirical combustion efficiency correlation

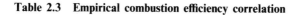

$$EC = C_\eta \left(\frac{W_3/P_3A}{4.0}\right)^A \left(\frac{T_3}{1000}\right)^B \left(\frac{\Delta T}{1000}\right)^C \left(\frac{L_c}{12}\right)^D \left(\frac{\Delta P/Q_R}{10}\right)^E \left(\frac{N_F/A_R}{10}\right)^F \left(\frac{W_1/W_3}{0.5}\right)^G \left(\frac{H_R}{6}\right)^H$$

Parameter	Combustion efficiency	
	$\Delta T = 600{-}1600°F$	$\Delta T = 1600{-}2500°F$
C_η	0.88	0.79
A	0.0	0.04
B	0.22	−0.04
C	0.11	−0.09
D	0.14	0.16
E	0.07	0.07
F	−0.07	0.0
G	−0.10	0.23
H	0.0	0.0

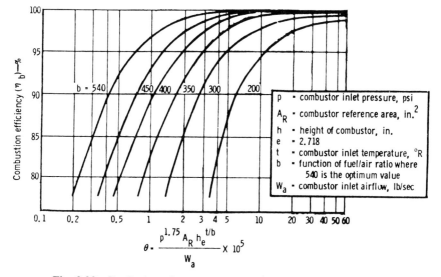

Fig. 2.23   Prediction of combustion efficiency using θ parameter.

depends on combustion zone fuel-air ratio. Herbert[14] estimated the effect of equivalence ratio on reaction rate by the following equation:

$$b = 220(\sqrt{2} \pm \ell n\phi/1.03) \qquad (2.11)$$

where ϕ is primary zone equivalence ratio. A graphic illustration of this expression (Fig. 2.24) describes the variation of b with primary zone

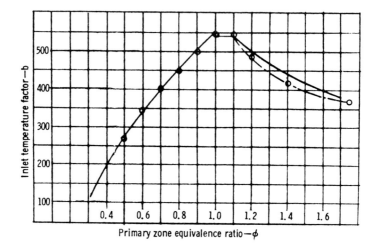

Fig. 2.24   Effect of primary zone ϕ on inlet temperature factor.

equivalence ratio. Hence, to achieve maximum efficiency, a primary zone fuel-air ratio of 0.067 ($\phi = 1$) should be used.

Combined empirical/theoretical models.   More complex theoretical descriptions of combustion must be incorporated into combustor models to allow predictions based on hardware design details. Physical models of important combustion processes (i.e., a "perfectly stirred reactor" primary zone) combined with available empirical analyses result in hybrid models. A number of two-dimensional models have been developed in recent years ranging from the perfectly stirred reactor plug flow models[15] to the more complex axisymmetric combustor flowfield calculations that account for heat, mass, and momentum transfer between fluid streams and include chemical reaction kinetic effects.[16,17] One such axisymmetric stream tube model was developed by Pratt & Whitney Aircraft in support of an Air Force-sponsored combustor program.[18] Figure 2.25 is the flowchart of computational steps for this analytical model, illustrating the increased complexity of this procedure in which both theory and experiment have been integrated. It is currently being used as an engineering tool for both current and advanced combustor design and exhaust emission analyses.

Another more recently developed model leans even more heavily on combustion theory and advanced numerical procedures. Anasoulis et al.[19] developed a two-dimensional computational procedure for calculating the coupled flow and chemistry within both cannular and annular combustors. A field relaxation method is used to solve the time-average Navier-Stokes equations with coupled chemistry including the effects of turbulence, droplet vaporization, and burning. Extensions to the three-dimensional case are also under development, permitting more accurate descriptions of the recirculation and mixing zones of the combustor.

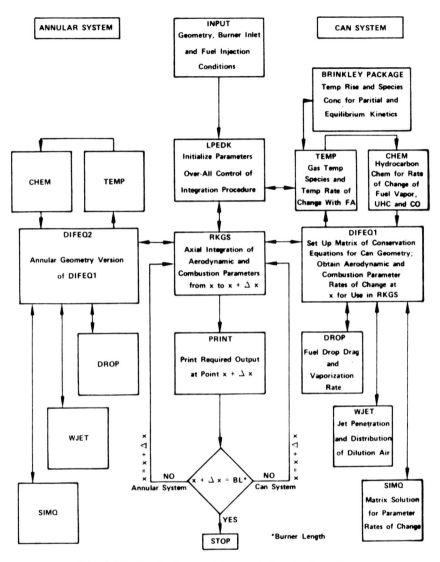

Fig. 2.25   Combustor model computational schematic.

Theoretical models. While Ref. 19 represents a significant advancement, the fully theoretical computerized design procedure has not yet evolved. A building block approach, however, has been envisioned in which a number of submodels (e.g., chemistry, turbulence, droplet vaporization, mixing) are developed independently, verified experimentally, and then coupled with the other submodels. Each coupled submodel set would undergo an experimental validation process before the next submodel is added to the overall analytical procedure. The ultimate is a fully theoretical description of all reacting flow processes, enabling prediction of overall combustor performance as a function of basic design parameters. Hence, prior to any hardware fabrication, the model will serve to evaluate the proposed combustion system design such that the desired objectives will be met with a minimum of subsequent hardware iterations and fine tuning.

Combustion Diagnostics

Development of valid combustor models is hampered by the difficulty of acquiring data for use in comparing and refining analytical solutions. Different types of measurements are required to validate various aspects of the important submodels, i.e., droplet size distribution, turbulence intensity, etc. In addition, improved instrumentation will be required to aid the combustor development process. Such equipment will provide the information on which the engineer will base subsequent design improvements and eventually the final version.

The rapidly growing field of combustion diagnostics will play an increasingly important role in satisfying the above objectives. Conventional thermocouple and sampling probes have been previously utilized to study combustion processes in practical systems. However, additional application and technique refinements are necessary. Furthermore, new, laser-based combustion diagnostic measurement equipment can be expected to play an important role in the future. Techniques such as laser Raman scattering and coherent anti-Stokes Raman scattering hold new promise for fundamental studies of combustion processes requiring real-time "point" ($\simeq 1 \text{ mm}^3$) measurements of temperature, concentration, and velocity. Other simpler methods may find application in measurement of combustor exit temperature profiles during combustion system development.

2.5 FUTURE REQUIREMENTS

The aeropropulsion combustion community is confronted with two additional future challenges: reduction of exhaust pollutant emissions and accommodation of new fuels that will reduce cost while increasing availability. The first two of the following three subsections summarize the problems and technology in each of these two areas. Furthermore, projected engine technology requirements necessitate advancements in combustion system design techniques and performance; the third subsection discusses the combustion engineer's task in this area.

Exhaust Emissions

Problem definition. In the 1960's, increased citizen concern over environmental issues coupled with the obvious visible smoke emissions from jet aircraft brought substantial public attention to aircraft-contributed pollution. As airport traffic increased, it became evident that at least the possibility existed that pollutant emissions, when concentrated in the local airport environment, could result in ambient levels in excess of allowable limits. Concern within the United States culminated in the inclusion of exhaust emissions from aircraft engines in the considerations of the Clean Air Act Amendments of 1970.[20] This legislation required that the Environmental Protection Agency (EPA) assess the extent to which aircraft emissions affect air quality, determine the technological feasibility of controlling such emissions, and establish aircraft emissions standards, if necessary.

The resulting EPA assessment[21] has indicated the necessity to regulate commercial aircraft emissions. Currently, EPA standards[22] apply to commercial and general aviation but not to military aircraft. The following excerpt from EPA's discussion accompanying the final announcement of the aircraft emissions standards[22] summarizes this policy.

In judging the need for the regulations, the Administrator has determined:

(1) that the public health and welfare is endangered in several air quality control regions by violation of one or more of the national ambient air quality standards for carbon monoxide, hydrocarons, nitrogen oxides, and photochemical oxidants, and that the public welfare is likely to be endangered by smoke emissions; (2) that airports and aircraft are now, or are projected to be significant sources of emissions of carbon monoxide, hydrocarbons, and nitrogen oxides in some of the air quality control regions in which the national ambient air quality standards are being violated, as well as being significant sources of smoke, and therefore (3) that maintenance of the national ambient air quality standards and reduced impact of smoke emissions requires that aircraft and aircraft engines be subject to a program of control compatible with their significance as pollution sources. Accordingly, the Administrator has determined that emissions from aircraft and aircraft engines should be reduced to the extent practicable with present and developing technology. The standards proposed herein are not quantitatively derived from the air quality considerations . . . but, instead, reflect EPA's judgment as to what reduced emission levels are or will be practicable to achieve for turbine and piston engines.

Current EPA regulations are based on reducing aircraft engine emissions during their operation below 3000 ft. However, an additional potential problem has been associated with aircraft—the possible environmental impact of high-altitude emissions.[23] There are many mechanisms by which this might arise: (1) emission of water vapor and carbon dioxide into the stratosphere may cause a "greenhouse effect"; (2) sulfur compound emissions can cause particulate formation that in turn cause solar radiation to be diverted away from the Earth's surface, reducing the equilibrium atmospheric temperature; and (3) increased concentrations of oxides of nitrogen due to emissions into the stratosphere may detrimentally react with the ozone layer and allow increased penetration of solar ultraviolet radiation. Potential problem 1 has been shown not to be significant. Much more investigation is needed concerning problems 2 and 3, however, before

Table 2.4 Engine combustion products

Group	Type	Species	Approximate concentrations		
			Low-power (idle)	High-power (non-AB)	Cruise (with afterburner)
1	Air	N_2	77%	77%	73–76%
		O_2	17.3–19%	13–16.3%	0–13%
		Ar	0.9%	0.9%	0.9%
2	Products of complete combustion	H_2O	1.4–2.4%	3–5%	5–13%
		CO_2	1.4–2.4%	3–5%	5–13%
3	Products of incomplete combustion	CO	50–2000 ppmv	1–50 ppmv	100–2000 ppmv
		Total HC	50–1000 ppmC	1–20 ppmC	100–1000 ppmC
		Partially oxidized HC	25–500 ppmC	1–20 ppmC	?
		H_2	5–50 ppmv	5–100 ppmv	100–1000 ppmv
		Soot	0.5–25 ppmw	0.5–50 ppmw	0.50–50 ppmw
4	Nonhydrocarbon fuel components	SO_2, SO_3	1–5 ppmw	1–10 ppmw	1–30 ppmw
		Metals, metal oxides	5–20 ppbw	5–20 ppbw	5–20 ppbw
5	Oxides of nitrogen	NO, NO_2	5–50 ppmv	50–500 ppmv	100–600 ppmv

the extent of potential stratospheric environmental problems can be suitably defined.

The discussion that follows defines the exhaust gas content, presents engine emission characteristics, and reviews emissions control technology.

Exhaust content. Aircraft engine exhaust constituents usually considered to be pollutants are smoke, carbon monoxide (CO), hydrocarbons (HC), and oxides of nitrogen (NO_x).‡ The magnitude of emissions depends on operating mode and engine type. The combustion products are conveniently organized into five groups, as listed in Table 2.4. More than 99% of the exhaust products are in the first two categories, which include those species not generally considered to be objectionable. The last three categories contain small quantities of constituents and are dominated by the principal pollutants: hydrocarbons (HC), carbon monoxide (CO), oxides of nitrogen (NO_x), and smoke. Because emissions characteristics at engine idle, nonafterburning high power, and afterburning operation vary substantially, columns listing typical composition for each of these operating modes are given. Note that levels given in Table 2.4 correspond to the turbojet case or to core flow only in the case of a turbofan.

(1) *Group 1, air.* These species pass through the engine unaffected by the combustion process and unchanged in chemical composition, except for oxygen depletion due to fuel oxidation. Argon is clearly inert. Although molecular nitrogen is nearly inert, the less than 0.01% that undergoes "fixation" to its oxide form (group 5) is, of course, extremely important.

(2) *Group 2, products of complete combustion.* Water and carbon dioxide are the dominant combustion products and the fully oxidized forms of primary fuel elements, hydrogen and carbon. It is the formation of these species that releases maximum energy from the fuel. H_2O and CO_2 are not generally considered to be air pollutants.

(3) *Group 3, products of incomplete combustion.* Hydrogen and carbon not converted to water or carbon dioxide are found in compounds categorized as products of incomplete combustion. The important species in this group are carbon monoxide, unburned and partially oxidized hydrocarbons, molecular hydrogen, and soot.

CO and HC emissions contain the largest portion of unused chemical energy within the exhaust during idle operation. Combustion efficiency at this operating condition may be calculated from exhaust CO and HC concentration data. At higher power settings, especially with afterburner operation, H_2 levels may also contribute significantly to inefficiency. Exhaust hydrocarbons are usually measured as total hydrocarbons as specified by the SAE ARP 1256.[24] Although it is well known that the toxicological

---

‡Exhaust nitrogen oxides are in the form of both NO and NO_2. Collectively, they are expressed as NO_x.

and smog-producing potential of different hydrocarbon types varies widely, characterization of the distribution of hydrocarbon types in the exhaust is difficult and seldom done. Presently available analytic techniques to accomplish such a characterization are complex, time consuming, expensive, and of unconfirmed accuracy.

A similar problem exists in quantifying soot emissions. The measurement technique that has evolved, ARP 1179,[25] does not relate directly to exhaust visibility or soot concentration. However, Champagne[26] has developed a correlation between measured ARP 1179 smoke number (SN) dry and particulate emissions. Efforts to measure and characterize particulate emissions directly are currently in progress. Complications have developed because the contribution of condensed hydrocarbons in the exhaust to the particulate measurement varies greatly with sampling conditions. Although a technique to determine exhaust soot concentrations may eventually be developed, characterization of size distribution appears to require a longer-range effort.

(4) *Group 4, nonhydrocarbon fuel components.*   The elemental composition of petroleum-based fuel is predominantly hydrogen and carbon. Of its trace components, sulfur is the most abundant. Most of the sulfur in the exhaust is in an oxidized form, probably as sulfur dioxide. Giovanni and Hilt[27] and Slusher[28] have found that the ratio of SO_3 to SO_2 ranges 0.03–0.14 in the case of heavy-duty stationary and aircraft gas turbines. The total amount of sulfur in exhaust compounds is directly related to, and calculable from, fuel sulfur content. The second most abundant trace component of nonhydrocarbon fuel involves metals. It is expected that these elements, which may be either natural components of the fuel or additions to it, appear in the exhaust as metal oxides. Further, it is generally expected that these species are particulates and often found with the soot.

(5) *Group 5, oxides of nitrogen.*   Although the ratio of NO to NO_2 emitted by aircraft gas turbines may shift with operating conditions, NO will eventually be converted to NO_2 in the atmosphere and subsequently participate in smog formation chemistry. Some attention, however, must be paid to the influence of time delay required for atmospheric NO_2 formation and the subsequent effect on smog formation. Stratospheric NO and NO_2 emissions are generally thought to have equally detrimental effects.

As stated above, product species are usually measured in terms of their volume (or sometimes mass, especially for condensed phases) fraction in the product sample. Occasionally, the suffix "dry" or "wet" is appended, according to whether or not the water is removed before analysis. A more useful and unambiguous method of reporting exhaust emissions from gas turbines has proved to be the use of an emission index that represents the ratio of the pollutant mass to the fuel consumption. A commonly used dimension is grams of pollutant per kilogram of fuel. Conversion of volume fraction measurements to emission indices requires assignment of molecular weights, which is not difficult for a single-compound category, but may lead to confusion for categories consisting of more than one compound. It is conventional to report oxides of nitrogen (NO_x) as though they were entirely NO_2. Similarly, the oxides of sulfur (SO_x) are usually reported as SO_2. Hydrocarbon measurements usually lead to a volumetric fraction related to a single hydrocarbon compound, e.g., parts per million equivalent hexane (or methane, propane, carbon atom, etc.). In reducing these measurements to an emission index, a hydrogen-carbon ratio of two is usually assumed.

Engine emissions characteristics. Processes that influence pollutant formation occur within both the main burner and afterburner. Conditions under which combustion occurs in these two systems are extremely different and studies of emissions from the main burner and augmenter are generally treated separately.

(1) *Main burner emissions.* Carbon monoxide and hydrocarbon emissions are a strong function of the engine power setting. As thrust is increased, the combustion system experiences greater inlet temperature and pressure, as well as higher fuel-air ratio. The increased fuel flow results in improved fuel atomization and a higher combustor inlet temperature provides more rapid vaporization. Chemical reaction rates responsible for CO and HC consumption are sharply increased by higher flame temperatures resulting from the higher fuel-air ratio (see Sec. 1.1 of Ref. 2). Each of these changes tends to decrease the rates at which HC and CO are emitted. Consequently, the relationship between an engine's emission of HC and CO (or combustion inefficiency) and the power setting indicates a sharply decreasing trend. Idle CO and HC emissions far exceed those at other engine conditions. A correlation of idle CO and hydrocarbon emis-

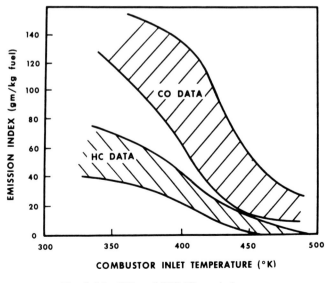

Fig. 2.26 CO and HC idle emissions.

sions for a number of engines can be established by plotting these values as a function of combustor inlet temperature. Figure 2.26 illustrates the trends that can be obtained.

As discussed in Chapter 1 of Ref. 2, combustion efficiency can be related to exhaust content. In the case of idle operation, the inefficiency is predominantly due to CO and HC. The idle CO and HC emission index values can be related to combustion inefficiency by

$$1 - \eta_b = [0.232(EI)_{CO} + (EI)_{HC}] \times 10^{-3} \qquad (2.12)$$

where η_b is the combustion efficiency, $(1 - \eta_b)$ the combustion inefficiency, and $(EI)_i$ the emission index of species i.

Oxides of nitrogen emission levels are greatest at high-power operating conditions. The predominant NO forming chemical reaction is

$$N_2 + O \rightarrow NO + N \qquad (2.13)$$

It is usually assumed that NO formation takes place in regions of the combustor where oxygen atoms are present at their equilibrium concentration. Reaction (2.13) and the oxygen atom concentration are extremely temperature sensitive; NO is produced only in the highest-temperature (near-stoichiometric) combustion zones. Since the stoichiometric flame temperature is dependent on combustor inlet temperature (a function of compressor pressure ratio and flight speed), oxide of nitrogen emissions can be expected to increase substantially with the power setting.

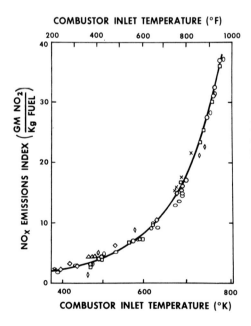

Fig. 2.27 Correlation of current engine NO_x emissions with combustor inlet temperature.

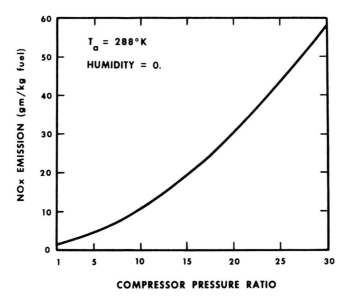

Fig. 2.28   Dependence of sea-level standard day NO_x emissions on compressor pressure ratio of engines developed in the 1970's.

An excellent correlation of NO_x emissions from a large number of engines has been established by Lipfert.[29] This correlation, reproduced in Fig. 2.27, relates the NO_x emission index to combustor inlet temperature. Note the strong temperature dependence previously discussed. Moreover, the fact that combustor design has little apparent effect on NO_x emission from these engines is noteworthy. This implies that the mixing and quenching processes within combustors operating with rich ($\phi > 1.0$) primary zones are strikingly similar; the temperature effect alone controls the NO_x emission rate. These combustors were not designed with the intent of controlling NO_x emissions.

No strong fuel effects are apparent from existing data. However, it has been shown that fuel-bound nitrogen is readily converted (50–100%) to NO_x in both stationary and aircraft turbine combustors.[30-33] Should aircraft fuel-bound nitrogen levels be increased in the future because of changing fuel requirements, fuel-bound nitrogen could become a significant problem. See subsection below on future jet fuels.

The dependence of NO_x emission on combustor inlet temperature is reflected in a strong relationship with cycle pressure ratio at sea-level conditions and with cycle pressure ratio and flight Mach number at altitude. Figure 2.28 illustrates the relationship between NO_x emission and cycle pressure ratio at sea-level static conditions. Figure 2.29 presents the dependence of NO_x emission on cycle pressure ratio and flight Mach number. Since the optimum pressure ratio for each flight Mach number changes with calendar time as technology developments allow higher-temperature operation, a band of logical operating conditions at the 1970 technology level has been indicated in Fig. 2.29.

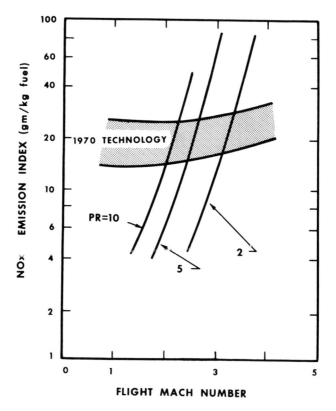

Fig. 2.29   Dependence of NO_x emission on flight Mach number.

Smoke formation is favored by high fuel-air ratio and pressure. Upon injection into the combustor, the heavy molecular weight fuel molecules are subjected to intense heating and molecular breakdown or pyrolysis occurs. If this process occurs in the absence of sufficient oxygen (i.e., high fuel-air ratio), the small hydrogen fragments can form carbon particulates that eventually result in smoke emission. The process by which carbon particulates are formed is known to be very pressure sensitive. Combustor designers have been successful in tailoring the burner to avoid fuel-rich zones, thus substantially reducing smoke levels. The current generation of engines has nearly invisible exhaust trails. The techniques resulting in these improvements will be highlighted in a subsequent subsection.

(2) *Afterburning engines.*   Relatively little information is available regarding emission during afterburner operation. General trends in existing data indicate possibly significant levels of CO and HC at the exhaust plane, especially at the lower afterburner power settings.[34-38] However, Lyon et al.[38] have confirmed that, at sea level, much of the CO and HC is chemically reacted to CO_2 and H_2O in the exhaust plume downstream of

the exhaust plane. These downstream reactions have been shown to consume up to 93% of the pollutants present at the exhaust plane. The extent of these plume reactions at altitude is uncertain. Although lower ambient pressures tend to reduce chemical reaction rates, reduced viscous mixing and the exhaust plume shock field tend to increase the exhaust gas time at high temperature and thus reduce the final emission of incomplete combustion products.

NO_x emission during afterburner operation expressed on an *EI* basis is lower than during nonafterburning operation because of reduced peak flame temperatures in the afterburner. The value, under sea-level conditions, is approximately 2–5 g/kg fuel.[39] At altitude, it is expected that the emission index would be 3.0 or less. Duct burners are expected to have an NO_x *EI* of about 5.0 during altitude operation.[39] While the total NO_x emission is not significantly influenced by plume reactions, there is speculation that conversion of NO to NO_2 occurs both within the afterburner and in the plume.[38]

Smoke or carbon particle emissions are reduced by the use of an afterburner.[38] Conditions within an afterburner are not conducive to carbon particle formation; furthermore, soot from the main burner may be oxidized in the afterburner, resulting in a net reduction.

Minimization of emissions.   Previous discussions of emissions levels concerned existing engines. Control technology may reduce emissions from these baseline levels by varying degrees. The fundamental means by which emissions may be reduced are discussed in this section.

(1) *Smoke emission.*   Technology to control smoke emission is well in hand and it would appear that future engines will continue to be capable of satisfying the future requirement of exhaust invisibility. The main design approach used is to reduce the primary zone equivalence ratio to a level where particulate formation will be minimized. Thorough mixing must be accomplished to prevent fuel-rich pockets that would otherwise preserve the smoke problem, even with overall lean primary zone operation. This must be done while maintaining other combustor performance characteristics. Airflow modification to allow leaner operation and air blast fuel atomization and mixing have been employed to accomplish these objectives. However, ignition and flame stabilization are the most sensitive parameters affected by making the primary zone leaner and must be closely observed during the development of low-smoke combustors.

(2) *HC and CO emission.*   To prevent smoke formation, the primary zone equivalence ratio of conventional combustors at higher-power operation must not be much above stoichiometric—this leads to much lower than stoichiometric operation at idle where overall fuel-air ratios are roughly one-third of the full-power value. Inefficient idle operation may be improved by numerous methods. The objectives in each technique are to provide a near-stoichiometric zone for maximum consumption of hydrocar-

bons while allowing sufficient time within the intermediate zone (where $\phi \simeq 0.5$) to allow for CO consumption.

To achieve an increased localized fuel-air ratio, dual orifice nozzles are frequently applied to modify fuel spray patterns at idle. Attempts to improve fuel atomization also provide decreased idle HC and CO through more rapid vaporization.[40] Greater local fuel-air ratios at idle can also be achieved by increased compressor air bleed or fuel nozzle sectoring. In the latter case, a limited number of nozzles are fueled at a greater fuel flow rate. Schemes where one 180 deg sector or two 90 deg sectors fueled have shown significant HC and CO reduction.[41]

Advanced approaches make use of staged combustion. The first stage, being the only one fueled at idle, is designed for peak idle combustion efficiency. The second stage is utilized only at higher-power conditions. This main combustion zone is designed with a primary motivation toward NO_x reduction. Significant HC and CO reductions have been demonstrated using the staged approach.[42–47] An example of such a design is shown in Fig. 2.30.

(3) *NO_x emission.* NO_x has been the most difficult aircraft engine pollutant to reduce in an acceptable manner. Currently available technology for reducing NO_x emissions consists of the two techniques discussed briefly below.

Water injection into the combustor primary zone has been found to reduce oxide-of-nitrogen emissions significantly (up to 80%). Peak flame temperatures are substantially reduced by the water injection, resulting in a sharp reduction in the NO_x formation rate. In a number of cases in which this technique has been attempted, however, CO emissions have increased, although not prohibitively. Figure 2.31 shows the relation between water

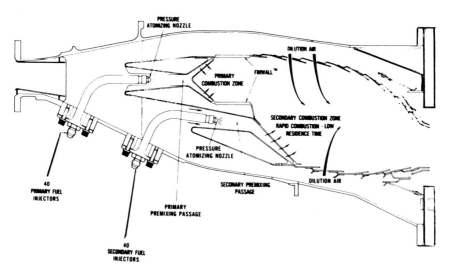

Fig. 2.30   Staged premix combustor (JT9D engine).

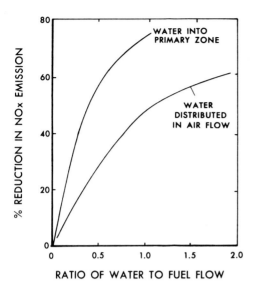

Fig. 2.31   Ideal effectiveness of water injection for NO_x control.

injection rate and NO_x reduction.[48] This method is not feasible for reducing cruise NO_x because the water flow required to attain significant abatement is of the order of the fuel-flow rate. In addition, there are difficulties with engine durability, performance, logistics, and economics associated with the cost of providing the necessary demineralized water. Consequently, these factors have caused this technique to receive negative evaluation as an approach toward reduction of ground-level NO_x for aircraft gas turbines.

A second method that has been used to reduce NO_x emissions involves air blast atomization and rapid mixing of the fuel with the primary zone airflow. Much has been written about this technique (notably the NASA swirl-can technology[49-51]). One engine, the F101, used in the B-1 aircraft, employs this principle. In the case of the F101, the overall combustor length was shortened from typical designs because of improved fuel-air mixture preparation. As a result, this method reduced both ground-level and altitude NO_x emissions. Reductions of approximately 50% below the uncontrolled case (Fig. 2.28) have been measured.

Advanced approaches to the reduction of NO_x can be divided into two levels of sophistication. The first level involves staged combustors like that shown in Fig. 2.30. In this case, fuel is injected upstream of the main combustion zone, which may be stabilized by a system of struts (or flameholders). Residence time in the premixing zone is short (i.e., high velocity and short length) because of the possibility of preignition or flame propagation upstream. It is known that these designs provide a fuel-air mixture far from ideal premix/prevaporization. In fact, the turbulence and nonuniformity characteristics of this sytem are probably not unlike those of conventional combustors. However, since the mixture ratio is only 0.6 stoichiometric, reduced NO_x levels result. Reductions of up to a factor of three have been achieved.[42-47]

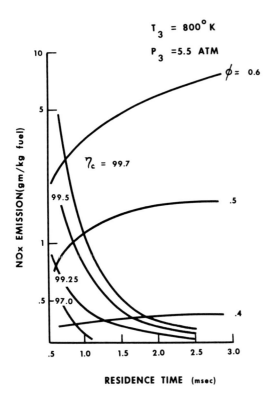

Fig. 2.32   Effect of residence time and ϕ on nitrogen oxide emissions.

Testing the advanced techniques of the second level—combustors utilizing premixing and prevaporization operating at equivalence ratios below 0.6—has aimed at ultra-low NO_x emission levels. These fundamental studies have been motivated by efforts to reduce automotive gas turbine emissions, as well as attempts to reduce stratospheric aircraft gas turbine emissions—the pressure ratio of the automotive engine is low with an absolute pressure level similar to the aircraft case at altitude. Nevertheless, inlet temperatures are high in the automotive case because of the use of regenerators. Ferri,[52] Verkamp et al.,[53] Anderson,[54] Wade et al.,[55] Azelborn et al.,[56] Collman et al.,[57] and Roberts et al.[58] have published results that indicate that levels below 1 g NO_2/kg fuel can be approached. Anderson's work is particularly thorough in discussing fundamental tradeoffs with combustion efficiency. His results are shown in Fig. 2.32. Good agreement with analytical model results indicates that useful conclusions may be drawn from the model predictions. These predictions all indicate that an "emissions floor" of approximately 0.3–0.5 g/kg fuel is the limit of NO_x emissions reduction.

A number of difficulties associated with the combustion of premixed/prevaporized lean mixtures can be alleviated by the use of a solid catalyst in the reaction zone. Recent developments to construct catalytic convertors to eliminate automotive CO and HC emissions have increased the temperature range within which such a device might operate. Test results have now been published that apply the concept of catalytic combustion to aircraft gas turbine combustors.[59-66] The presence of the catalyst in the combustion region provides stability at lower equivalence ratios than possible in gas-phase combustion. This is due to the combined effect of heterogeneous chemical reactions and the thermal inertia of the solid mass within the combustion zone. The thermal inertia of the catalytic combustor system has been calculated to be more than two orders of magnitude greater than in the gas-phase combustion system. NO_x reductions up to factors of 100 seem to be possible using the catalytic combustor approach.

Future Jet Fuels

Since 1973, the cost and availability of aircraft jet fuels have drastically changed. Jet fuel costs per gallon have increased tenfold for both commercial and military consumers. At times, fuel procurement actions have met with difficulties in obtaining desired quantities of fuel. These developments have encouraged examinations of the feasibility of expanding specifications to improve availability in the near term and, eventually, of producing jet fuels from nonpetroleum resources.[67-69]

Although economics and supply are primarily responsible for this recent interest in new fuel sources, projections of available worldwide petroleum resources also indicate the necessity for seeking new means of obtaining jet fuel. Regardless of current problems, the dependence on petroleum as the primary source of jet fuel can be expected to cease sometime within the next half-century.[70]

If the general nature of future aircraft (size, weight, flight speed, etc.) is to remain similar to today's designs, liquid hydrocarbons can be expected to continue as the primary propulsion fuel. Liquefied hydrogen and methane have been extensively studied as alternatives, but seem to be practical only for very large aircraft. The basic nonpetroleum resources from which future liquid hydrocarbon fuels might be produced are numerous. They range from the more familiar energy sources of coal, oil shale, and tar sands to possible future organic materials derived from energy farming. Some of the basic synthetic crudes, especially those produced from coal, will be appreciably different than petroleum crude. Reduced fuel hydrogen content would be anticipated in jet fuels produced from these alternate sources.

Because of the global nature of aircraft operations, jet fuels of the future could be produced from a combination of these basic sources. Production of fuels from blends of synthetic and natural crudes may also be expected. In light of the wide variations in materials from which worldwide jet fuel production can draw, it is anticipated that economics will dictate the acceptance of future fuels with properties other than those of the currently used JP-4, JP-5, and Jet A. Much additional technical information will be required to identify the fuel characteristics that meet the following objectives:

(1) Allow usage of key worldwide resources to assure availability.

(2) Minimize the total cost of aircraft system operation.

(3) Avoid major sacrifice of engine performance, flight safety, or environmental impact.

A complex program is necessary to establish the information base from which future fuel specifications can be derived. Figure 2.33 depicts the overall nature of the required effort. Fuel processing technology will naturally be of primary importance to per gallon fuel costs. The impact of reduced levels of refining (lower fuel costs) on all aircraft system components must be determined. These include fuel system (pumps, filters, heat exchangers, seals, etc.) and airframe (fuel tank size and design, impact on range, etc.) considerations, as well as main burner and afterburner impacts. In addition, handling difficulties (fuel toxicity) and environmental impact (exhaust emissions) require evaluation. The overall program must be integrated by a system optimization study intended to identify the best solution to the stated objective.

Fuel effects on combustion systems.   Future fuels may affect combustion system/engine performance through changes in hydrogen content, volatility, viscosity, olefin content, fuel nitrogen, sulfur, and trace metal content.

Fuel hydrogen content is the most important parameter anticipated to change significantly with the use of alternate fuels. In particular, fuels produced from coal can be expected to have significantly reduced hydrogen content. In most cases, the reduction in fuel hydrogen content will be due to increased concentrations of aromatic-type hydrocarbons in the fuel. These may be either single-ring or polycyclic in structure. Experience has shown that decreased hydrogen content significantly influences the fuel pyrolysis process in a manner that results in increased rates of carbon particle formation. In addition to increased smoke emission, the particulates are responsible for formation of a luminous flame in which radiation from the particles is a predominant mode of heat transfer.

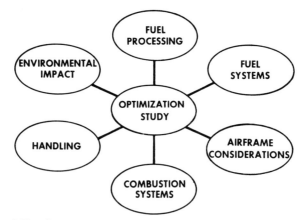

Fig. 2.33   Overall scheme for alternate jet fuel development program.

Significantly increased radiative loading on combustor liners can result from decreased fuel hydrogen content. Increases in liner temperature translate into decreases in hardware life and durability. Figure 2.34 illustrates the sensitivity of combustor liner temperature to hydrogen content. The following nondimensional temperature parameter is used to correlate these data,[31,71-74] which are representative of *older engine designs*:

$$\frac{T_L - T_{L0}}{T_{L0} - T_3}$$

The numerator of this expression represents the increase in combustor liner temperature T_L over that obtained using the baseline fuel (14.5% hydrogen JP-4) T_{L0}. This is normalized by the difference between T_{L0} and combustor inlet temperature T_3. It was found that data obtained using different combustors could be correlated using this parameter. It should also be noted that the parameter is representative of the fractional increase (over the baseline fuel) in heat transfer to the combustor liner.

Because combustor design differences play an important part in determining engine smoke characteristics, differences in emission are not correlatable in the same manner as combustor liner temperature. However, results obtained using a T56 single combustor rig[31] are illustrative of the important trends (see Fig. 2.35). Significantly increased smoke emission was determined with decreased hydrogen content for each condition tested. Trends between smoke emission and hydrogen content are similar for each

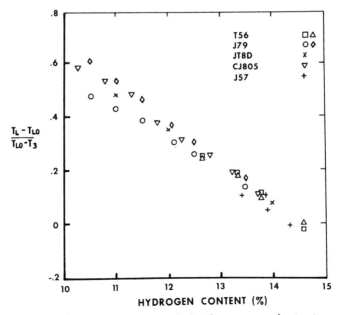

Fig. 2.34   Liner temperature correlation for many combustor types.

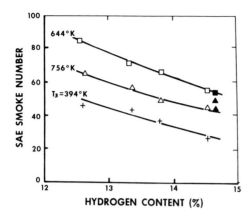

Fig. 2.35 Smoke emission dependence on hydrogen content.

combustion condition. Increased absolute smoke emission between the 394 and 644 K combustor inlet temperature conditions is attributable to increased pressure and fuel-air ratio. Although a further small increase might be expected for the 756 K condition because of higher pressure, the lower fuel-air ratio required to maintain the 1200 K exhaust temperature results in a lower absolute smoke emission.

Volatility affects the rate at which liquid fuel introduced into the combustor can vaporize. Since important heat release processes do not occur until gas-phase reactions take place, reduction of volatility shortens the time for chemical reaction within the combustion system. In the aircraft engine, this can result in reduced ground or altitude ignition capability, reduced combustor stability, increased emissions of carbon monoxide (CO) and hydrocarbons (HC), and the associated loss in combustion efficiency. Moreover, carbon particle formation is aided by the formation and maintenance of fuel-rich pockets in the hot combustion zone. Low volatility allows rich pockets to persist because of the reduced vaporization rate. Again, increased particulates can cause additional radiative loading to the combustor liners and more substantial smoke emissions.

The desired formation of a finely dispersed spray of small fuel droplets is adversely affected by viscosity. Consequently, the shortened time for gas-phase combustion reactions and the prolonging of fuel-rich pockets experienced with low volatility can also occur with increased viscosity. The ignition, stability, emissions, and smoke problems previously mentioned also increase for higher-viscosity fuels.

Olefin content is known to influence fuel thermal stability. Potential problems resulting from reduced thermal stability include fouling of oil fuel heat exchangers and filters and plugging of fuel metering valves and nozzles. No negative effect of fuel olefin content on gas-phase combustion processes would be expected.

The effect of increased fuel-bound nitrogen is evaluated by determining the additional NO_x emission occurring when nitrogen is present in the fuel and by calculating the percentage of fuel nitrogen conversion to NO_x necessary to cause this increase. Petroleum fuels have near-zero

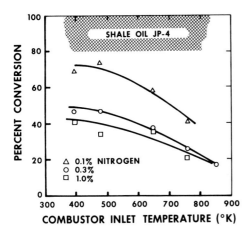

Fig. 2.36 Fuel-bound nitrogen conversion to NO_x in an aircraft gas turbine combustor.

(< 10 ppmw) fuel-bound nitrogen. Data presented in Fig. 2.36 for 0.1, 0.3, and 1.0% fuel nitrogen were obtained by doping a petroleum fuel with pyridine. The shale fuel was refined from a retorted Colorado oil shale. The results indicate the importance of two variables. First, as the combustor inlet temperature is increased, conversion is reduced. Second, as fuel-bound nitrogen concentrations are increased, conversion decreases. This second trend is consistent with the available results for oil shale JP-4, which has less than 0.08% nitrogen. The shale results are shown as a band in Fig. 2.36 because of the difficulties in accurately measuring small increases in NO_x emissions during that test.

Both sulfur and trace metals are at very low concentrations in current jet fuels. Sulfur is typically less than 0.1% because the petroleum fraction used for jet fuel production is nearly void of sulfur-containing compounds. Although syncrudes from coal or oil shale could be expected to contain higher sulfur levels, it is not likely that the current specification limit of

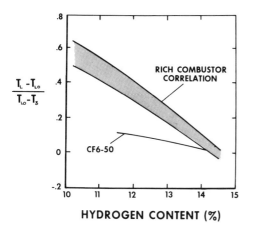

Fig. 2.37 Effect of lean operation on combustor fuel sensitivity.

0.4% would be exceeded with the processed jet fuel. Because of the way in which future jet fuels are expected to be produced, trace metals are also expected to continue to be present at low concentrations (less than 1 ppmw). Should higher levels appear possible, the serious consequences (deleterious effects on turbine blades) would probably justify the additional expense of their removal.

Combustion system design impact. Although assessment is still in the early stages, it appears certain that future combustion system designs will be significantly influenced by the changing character of fuel properties as alternate energy sources are tapped. It is essential to develop designs that accommodate lower-hydrogen-content fuels with good combustor liner durability and low-smoke emission and, at the same time, to maintain the customary level of combustion system performance.

Lean primary zone combustion systems, which are much less sensitive to fuel hydrogen content, will comprise a major approach to utilizing new fuels. Low-smoke combustor designs have been shown to be much less sensitive to variations in fuel hydrogen content. Figure 3.37 compares the correlation for older designs (Fig. 2.34) with results for a newer, smokeless combustor design, the CF6.[75] Current research on staged combustion systems will further contribute to the goal of achieving leaner burning while maintaining the desired system performance. Some of these designs have demonstrated very low sensitivity to fuel type.[75] These extremely important developments provide hope that future fuels having a lower hydrogen content can be accommodated while maintaining acceptable emissions characteristics.

Design and Performance Advancements

This subsection briefly addresses three new design concepts currently under consideration intended to address future turbopropulsion performance requirements. The variable-geometry combustor (VGC) is an advanced concept that resolves conflicting design requirements through control of the primary zone airflow. The vortex-controlled diffuser (VCD) is an improved, low-loss boundary-layer bleed diffuser that supports the needs of both current and future combustion systems. The shingle liner is an advanced concept combining new design features for both improved structural and thermal durability.

Variable-geometry combustor (VGC). For modern high-performance turbine engines, conflicting requirements are placed on the combustor design, including rapid deceleration without blowout, good altitude ignition characteristics, low idle emissions, and a lean primary zone for low smoke. As combustor design temperature rises have increased, it has become increasingly difficult to achieve a satisfactory design compromise. One solution, which may find application in future engines, is to vary the combustor geometry to control the distribution of air between the primary,

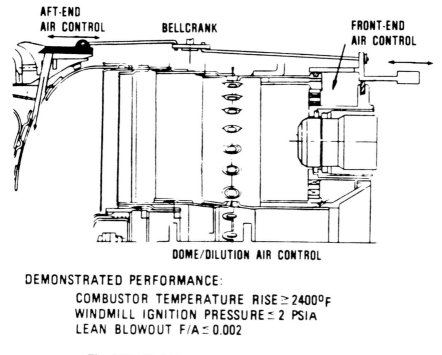

DEMONSTRATED PERFORMANCE:
COMBUSTOR TEMPERATURE RISE $\geq$ 2400°F
WINDMILL IGNITION PRESSURE $\leq$ 2 PSIA
LEAN BLOWOUT F/A $\leq$ 0.002

Fig. 2.38   Variable-geometry combustor (VGC).

intermediate, and dilution zones. Such a VGC has been developed by the Garrett Turbine Engine Company under joint Air Force and Navy sponsorship. A schematic cross section of this VGC, an annular reverse flow combustor, is shown in Fig. 2.38. A bell-crank arrangement is used to control both front- and aft-end airflow simultaneously. This is done in order to hold the liner pressure drop constant as the front-end airflow is varied. In the open position, the combustor primary zone operates lean, which provides good high-power efficiency, low smoke, and low unburned hydrocarbons. In the closed position, primary airflow is greatly reduced, providing good idle, altitude relight, and rapid deceleration performance. The Garrett VGC has been rig-tested using both the variable geometry described above and fuel staging. (Fuel staging involves turning some of the fuel nozzles off during low-power operation or rapid decelerations so that the fuel-air ratios for the remaining nozzles may be kept at a level that will ensure stable combustion.) The deceleration lean blowout performance is shown in Fig. 2.39. About half the decrease in lean blowout fuel-air ratio is due to the variable geometry and half to the fuel staging. The altitude ignition/blowout performance is equally good.

Vortex-controlled diffuser (VCD).   The VCD is a compact boundary-layer-bleed combustor inlet diffuser designed to effectively diffuse both conventional and high Mach number flowfields while providing good

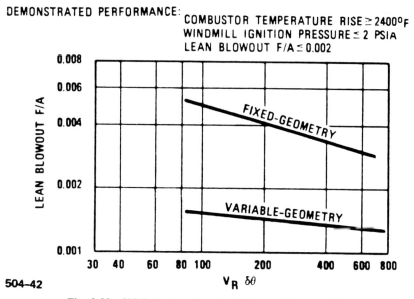

Fig. 2.39   VGC deceleration lean blowout characteristics.

pressure loss and flow stability in a very short length (relative to contemporary diffusion systems). The VCD was initially investigated at Cranfield Institute of Technology by Adkins.[76] The basic VCD geometry and nomenclature are defined in Fig. 2.40. Inner and outer VCD bleeds flow from the primary duct exit providing high-pressure recovery and low-pressure loss. The VCD advantages are principally short diffuser length, high-pressure recovery, design simplicity, and stable flow provided by the vortex-retaining fences. This concept offers considerable promise and is expected to find its way into a wide range of future propulsion system applications. Extended development of the VCD has been sponsored by the Air Force and conducted at the Allison Gas Turbine Division of General Motors Corporation.

Shingle liner.   The shingle liner design concept is an advanced combustor cooling technique featuring an innovation in which the thermal and mechanical stress loads of the combustor are isolated and controlled by independent means. The liner is basically an impingement-cooled segmented design as illustrated in Fig. 2.41. The outer shell serves as the structural or load-carrying portion of the combustion system and provides impingement cooling for the inner segments or shingles. As a result, the shingles provide an effective thermal barrier, protecting the highly stressed outer shell. The shingle liner is particularly well suited to high-temperature-rise combustor operation where cooling airflow is at a premium. Additionally, the shingle concept offers improved liner life due to its thermally relieved mechanical design aspect and the possibility of improved maintain-

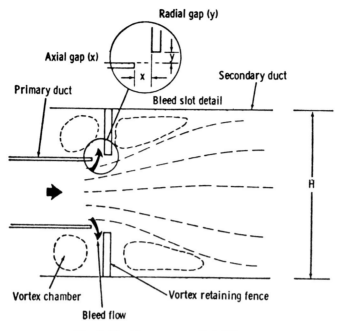

Fig. 2.40   Vortex-controlled diffuser.

ability, because low-cost "throwaway" segments may be employed. The shingle liner concept also permits the combustor designer a greatly increased latitude in selecting liner materials. Since the support shell remains relatively cool, lightweight lower-temperature alloys may be used for it. And, since the thermally induced stresses are reduced in the shingles, materials ranging from conventional liner alloys to cast turbine alloys to ceramics may be used.

The shingle liner development was sponsored by the Navy and conducted by the General Electric Company. It is presently being considered for both near-term and future propulsion system application. A similar concept, trade named Floatwall, has also been developed at Pratt and Whitney Aircraft.

2.6 CONCLUSIONS

As discussed earlier, the turbine engine combustion system has undergone an evolutionary development process over the past 50 years, beginning with long, bulky, can-type combustors (i.e., the J33 shown in Fig. 2.3) and progressing to the compact, high-temperature-rise annular combustors of today's newest turbopropulsion systems. In recent years, significant technological advancements have been realized in both combustion system design and performance. With respect to the important design parameters of combustion efficiency and stability, pressure loss, combustor size, and pattern factor, the annular combustors recently developed have provided

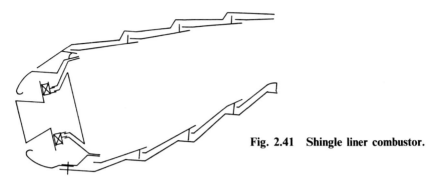

Fig. 2.41 Shingle liner combustor.

substantial improvements. Further improvements in these parameters will be required, however, if propulsion system demands of the future are to be met.

In the vital area of durability, improvements in liner design and cooling have added substantially to the maintainability and durability aspects of the combustor at a time when system operating pressures and temperatures are on the rise. Figure 2.42 illustrates the technological improvements realized in the 20–25 years since annular combustors were introduced. For example, the continued drive for reduced cost, improved fuel economy, and design compactness and simplicity has led to the compact, high-temperature combustor of the F101 engine (developed for the B-1 bomber) illustrated in Fig. 2.43. This combustor is a low pressure loss (5.1%), high heat release (7.5×10^6 Btu/h/atm/ft^3) design employing an improved low-

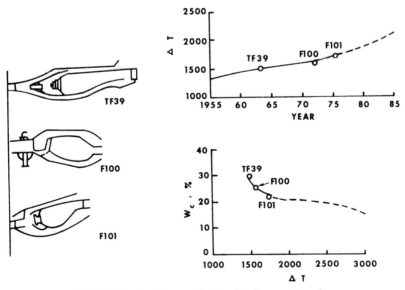

Fig. 2.42 Annular combustor development trends.

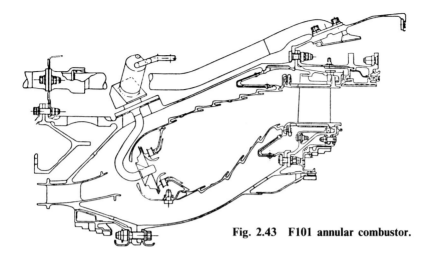

Fig. 2.43   F101 annular combustor.

pressure fuel injection system, a machined-ring high-durability, liner and a simple, cast, low-loss inlet dump diffuser. Relative to other contemporary combustion systems, the F101 is the most advanced annular design developed to date and introduces a new family of compact, high-temperature systems for fighter, bomber, and transport applications.

Future aircraft propulsion requirements call for primary combustors capable of (1) accepting greater variations in compressor discharge pressure, temperature, and airflow; (2) producing heat release rates and temperature rise that will ultimately approach stoichiometric levels; and (3) providing high operational reliability and improved component durability, maintainability, and repairability. In addition, the new requirements discussed in Sec. 2.5—exhaust emissions and fuel flexibility—must be addressed.

It is possible only to speculate on new concepts that might be employed in the next quarter century in aeropropulsion combustion. Nevertheless, such an effort is worthwhile as the reader may gain an appreciation for the wide range of opportunity and flexibility that remains available to the combustion system designer. Two concepts—catalytic combustion and photochemically assisted combustion—will be highlighted here.

Catalytic combustion involves the use of a heterogeneous catalyst within the combustion zone to increase the energy release rate.[59-66] Fuel and air are premixed at low equivalence ratios, often below the lean flammability limit, and passed through a catalytically coated, ceramic honeycomb structure. Due to the combined effect of heterogeneous chemical reactions and the thermal inertia of the solid structure, this concept can be utilized to achieve stable, efficient combustion outside the normal flammability limits of gas-phase systems.[59-66] Typical experimental results are shown in Fig. 2.44. The benefits of lean combustion (low radiative emission, decreased tendency for turbine inlet temperature nonuniformities, reduced smoke and

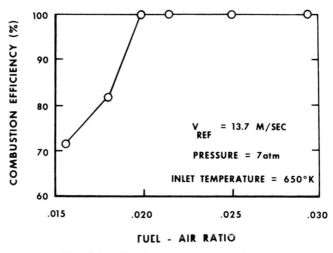

Fig. 2.44   Catalytic combustor efficiency.

NO_x formation) provide significant potential payoff. Applications in both main burners and afterburners are possible. The promising concept of a porous flameholder with low-pressure drop has been investigated. Such a device would provide a means of stabilizing combustion flows that reach final flame temperatures in excess of flameholder material limitations.

A second concept utilizes either ultraviolet light sources or plasmas to stabilize or promote combustion processes. Studies have confirmed the possibility of photochemical ignition where a pulse of ultraviolet light (from an arc discharge) is used to dissociate molecular oxygen and provide ignition at lower temperatures and with lower energy input than in the case of a spark ignition source.[77,78] Other investigations have confirmed the capability of plasma jets in supplying reactive species to stabilize and improve the efficiency of combustion flows.[79] The payoff of these two concepts can be flame stabilization without flameholding devices—an "optical flameholder" with zero pressure drop. Such a device would further promote the feasibility of practical lean combustion.

While the example concepts just described are in their infancy and may not find eventual practical application in the turbopropulsion field, they are illustrative of future developments in combustion technology. As the combustor designer is confronted with the new requirements of the future, especially exhaust emissions and fuel flexibility, new concepts like these will play an important role in problem solution. Engineers in the aeropropulsion combustion community will certainly enjoy challenges with the possibility for imaginative solutions as the next quarter century unfolds.

References

[1]Whittle, F., *Jet, The Story of a Pioneer*, Frederich Muller, London, 1953.

[2]Oates, G. C. (ed.), *Aerothermodynamics of Aircraft Engine Components*, AIAA Education Series, New York, 1985.

[3]Smith, I. E., *Combustion in Advanced Gas Turbine Systems*, Pergamon, New York, 1967.

[4]Cornelius, W. and Agnew, W. G., *Emissions From Continuous Combustion Systems*, Plenum, New York, 1972.

[5]Lefebvre, A. H., *Gas Turbine Combustors*, McGraw-Hill, New York, 1983.

[6]Swithenbank, J., *Combustion Fundamentals*, Air Force Office of Scientific Research, Washington, DC, Feb. 1970.

[7]Barnett, H. C. and Hibbard, R. R. (eds.), "Basic Consideration in the Combustion of Hydrocarbon Fuels with Air," NACA Rept. 1300, 1957.

[8]"The Design and Performance Analysis of Gas-Turbine Combustion Chambers," Vol. I, Northern Research and Engineering Corp., Cambridge, MA, 1964.

[9]Reneau, L. R., Johnston, J. P., and Kline, S. J., "Performance and Design of Straight Two-Dimensional Diffusers," Thermosciences Div., Dept. of Mechanical Engineering, Stanford Univ., Stanford, CA, Rept. PD-8, 1964.

[10]Rizkalla, A. A. and Lefebvre, A. H., "The Influence of Air and Liquid Properties on Airblast Atomization," *Journal of Fluids Engineering*, Vol. 97, No. 3, Sept 1975, pp. 316–320.

[11]Rizkalla, A. A. and Lefebvre, A. H., "Influence of Liquid Droplets on Airblast Atomizer Spray Characteristics," *Journal of Engineering for Power*, Vol. 97, April 1975.

[12]Lefebvre, A. H. and Miller, D., "The Development of Airblast Atomizer for Gas Turbine Application, College of Aeronautics," Cranfield, England, Aero Note 193, 1967.

[13]Lefebvre, A. H., *Theoretical Aspects of Gas Turbine Combustion Performance*, College of Aeronautics, Cranfield, England, Aero Note 163, Aug. 1966.

[14]Herbert, J. D., "Theoretical Analysis of Reaction Rate Controlled Systems— Part I," *AGARD Combustion Research and Reviews*, 1957, Chap. 6.

[15]Swithenbank, J., Poll, I., Wright, D. D., and Vincent M. W., "Combustion Design Fundamentals," *14th Symposium (International) on Combustion*, The Combustion Institute, Pittsburgh, PA, Aug. 1972.

[16]Spaulding, D. B., "Mathematical Models of Continuous Combustion," *Emissions From Continuous Combustion Systems*, edited by W. Correlius and W. G. Agnew, Plenum, New York, 1972.

[17]Gosman, A. O., Pun, W. M., Runchal, A. K., Spaulding, D. B., and Wolfshtein, M., *Heat and Mass Transfer in Recirculating Flows*, Academic, Orlando, FL, 1969.

[18]Mosier, S. A. and Roberts, R. "Low Lower Turbopropulsion Combustor Exhaust Emissions," Vols. I–III, Air Force Aero Propulsion Laboratory, AFAPL-TR-73-36, June 1973.

[19]Anasoulis, R. F., McDonald, H., and Buggelin, R. C., "Development of a Combustor Flow Analysis," Vols. I and II, Air Force Aero-Propulsion Laboratory, AFAPL-TR-73-98, Jan. 1974.

[20]Clean Air Act Amendments of 1970, 40 USC 1857 as amended by PL 91-604.

[21]"Aircraft Emissions: Impact on Air Quality and Feasibility of Control," U.S. Environmental Protection Agency, Washington, DC, Feb. 1972.

[22]U.S. Environmental Protection Agency, "Control of Air Pollution from Aircraft and Aircraft Engines," *Federal Register*, Vol. 47, No. 251, Dec. 30, 1982.

[23]Grobecker, A. J., Coroniti, S. C., and Cannon, R. H., Jr., "Report of Findings—Executive Summary, The Effects of Stratospheric Pollution by Aircraft,"

Climatic Impact Assessment Program, U.S. Department of Transportation, Washington, DC, Dec. 1974.

[24]"Procedure for the Continuous Sampling and Measurement of Gaseous Emissions from Aircraft Turbine Engines," Aerospace Recommended Practice 1256, Society of Automotive Engineers, Warrendale, PA, 1971.

[25]"Aircraft Gas Turbine Engine Exhaust Smoke Measurement," Aerospace Recommended Practice 1179, Society of Automotive Engineers, Warrendale, PA, 1970.

[26]Champagne, D. L., "Standard Measurement of Aircraft Gas Turbine Exhaust Smoke," ASME Paper 71-GT-88, March 1971.

[27]Giovanni, D. V. and Hilt, M. B., "Particulate Matter Emission Measurements from Stationary Gas Turbines," ASME Paper 73-PWR-17, Sept. 1973.

[28]Slusher, G. R., "Sulfur Oxide Measurement in Aircraft Turbine Engine Exhaust," Federal Aviation Administration, Washington, DC, Rept. FAA-RD-75-101, Sept. 1975.

[29]Lipfert, F. W., "Correlations of Gas Turbine Emissions Data," ASME Paper 72-GT-60, March 1972.

[30]Blazowski, W. S., Fahrenbruck, F. S., and Tackett, L. P., "Combustion Characteristics of Oil Shale Derived Jet Fuels," presented at Western States Section of The Combustion Institute, Paper 75-13, Oct. 1975.

[31]Blazowski, W. S., "Combustion of Considerations for Future Jet Fuels," *16th Symposium (International) on Combustion*, The Combustion Institute, Pittsburgh, PA, 1976.

[32]Butze, H. F. and Ehlers, R. C., "Effect of Fuel Properties on Performance of a Single Aircraft Turbojet Combustor," NASA-TM-X-71789, Oct. 1975.

[33]Wilkes, C. A. and Johnson, R. H., "Fuel Property Effects on Gas Turbine Emissions Control," ASME Paper, Sept. 1974.

[34]Lazalier, G. R. and Gearhart, J. W., "Measurement of Pollution Emissions from an Afterburning Turbojet Engine at Ground Level, Part 2, Gaseous Emissions," Arnold Engineering Development Center, Tullahoma, TN, AEDC-TR-72-70,

[35]Palcza, J. L., "Study of Altitude and Mach Number Effects on Exhaust Gas Emissions of an Afterburning Turbofan Engine," Federal Aviation Administration, Washington, DC, Rept. FAA-RD-72-31, 1971.

[36]German, R. C., High, J. D., and Robinson, C. E., "Measurement of Exhaust Emissions from a J85-GE-5B Engine at Simulated High-Altitude, Supersonic, Free-Stream Flight Conditions," Arnold Engineering Development Center, Tullahoma, TN, Rept. AEDC-TR-73-103, 1973.

[37]Diehl, L. A., "Measurement of Gaseous Emissions from an Afterburning Turbojet Engine at Simulated Altitude Conditions," NASA TM-X-2726, 1973.

[38]Lyon, T. F., Colley, W. C., Kenworthy, M. J., and Bahr, D. W., "Development of Emissions Measurement Techniques for Afterburning Turbine Engines," Air Force Aero Propulsion Laboratory, Wright-Patterson AFB, OH, Rept. AFAPL-TR-75-52, Oct. 1975.

[39]Grobman, J. and Ingebo, R. D., "Forecast of Jet Engine Exhaust Emissions of High Altitude Commercial Aircraft Projected to 1990," Propulsion Effluents in the Stratosphere, CIAP Monograph 2, Rept. DOT-TST-75-52, Sept. 1975.

[40]Norgren, C. T. and Ingebo, R. D., "Effect of Fuel Vapor Concentrations on Combuster Emissions and Performance," NASA TM-X-2800, 1973.

[41]Bahr, D. W., "Control and Reduction of Aircraft Turbine Engine Exhaust Emissions," *Emissions from Continuous Combustion Systems*, Plenum, New York, 1972, pp. 345–373.

[42]Mularz, E., "Results of the Pollution Reduction Technology Program for Turboprop Engines," AIAA Paper 76-760, July 1976.

[43]Roberts, R., Fiorentino, A., and Diehl, L., "Pollution Reduction Technology for Turbofan Engine Can Annular Combustors," AIAA Paper 76-761, July 1976.

[44]Niedzwiecki, R. and Roberts, R., "Low Pollution Combustors for CTOL Engines—Experimental Clean Combustor Program (JT9D)," AIAA Paper 76-762, July 1976.

[45]Niedzwiecki, R. and Gleason, C., "Low Pollution Combustor for CTOL Engines—Experimental Clean Combustor Program (CF6)," AIAA Paper 76-763, July 1976.

[46]Roberts, P. B., White, D. J., Shekleton, J. R., and Butze, H. F., "Advanced Low NO_x Combustors for Aircraft Gas Turbines," AIAA Paper 76-764, July 1976.

[47]Verdouw, A. J., "Evaluation of a Staged Fuel Combustor for Turboprop Engines," ASME Paper 76-WA/GT-5, July 1976.

[48]Blazowski, W. S. and Henderson, R. E., "Aircraft Exhaust Pollution and Its Effects on the U.S. Air Force," Air Force Aero Propulsion Laboratory, Wright-Patterson AFB, OH, Rept. AFAPL-TR-74-64, 1974.

[49]Niedzwiecki, R. W. and Jones, R. E., "Parametric Test Results of Swirl Can Combustor," NASA TM-X-68247, June 1973.

[50]Jones, R. E., "Advanced Technology for Reducing Aircraft Engine Pollution," NASA TM-X-68256, Nov. 1973.

[51]Mularz, E. J., Wear, I. D., and Verbulecy, P. W., "Pollution Emissions from Single Swirl Can Combustor Modules at Parametric Test Conditions," NASA TM-X-3167, Jan. 1975.

[52]Ferri, A., "Reduction of NO Formation by Premixing in Atmospheric Pollution by Aircraft Engines," AGARD CP 125, April 1973, A-1–A-9.

[53]Verkamp, F. J., Verdouw, A. J., and Tomlinson, J. G., "Impact of Emissions Regulations on Future Gas Turbine Engine Combustors," AIAA Paper 73-1277, Nov. 1973.

[54]Anderson, D. N., "Effects of Equivalence Ratio and Dwell Time on Exhaust Emissions from an Experimental Premixing Prevaporizing Burner," NASA TM-X-71592, 1974.

[55]Wade, W. R., Shen, P. I., Owens, C. W., and McLean, A. F., "Low Emissions Combustion for the Regenerative Gas Turbine, Part I—Theoretical and Design Considerations," ASME Paper 73-GT-11, Dec. 1972.

[56]Azelborn, N. A., Wade, W. R., Secord, J. R., and McLean, A. F., "Low Emissions Combustion for the Regenerative Gas Turbine, Part II—Experimental Techniques, Results, and Assessment," ASME Paper 73-GT-12, Dec. 1972.

[57]Collman, J. S., Amann, C. A., Mathews, C. C., Stettler, R. J., and Verkamp, F. J., "The GT-255—An Engine for Passenger Car Gas Turbine Research," SAE Paper 750167, Feb. 1975.

[58]Roberts, P. B., Shekleton, J. R., White, D. J., and Butze, H. F., "Advanced Low NO_x Combustor for Supersonic High-Altitude Aircraft Gas Turbines," ASME Paper 76-GT-12, 1976.

[59]Blazowski, W. S. and Bresowar, G. E., "Preliminary Study of the Catalytic Combustor Concept as Applied to Aircraft Gas Turbines," Air Force Aero Propulsion Laboratory, Wright-Patterson AFB, OH, Rept. AFAPL-TR-74-32, 1974.

[60]Wampler, F. P., Clark, D. W., and Gaines, F. A., "Catalytic Combustion of C_3H_8 on Pt Coated Monolith," Western States Section of the Combustion Institute, Paper 74-36, Oct. 1974.

[61]Blazowski, W. S. and Walsh, D. E., "Catalytic Combustion: An Important Consideration for Future Applications," Combustion Science and Technology, Vol. 10, 1975, pp. 233–244.

[62]Rosfjord, T. J., "Catalytic Combustors for Gas Turbine Engines," AIAA Paper 76-46, Jan. 1976.

[63]Anderson, D. N., Tacina, R. R., and Mroz, T. S., "Performance of a Catalytic Reactor at Simulated Gas Turbine Operating Conditions," NASA TM-X-71747, 1975.

[64]Pfefferle, W. C. et al., "Catathermal Combustion: A New Process for Low Emissions Fuel Conversion," ASME Paper 75-WA/Fu-1, Dec. 1975.

[65]DeCorso, S. M., et al., "Catalysts for Gas Turbine Combustors—Experimental Test Results," ASME Paper 76-GT-4, March 1976.

[66]Anderson, D. N., "Preliminary Results from Screening Tests of Commercial Catalysts with Potential Use in Gas Turbine Combustors, Part I: Furnace Studies of Catalyst Activity," NASA TM-X-73410 and "Preliminary Results from Screening Tests of Commercial Catalysts with Potential Use in Gas Turbine Combustors, Part II: Combustion Test Rig Evaluation," NASA TM-X-73412, May 1976.

[67]Goen, R. L., Clark, C. F., and Moore, M. A., "Synthetic Petroleum for Department of Defense Use," Air Force Aero Propulsion Laboratory, Wright-Patterson AFB, OH, AFAPL-TR-74-115, Nov. 1974.

[68]Shaw, H., Kalfadelis, C. D., and Jahnig, C. E., "Evaluation of Methods to Produce Aviation Turbine Fuels from Synthetic Crude Oils, Phase I," Air Force Aero Propulsion Laboratory Wright-Patterson AFB, OH, Rept. AFAPL-TR-75-10, March 1975.

[69]Bartick, H., Kunchal, K., Switzer, D., Bowen, R., and Edwards, R., "The Production and Refining of Crude Shale Oil into Military Fuels," Office of Naval Research, Arlington, VA, Final Report for Contract N00014-75-C-0055, Aug. 1975.

[70]Pinkel, I. I., "Future Fuels for Aviation," AGARD Advisory Rept. 93, Jan. 1976.

[71]McClelland, C. C., "Effects of Jet Fuel Constituents on Combustor Durability," Naval Air Propulsion Test Center, Rept. NAEC-AEL-1736, May 1963.

[72]Butze, H. F. and Ehlers, R. C., "Effect of Fuel Properties on Performance of A Single Aircraft Turbojet Combustor," NASA TM-X-71789, Oct. 1975.

[73]Macaulay, R. W. and Shayeson, M. W., "Effects of Fuel Properties on Linear Temperatures and Carbon Deposition in the CJ805 Combustor for Long Life Applications," ASME Paper 61-WA-304, Oct. 1961.

[74]Schirmer, R. M., McReynolds, L. A., and Daley, J. A., "Radiation from Flames in Gas Turbine Combustors," SAE Transactions, Vol. 68, 1960, pp. 554–561.

[75]Gleason, C. C. and Bahr, D. W., "Experimental Clean Combustor Program Alternate Fuels Addendum Phase II Final Report," NASA CR-134972, Jan. 1976.

[76]Adkins, R. C., "A Short Diffuser with Low Pressure Loss," ASME Paper, May 1974.

[77]Cerkanowicz, A. E., Levy, M. E., and McAlevy, R. F., "The Photochemical Ignition Mechanism of Unsensitized Fuel-Air Mixtures," AIAA Paper 70-149, Jan. 1970.

[78]Cerkanowicz, A. E., "Photochemical Enhancement of Combustion and Mixing in Supersonic Flows," Air Force Office of Scientific Research, Washington, DC, Rept. AFOSR-TR-74-0153, Nov. 1973.

[79]Weinberg, F. J., "The First Half Million Years of Combustion Research and Today's Burning Problems," *15th Symposium (International) on Combustion*, The Combustion Institute, Pittsburgh, PA, 1975.

CHAPTER 3. ENGINE/AIRFRAME PERFORMANCE MATCHING

D. B. Morden
Boeing Commercial Airplanes, Seattle, Washington

3
ENGINE/AIRFRAME
PERFORMANCE MATCHING

NOMENCLATURE

Throughout this chapter, the symbols are defined in the text where they are used. Commonly accepted symbols are used whenever possible.

3.1 INTRODUCTION

When engine cycles and airframe concepts are analytically integrated to define an airplane, the engine and airframe subsystems interact with one another. Engine physical characteristics affect the size, shape, weight, and balance of the airframe. Airframe characteristics affect the installed performance of the propulsion system. The "best" engines for a given airplane cannot be identified until the required installed thrust has been established for the desired mission. Analytical and empirical tools can be used to estimate the required engine size and airplane size to perform any desired mission. Further, the calculation can be used to optimize the engine cycles and the aircraft geometry for a mission and thus provide a quantitative measure of the sensitivity of the system's performance to any changes in engine, airplane, or mission description.

System matching requires that the installed engine/airframe performance be evaluated over the entire operating range of Mach numbers and altitudes. The specified flight path elements and the corresponding performance requirements for the aircraft are referred to as the design mission. Mission analysis is introduced to evaluate engine/airframe performance by summing the increments in fuel consumed and/or distance covered for the specified mission segments or legs. The performance is also computed at constraint points where, for example, a particular acceleration, maneuverability, or noise level is required. If the design objectives are not met, the candidate engine cycle and airframe are resized and the process is repeated. Several iterations are usually necessary to size a candidate engine and airplane to meet the mission. The objective of the design process, referred to as optimization, is to compare satisfactory candidate configurations and identify the one that *best* satisfies the mission requirements. The digital computer provides the means of quickly and efficiently assessing large numbers of configurations. While the computer is a powerful tool, the quality of the final airplane is dependent on the engineering judgment applied in choosing candidates and by the accuracy of the analytical methods and empirical expressions used to describe the engine and airframe.

Over a period of years, airframe and engine manufacturers have compiled a large inventory of experience that has been used to carry out these simulations. Routines vary within industry and on-going efforts keep changing the program formats and complexity in a continual quest for more accurate and less costly computation techniques. It is not the purpose of this text to describe specific current programs. Rather, this chapter presents some examples of the basic procedures that are fundamental to computer simulations of mission analysis, design optimization and sensitivity, and gas generator performance calculation.

3.2 MISSION ANALYSIS

Philosophy and Logic

The mission that an aircraft is designed to perform consists of a flight path with several distinct segments where specific performance requirements must be met. Each of the mission segments can be analyzed using Newton's laws of motion. The final condition resulting from calculation of one segment becomes the initial condition for the next segment. By combining the various segments into an appropriate sequence, any mission can be described. Relationships must be provided to describe the following for each segment: (1) lift and drag for the airframe, (2) engine cycle characteristics, (3) weight buildup and scaling rules, (4) inlet and nozzle performance, and (5) interaction or installation effects.

To begin the analysis, an initial airplane configuration must be specified based on a perceived market requirement or knowledge of a customer's requirements. Usually, the various disciplines contributing components to the airplane supply initial weight and volume requirements. A configuration is produced that incorporates the various constraints into an aerodynamic shape. This stage relies very heavily on past experience and perceived benefits. The result constitutes the "candidate airplane." The wetted area, volume distribution, weight, and drag are then estimated. The design point for the engine cycle usually corresponds to the most demanding performance point or segment in a mission. The engine operation is simulated for appropriate altitude and Mach number conditions characteristic of the mission with engine size (i.e., airflow rate) determined from the required thrust. The paper engine can then be resized if necessary by comparing the thrust obtained to the thrust required at other operating conditions. Thus, the mission analysis also requires scaling rules for changes in engine dimensions and weight as a function of airflow.

Mission analysis is used to iterate on engine and airplane size to satisfy mission requirements for a fixed payload or to iterate on range with trades in fuel and payload for a fixed airplane size. Normally, the most demanding mission or the mission resulting in the largest takeoff gross weight (TOGW) or carrying a desired payload a specified distance is referred to as the design mission. Once an engine and airplane are matched for the design mission, off-design missions can be run using the fixed airplane to establish performance or payload on other missions of interest. The designer will usually choose an efficient computer program that uses a minimum number of inputs to simulate the aircraft system to the desired accuracy. The required accuracy and complication in system description increases as an airplane concept proceeds from initial concept evaluation through final design.

The fundamentals of the process of combining subsystems and the assessment of the resulting systems are the subject of the remainder of this section. Increases in accuracy do not change the basic approach, but only increase the complexity of the component and interaction descriptions.

Fixed missions, variable airplane size. Missions may be specified partially or completely; i.e., while the sequence of mission segments is always specified, either the distance traveled or the elapsed time may be taken as a variable determined by the amount of fuel available. For a given set of performance requirements, an aircraft can be sized to fly any desired distance up to the point of design divergence (zero payload) if all major components (including the engine) are appropriately scaled. A mission is said to be fixed when the total distance and altitude/Mach number requirements, together with either segment distance or segment time, are specified for all phases of flight.

The weight of the airplane is the sum of all the component weights, including payload and fuel, as represented in Fig. 3.1. The takeoff gross weight is initially estimated for the airplane. The engine is initially sized to produce the required thrust-to-weight ratio at each of several points where a specific level of performance is required. Typical engine sizing points include: maximum Mach number at specified (or best) altitude, maximum speed at sea level, takeoff field length, landing distance, excess power, and acceleration. The airframe size and weight must be chosen to provide necessary lift, maneuverability, and fuel. Starting with this initial engine and airplane weight estimate, the fixed mission is "flown" to see if the performance requirements are met and to compute the amount of fuel consumed. An iterative process is then used until the airplane gross takeoff weight is appropriate for the entire mission, consuming all the fuel except for specified reserves.

Fixed airplane size, variable mission. In this mode of operation, the airplane size and weight are held constant while trading fuel and payload weight. Usually, the capability of the airplane is evaluated on several different missions with different payloads to develop a payload/range curve. This approach is used to evaluate candidate engines using the criteria of best range or largest payload. Occasionally, this form of mission analysis is also used to evaluate new or modified airframes using existing engines.

Overview of Aerodynamic Force Determination

Engines must produce enough thrust to overcome drag and the additional thrust to provide the desired acceleration at each point in a mission. Traditionally, the aircraft industry employs separate staffs to evaluate the propulsion system and the aerodynamics of the airplane. Although it is

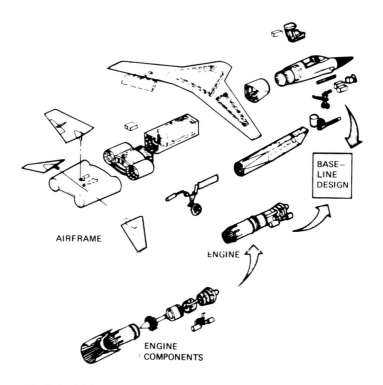

BASE–
LINE
DESIGN

AIRFRAME

ENGINE

ENGINE
COMPONENTS

Fig. 3.1   Multitechnology computer program design representation.

seldom necessary for the propulsion engineer to become involved in the detailed analysis of the airplane aerodynamics, it is important that he understand the principal drag sources. This is becoming increasingly important with the advent of multimission aircraft, fighters, and supersonic transports because of the substantial interactions between the propulsion system and airframe.

This section presents a brief overview of airplane lift and drag and shows how to build up the drag polar that will be used to relate the airplane weight and thrust requirements.

Lift and drag. Dimensional analysis can be conveniently exploited to show that the dimensionless force C_F acting on a body immersed in a fluid can be expressed as a function of three dimensionless parameters: the Reynolds number Re, Mach number M, and ratio of specific heats of the fluid γ. We consider here only bodies interacting with air, so γ remains approximately constant. Thus, if geometrically similar bodies are considered, the force relationship may be written in the form

$$C_F = f(Re, M) \tag{3.1}$$

where

$C_F = F/qA \equiv$ force coefficient
F = force (component)
q = dynamic pressure, $= \frac{1}{2}\rho V^2$
A = representative area, usually the wing area
Re = Reynolds number, $= \rho V \ell / \mu$
ℓ = representative length; for wings, the average wing chord

The most commonly used forms of Eq. (3.1) are those in which the forces considered are lift and drag. In these cases, the coefficients are denoted C_L and C_D, respectively. Experiments demonstrate that Mach number has much more influence on C_F than does the Reynolds number in the cruise flight regime of both subsonic and supersonic aircraft.

The vector integral of the static pressure distribution over a surface is the net pressure force. It is convenient to express this force as acting at a point about which moments generated by the force are zero (referred to as the center of pressure). Figure 3.2 shows the pressure distribution and net pressure force for a typical airfoil section. The force is resolved into lift and drag components, which are perpendicular and parallel to the flight velocity vector.

Airfoils usually show a linear variation of lift coefficient within a range of angle of attack. At large airflow, separation is experienced and the lift curve becomes nonlinear (see Fig. 3.3) until a maximum lift coefficient value is reached at $C_{L\,max}$. Beyond this point, C_L decreases, either abruptly or gradually depending on the particular airfoil (except for a low-aspect-ratio delta wing). This phenomenon is called "stalling" and the angle corresponding to $C_{L\,max}$ is called the stall angle. The angle of attack

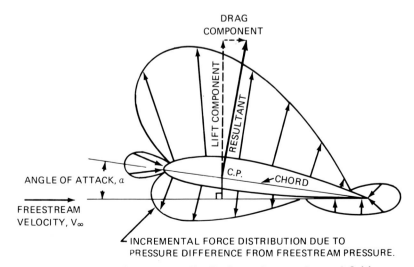

Fig. 3.2   Airfoil section pressure distribution and nomenclature definition.

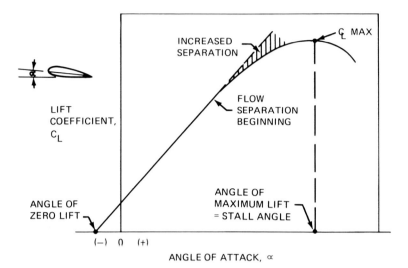

Fig. 3.3   Variation of lift coefficient with angle of attack for a typical airfoil.

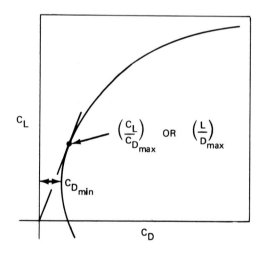

Fig. 3.4   Typical wing drag polar showing points of minimum drag and maximum L/D.

resulting is zero; lift is zero for airfoils with symmetrical cross sections and negative for typical wing cross sections having camber.

For a wing, the relationship between the drag coefficient and lift coefficient is approximately parabolic in shape, up to $C_{L_{max}}$, and the plot of C_L vs C_D is called a drag polar (Fig. 3.4). The angle of attack associated with such a curve increases as C_L increases below stall; for wings with camber, the minimum C_D value does not occur at zero C_L. The drag polar for a complete airplane includes the drag of all exposed surfaces and has a similar shape to that of the wing alone.

From the definitions of C_L and C_D,

$$\frac{C_L}{C_D} = \frac{C_L q S}{C_D q S} = \frac{L}{D} \tag{3.2}$$

A tangent line through the origin on Fig. 3.4 defines the C_L and C_D values for the maximum lift-to-drag ratio. The tangent point locates the angle of attack that produces the most lift for the least drag.

By extending the idea of a wing drag polar to represent the entire airplane, the relationship between lift and drag combined with the equation of motion for each flight segment is sufficient for computing the thrust requirement for each increment of a mission.

Drag polar buildup. The thrust the engines must produce for each increment of a given mission is dependent on the following factors: atmospheric conditions (altitude and temperature), flight Mach number M, average gross weight during the increment (GW), and type of segment. These conditions, together with the aircraft drag polar, are sufficient for mission analysis.

A drag polar for the entire airplane can be developed using the lift/drag relationships for each of the components including the components' interactions. In an effort to understand drag and to produce equations suitable for determining drag polars, a methodology has been developed that combines theoretical and empirical drag contributions into a drag "buildup." Traditionally, subsonic drag is represented as the sum of parasite (friction plus pressure) drag, induced (drag due to lift), and compressibility drag terms.

The parasite drag term includes the skin-friction drag on the entire wetted area of the airplane and the profile or form drag due to the net force in the drag direction arising from the zero lift pressure distribution on the wing, body, and appendages. Another factor of the pressure drag, traditionally grouped with the parasite drag, is interference drag due to the changes in the flow pattern that accompanies placing two bodies in close proximity. (The total drag on the two bodies generally differs from the sum of the isolated drags on each body. Depending on the configuration, the inferred interference force can act in the thrust or drag direction. The parasite drag C_{D_P} is, by convention, defined as the minimum drag experienced by an airplane.) Some changes in drag also occur as a result of the change of the aircraft's geometrical orientation in the flowfield due to changes in angle of attack. Such (small) changes that are not attributable to the changes induced in the flowfield due to lift are traditionally accounted for as an attitude dependent term ΔC_{D_P} and are usually obtained from wind-tunnel tests and adjusted based on actual flight experience. As defined for any given configuration, C_{D_P} is a constant that does not depend on lift or Mach number. C_{D_P} changes only with Reynolds number and alterations to the configuration such as landing gear, flaps, and changes in external stores.

The portion of drag resulting from lift is referred to as induced drag. In practice, induced drag from components other than the wing is usually small and is lumped into the ΔC_{D_p}. A physical feeling for induced drag may be obtained by considering the flow over a wing with positive lift in subsonic flow. For this condition, the air pressure on the upper surface is less than the air pressure on the lower surface. This pressure differential causes air to flow around the ends of the wing, from the high pressure below the wing toward the low-pressure area above the wing and thus form vortices that concentrate near the wing tips. For the net effect of this vortex system is to create a downward inclination to the air behind the wing. The influence of the transverse flow on the wing depends primarily on its aspect ratio (AR = wing span^2/wing planform area). In the limit of infinite aspect ratio, the transverse flow would be absent. For a finite aspect ratio, the average downwash angle is finite. It can be shown that the effective angle of attack is reduced by half the average downwash angle.[1] This angle of the incoming flow results in an additional component in the drag direction referred to as induced drag. The planform of the wing also influences the induced drag because of its effect on the average downwash angle. An elliptical lift distribution results in the theoretical minimum of induced drag.[2] This distribution can be obtained by (1) varying airfoil sections along the span, (2) using the same airfoil section and an elliptical planform, (3) twisting the wing to vary the geometric angle of attack along the span, or (4) using a combination of these on a tapered planform. The effect of the nonelliptical loading is accounted for with a span efficiency factor e (≤ 1.0) which should be maximized within structural and design constraints. The use of winglets (mainly vertical fins on the wing tips, as on the 747-400) can result in values of e greater than one.

The relationship between lift and induced drag is

$$C_{D_{\text{induced}}} \equiv C_{D_L} = \frac{(C_L - C_{L_{\min}})^2}{\pi \cdot AR \cdot e} \tag{3.3}$$

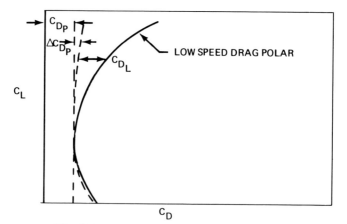

Fig. 3.5   Low-speed drag polar components.

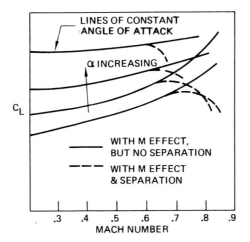

Fig. 3.6   Lift coefficient variation with Mach number.

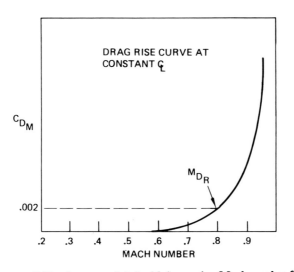

Fig. 3.7   Compressibility drag associated with increasing Mach number for a constant lift coefficient.

where $C_{L_{min}}$ is the value of the lift coefficient when the induced drag equals zero. The induced drag is treated as independent of Reynolds number.

The sum of the parasite and induced drag completely accounts for the drag in low-speed flight. A typical low-speed polar is shown in Fig. 3.5. Notice that the parabolic form of the drag polar is primarily due to the induced drag term. Transport airplanes are usually designed so that the minimum value of C_D corresponds to the required C_L for a cruise condition.

As flight speed increases, compressibility effects become significant when the air reaches local sonic speeds. The sudden rise in pressure through the resulting shock waves necessary for recompressing the high-speed flow cause boundary-layer thickening and possibly separation, resulting in a general degradation of lift and an increase in drag. A typical C_L vs Mach number curve is shown in Fig. 3.6. An associated compressibility drag curve (Fig. 3.7) shows the onset of "drag rise" associated with shock-induced separation and wave drag from local supersonic regions.

The compressibility drag C_{D_M} onset may be delayed by sweeping the wing. The critical or initial drag rise Mach number M_{D_R} is defined as the Mach number at which the C_D increases by 0.002 above the low-speed value.

The compressibility drag is affected to some extent by the airfoil's thickness-to-chord ratio and C_L value. An empirical equation giving approximate compressibility drag for subsonic airplanes has the form

$$C_{D_M} = K_1(M - M_{D_R})^8 + K_2(M - M_{D_R})^2(C_L - C_{L_{\min}})^3 \qquad (3.4)$$

where K_1 and K_2 are constants that depend on the configuration (thickness to chord, sweep angle, etc.).

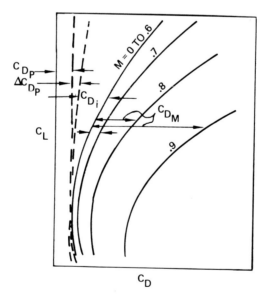

Fig. 3.8   Drag polar of a subsonic airplane showing contribution due to compressibility effects.

Combining the various drag terms gives an equation from which subsonic drag polar can be plotted, as

$$C_D = C_{D_P} + \Delta C_{D_P} + \frac{(C_L - C_{L_{min}})^2}{\pi \cdot AR \cdot e} + K_1(M - M_{D_R})^8$$
$$+ K_2(M - M_{D_R})^2(C_L - C_{L_{min}})^3 \qquad (3.5)$$

A typical subsonic drag polar is shown in Fig. 3.8.

The drag rise in high subsonic flow continues through transonic flow. Depending on the configuration, a maximum drag coefficient is observed in the transonic or low-supersonic Mach number region.

An additional term, wave drag, is required in supersonic flow. Wave drag replaces the subsonic terms of profile and compressibility drag. Approximate equations describing the wave drag on bodies and planar surfaces are

(1) Bodies:

$$C_{D_W} \approx 0.7 \left(\frac{\text{length}}{\text{maximum equivalent diameter}} \right)^{5/3} \qquad (3.6)$$

(2) Planar surfaces:

$$C_{D_W} \approx 3.05 \left(\frac{\text{thickness}}{\text{chord}} \right)^{5/3} \cos^{3/2}(\Lambda \text{ leading edge}) \qquad (3.7)$$

Notice that these approximations are not Mach number dependent. The drag coefficient vs Mach number for a typical supersonic aircraft in level flight is shown in Fig. 3.9.

More detailed analyses of high-speed flight indicate that C_D at $M = 3$ can be below the maximum value by as much as 25% depending on the configuration.

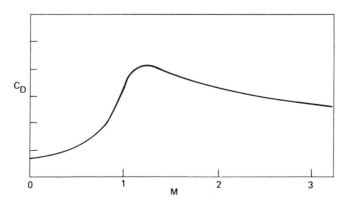

Fig. 3.9 Representative drag coefficient variation for a supersonic airplane in level flight.

The aerodynamicist depends largely on three sources for information pertaining to the drag polar: similar existing airplanes, aerodynamic component test, and scale wind-tunnel tests. Wind-tunnel models are used to verify analytic estimates and to investigate certain design features for which no data are available. The basic drag polar is usually obtained for the complete airplane from wind-tunnel tests, although the drag so obtained is not used directly because of the differences in scale. While the shape and spacing of the curves relative to each other are used directly, the intersection of the low Mach number polar with the C_D axis is obtained by other means. Experience has shown that wind-tunnel data will accurately predict the polar shape and drag rise characteristics, but will not accurately predict the basic drag level or the onset of drag size. Two major reasons for this are Reynolds number effects and the difficulty in accurately separating out the drag and interference effects of the model suspension system.

Equation (3.1) showed that the forces are a function of Mach and Reynolds numbers. Since the Reynolds number is determined by $\rho V \ell / \mu$, one way to preserve the Reynolds number for a scale model test is to use a fluid at conditions where ρ / μ for the model are enough different from the atmospheric conditions for actual full-scale flight that the dimensions of the model to full scale are offset (i.e., $Re = (\rho V \ell / \mu)_{model} = (\rho V \ell / \mu)_{full \ scale}$). Test facilities that produce this relationship are called high Reynolds number facilities.

With appropriate coefficients, the equations shown above are sufficiently accurate for most preliminary design work and initial engine cycle selection studies. As more detail is desired, the descriptive equations are improved through the use of additional terms and modified coefficients. Provision for the more accurate equations is usually the responsibility of the aerodynamicist and the propulsion designer need only assure himself that accuracy of the polar is compatible with that required for his study.

Because the interactions between the engine and airframe are very strong, particularly in fighter aircraft, detailed thrust/drag bookkeeping is a substantial problem. The present section avoids bookkeeping details in order to provide insight into the method of mission analysis.

Use of the drag polar for mission analysis.   In order to clarify the use of the drag polar for mission analysis, a routine airlift mission will be considered. It is common in this case to consider altitude changes within cruise mission segments to be negligible, to assume that the engine propulsive force and drag act along the same line, and to note that the lift acts at a right angle to the drag and to the aircraft's instantaneous velocity. As a result, in nonaccelerating level flight, the lift is equal to the weight. When the airplane is accelerating in the lift direction (as would occur when pulling out of a dive or in a level turn, for example), the lift must, of course, be greater than the aircraft weight. If the total lift is expressed as a fraction of standard gravity N, then the equation for the lift coefficient can be generalized as

$$C_L = \frac{L}{qS} = \frac{W \cdot N}{qS} \tag{3.8}$$

Mission analysis is an incremental progression through the following steps for the number of segments into which the mission has been divided:

(1) The weight, reference area, and flight conditions are used to compute the required C_L for the first mission segment.

(2) The drag polar and Mach number are then used to determine the drag coefficient C_D.

(3) The calculated C_D, flight condition, and reference area permit the airplane drag to be computed.

(4) The net thrust is calculated to be the drag plus additional thrust to provide linear acceleration requirements for the flight segment.

(5) The weight of fuel consumed by the engine to produce the net thrust during the increment is computed and the weight of the aircraft diminished by that amount. (This is done for each increment into which the segment has been divided.)

(6) The procedure is repeated by returning to step 1, with the newly computed weight and flight conditions for the next mission segment.

(7) This incremental process is continued until all fuel is used or the mission is completed. If the fuel aboard does not match the fuel required for the mission (including reserves), the airplane is resized and the computation repeated.

(8) If the engine size (usually estimated initially from a thrust-to-takeoff gross weight ratio) does not exactly provide the required thrust at the most demanding point of the segment, the engine size is altered and the entire computation is repeated.

The unknowns remaining in the above procedure are the characteristics of the mission segments and thrust required for each. These will be discussed for the segments that commonly provide the building blocks for a mission.

Mission Segments and Thrust Requirements

Mission segments can be combined to define any desired flight profile. A typical combat mission profile is shown in Fig. 3.10. The following subsections provide a brief description of each segment and standard assumption and the appropriate equations for use in mission analysis. For further details, see Refs. 1 and 3.

Computer programs for mission analysis break each major segment of the mission into increments. At each step, the program calculates the airplane weight change due to fuel consumption, refueling, and payload expenditure.

All of the methods of mission performance analysis are based on the same fundamental laws of motion and flight mechanics; however, one technique[4] is particularly well suited for the purpose of engine/airplane matching. This method will be emphasized in the following sections.

During mission analysis the range is not known a priori when calculating climb and descent segments. During cruise, required range may be specified or maximized by assuming exhaustion of fuel allotted for the segment. During loiter conditions, no range is credited and fuel consumption is computed from maximum endurance considerations (safe flight speed,

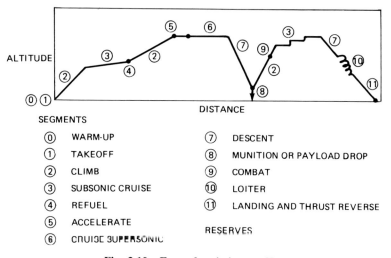

SEGMENTS

⓪	WARM-UP	⑦	DESCENT
①	TAKEOFF	⑧	MUNITION OR PAYLOAD DROP
②	CLIMB	⑨	COMBAT
③	SUBSONIC CRUISE	⑩	LOITER
④	REFUEL	⑪	LANDING AND THRUST REVERSE
⑤	ACCELERATE		
⑥	CRUISE SUPERSONIC		RESERVES

Fig. 3.10   Example mission profile.

minimum fuel consumption). The range for any other kind of segment cannot be specified (or independently varied) because range for other segments depend upon the flight path conditions specified.

This flexibility allows the representation of virtually any mission profile. In addition, at any point in the mission and for any segment, performance requirements and constraints can be specified. Typical examples of these requirements include: payload may be added or subtracted, external stores or tanks or weapons may be released, the g loading experienced by the aiirplane can be changed to any desired level (to simulate combat conditions), the thrust of any number of engines can be omitted to simulate engine-out, the wing sweep may be varied, the specific excess power required for acceleration and climb can be specified, and maximum noise levels can be prescribed.

Although several engine sizing criteria must be specified in one mission, the most demanding condition must be identified and used. Once the size of the engine has been established, most portions of the mission will require only part power. The fuel consumption is then deduced as a function of thrust for each altitude and Mach number specified in the mission analysis program as will be discussed in the following section on engine performance representation.

Warmup and takeoff.   Ground taxiing and holding before takeoff consumes fuel in an amount that can be calculated from the power settings and time requirements specified. Since the fuel consumed and resulting weight changes do not contribute to the flight portion of the mission, it is usually treated as an allowance and the required fuel weight is added to the takeoff gross weight of the airplane. The resulting weight allowance added to the takeoff (or brake release) gross weight is defined as the maximum taxi weight. The additional fuel tank volume must be accounted for in the aircraft size and shape.

The takeoff segment of the mission includes the ground run, rotation, and a climb to a standard height of 35 ft. The analysis of the takeoff segment yields an estimate of the fuel used during takeoff and the size of the engine to provide takeoff within the desired takeoff distance.

For the ground roll the relationship between acceleration a, velocity v, distance s, and time t is, of course, $ds = V\,dt$ and $a = dV/dt$. Thus,

$$ds = V\,dV/a \qquad (3.9)$$

If a steady headwind velocity of V_W exists, the ground roll distance S_G may be obtained as an integral over the velocity relative to the airplane V_r by noting $V_r = V + V_W$. Then,

$$S_G = \int_{V_W}^{V_{LO}} \frac{V_r - V_W}{a}\,dV_r \qquad (3.10)$$

where V_{LO} is the required relative velocity for liftoff.

The acceleration can be evaluated using Newton's law of motion $[\Sigma F = Ma = (W/g)a]$ and Fig. 3.11.

Solving for a and substituting into Eq. (3.10) yields an expression for the takeoff distance with the symbols defined as in Fig. 3.11,

$$S_G = \int_{V_W}^{V_{LO}} \frac{W}{g} \cdot \frac{(V_r - V_W)\,dV_r}{[(T - \mu W) - (C_D - \mu C_L)qS - W\Phi]} \qquad (3.11)$$

In order to carry out the integration, the variation of thrust, weight, drag, and lift with velocity must be determined. In general, thrust is a function of velocity, air temperature, and pressure; weight is nearly constant for the ground run; and drag and lift are functions of velocity and air density.

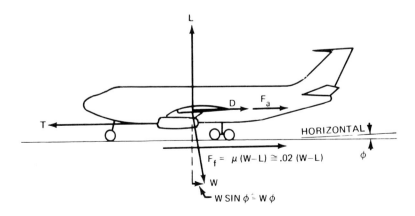

Fig. 3.11   Forces on airplane during ground run.

With a tricycle landing gear, the airplane attitude will remain constant during the ground operation and C_L and C_D will also remain constant. Figure 3.12 shows the approximate order of magnitude of the forces acting during a takeoff run.

The time required for the ground run is

$$t = \int_{V_W}^{V_{LO}} \frac{W}{g} \cdot \frac{dV_r}{[(T - \mu W) - (C_D - \mu C_L)Sq - W\Phi]} \tag{3.12}$$

Equations (3.11) and (3.12) can be solved by stepwise integration.

In preliminary design, the above degree of accuracy is not usually necessary and approximate values are often used. Wind speed is ignored and Eq. (3.10) is integrated to give

$$S_G = V_{LO}^2/2\bar{a} \tag{3.13}$$

where $\bar{a}$ is a representative constant acceleration. With the assumption of a horizontal runway and negligible aerodynamic drag during ground roll, an appropriate value for $\bar{a}$ is

$$\bar{a} = \frac{(F_N - 0.02W_0)}{W_0} \cdot g \tag{3.14}$$

where W_0 is the initial brake release weight (and usually equals the takeoff gross weight in mission analysis) of the airplane and F_N is the net thrust of the propulsion system at $0.707V_{LO}$. The time required is approximately

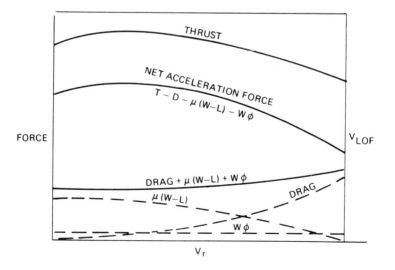

Fig. 3.12   Relative magnitude of forces during ground roll.

$V_{\text{LO}}/\bar{a}$. (Note: thrust T is the net thrust of the propulsion system and is interchangeable with F_N. T is commonly used by the aerodynamicist to mean the net thrust required for the airplane, while F_N is used by the propulsion specialist to mean either the net thrust per engine or total net thrust.)

Several rules governing speed requirements for airplane control and engine-out operation govern the liftoff velocity in actual practice. For preliminary design, a liftoff speed of 1.2 times the stall speed is a reasonable approximation. When conducting precise mission analysis, the appropriate military or FAA rules should be used.

Various field length definitions are in common usage. When field length is a critical segment in the airplane analysis, a consistent definition of that length must be used. For example, one criterion for evaluating a multiple-engine commercial aircraft is the balanced field length: a takeoff calculated according to Federal Aviation Regulations, FAR part 25, provides for a field length that satisfies both the takeoff and accelerate-stop requirements. If it is assumed that an engine failure occurs at a speed such that the distance to continue takeoff and climb to a stipulated height is equal to the distance required to stop, the total field length is termed "balanced." FAR part 25 stipulates using a field length that is the greatest of either the accelerate-and-go distance, the accelerate-and-stop distance, or 115% times the all-engine-operating distance to a 35 ft height.

Detailed takeoff calculations are generally performed in discrete segments that define the distinct parts of the takeoff, as shown in Fig. 3.13.

The drag polar used to compute the increment in time and distance required to clear the 35 ft obstacle is often different from that used for the majority of the mission. The appropriate polar must include the drag penalty and lift benefits due to flaps or other lift augmentation devices, as well as landing gear drag. Detailed calculations require estimating the effects of wind, flap setting, altitude, temperature, desired climbout rate, and liftoff gross weight. Preliminary mission studies use a nominal drag polar penalty and account for the temperature and altitude effects on engine performance, computed as changes from standard day (unless the engine is "flat rated" to the desired temperature, in which case the throttle is advanced until standard-day thrust is produced).

The weight of fuel consumed during takeoff is computed, knowing the takeoff power setting, fuel consumption per pound of thrust (TSFC), and time duration.

Thrust available and required. For an airplane in level unaccelerated flight, the net thrust F_N must equal drag D and the lift must equal weight W. The lift and drag coefficient equations can be rewritten for this flight condition as

$$\frac{D}{\delta} = \frac{\gamma}{2} p_{\text{STD}} C_D M^2 S \qquad (3.15)$$

$$\frac{W}{\delta} = \frac{\gamma}{2} p_{\text{STD}} C_L M^2 S \qquad (3.16)$$

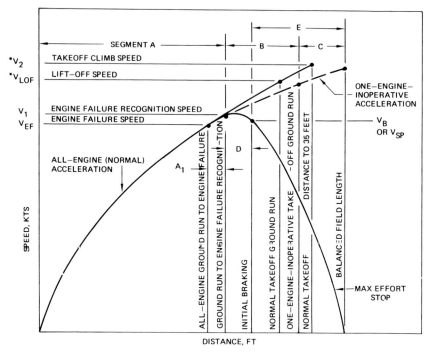

*ONE–ENGINE–INOPERATIVE V_2 AND V_{LOF} SPEEDS WILL BE SLIGHTLY LOWER THAN THE
ALL–ENGINES–OPERATING V_2 AND V_{LOF} SPEEDS FOR THE SAME ROTATION SPEED, V_R

Fig. 3.13   Balanced field length and performance segments.

$$\frac{F_N}{\delta} = \frac{D}{\delta} \tag{3.17}$$

where δ is here defined as the local freestream static pressure divided by the standard reference pressure ($p_{STD} = 1.013 \times 10^5$ Pa or 2116 lb_f/ft^2).

The equations describing the off-design performance of various engines are developed in Chapter 8 of Ref. 4 where it is shown that F/δ is independent of altitude if the ambient temperature, flight speed, and engine setting (turbine inlet temperature) remain constant. F/δ is hence a convenient thrust parameter and the ratio W/δ is used for consistency.

The above equations are applied to performance computations as follows. The value of C_L at a given W/δ and Mach number is computed from Eq. (3.16). The C_L and Mach number are used to find the appropriate C_D from the drag polar, which is then substituted into Eq. (3.17) to establish the required net thrust.

Selecting a range of Mach numbers for a given W/δ results in a series of points defining a curve that may be plotted on axes of F_N/δ and Mach number. By selecting several values of W/δ, a family of curves may be

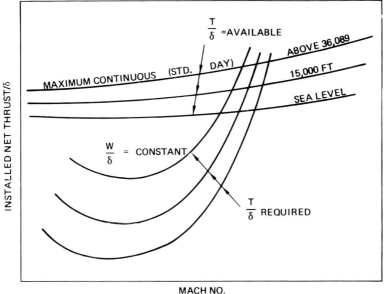

INSTALLED NET THRUST/δ

$\frac{T}{\delta}$ =AVAILABLE

ABOVE 36,089

MAXIMUM CONTINUOUS   (STD.   DAY)

15,000 FT

SEA LEVEL

$\frac{W}{\delta}$ = CONSTANT

$\frac{T}{\delta}$ REQUIRED

MACH NO.

Fig. 3.14   Thrust required and available for level unaccelerated flight.

obtained as in Fig. 3.14. These curves define the T/δ that must be provided by the engines to sustain level unaccelerated flight. Typical F_N/δ available curves from jet engines are superimposed on Fig. 3.14. The intersections show the maximum Mach number that can be maintained in level flight for each weight and altitude.

For well-designed high-speed aircraft, the decrease in dynamic pressure with altitude increases more than compensates for the increase in drag coefficient with Mach number, resulting in less drag at altitude for transonic and supersonic speeds than for high subsonic speeds flown at sea level.

At Mach numbers below the intersections in Fig. 3.14, the available net thrust exceeds that required for level unaccelerated flight and either the throttle setting must be reduced or the excess thrust will result in acceleration and/or climb.

Climb, acceleration, and descent.   If excess thrust is available, it can be used for climb and/or acceleration. Newton's laws of motion can be applied to this situation with the aid of Fig. 3.15.

The gravitational force is resolved into components perpendicular and parallel to the flight path. The inclination of the flight path to the horizontal is ϕ and the thrust vector is at an angle α_T to the flight path. The following analysis presumes that α_T is small enough (<15 deg) to assume that $T \cdot \sin\alpha_T \cong 0$ and $T \cdot \cos\alpha_T = T$. The small angle assumption is adequate for the majority of missions. (In configurations using vectored thrust or at high angle of attack, these contributions must be included.)

Resolving the forces along the flight path yields

$$T - D - W \cdot \sin\phi = \frac{W}{g}\frac{dV}{dt} \tag{3.18}$$

Perpendicular to the flight path, the sum of the forces is equal to the centripetal force necessary to change the flight path angle at a rate $d\phi/dt$,

$$L - W\cos\phi = \frac{W}{g}V\frac{d\phi}{dt} \tag{3.19}$$

In a climb, $d\phi/dt$ is essentially zero and thus

$$L = W\cos\phi \tag{3.20}$$

This equation shows that, in a climb, the lift will be less than the weight (and thus the level flight lift) by the factor $\cos\phi$; the balance of the weight is supported by a thrust vector component. This means that at a particular speed, the induced drag will be less than at the same speed in level flight. The reduction in lift (and therefore drag) will be greater for larger climb angles. For transport airplanes, under most conditions, the climb angles are small so that $\cos\phi$ is very close to unity. Thus, the lift and drag during climb are practically identical to those existing under level flight conditions and the level flight drag polar data for the airplane can be used. It is consistent for this approximation to use

$$C_L = W/qS \tag{3.21}$$

For fighter-type aircraft missions with segments calling for climb angles greater than 15 deg, the effect of climb angle on weight and hence drag must be considered. Occasionally, fighter and aerobatic-type aircraft missions will include segments where $d\phi/dt$ is large and Eq. (3.19) must be used without simplification.

Equation (3.18) can be rewritten as

$$\sin\phi = \frac{1}{W}\left(T - D - \frac{W}{g}\frac{dV}{dt}\right) = \frac{T-D}{W} - \frac{1}{g}\frac{dV}{dt} \tag{3.22}$$

The rate of climb dh/dt is equal to the vertical component of the flight velocity,

$$\frac{dh}{dt} = V\sin\phi \tag{3.23}$$

or

$$\frac{dh}{dt} = \frac{(T-D)V}{W} - \frac{V}{g}\frac{dV}{dt} \tag{3.24}$$

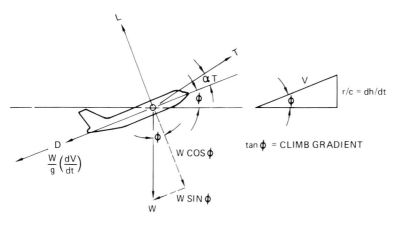

Fig. 3.15   Free-body diagram of airplane with acceleration and climb.

and noting that $dV/dt = (dV/dh)(dh/dt)$, it follows that

$$\frac{dh}{dt}\left(1 + \frac{V}{g}\frac{dV}{dh}\right) = \frac{(T-D)V}{W} \tag{3.25}$$

The right side of this equation is the excess thrust times velocity (excess power) divided by the weight and is defined as the specific excess power P_s. The dimensionless term $(V/g)(dV/dh)$ is an acceleration factor. When a climb is executed at a constant true airspeed, the acceleration factor is zero and Eq. (3.25) becomes

$$\text{Rate of unaccelerated climb} = \frac{(T-D)V}{W} \tag{3.26}$$

The specific excess power is one of the criteria by which fighter aircraft are evaluated and mission constraints usually specify a minimum level.

For commercial transports, the climb gradients are specified by rules (FAR part 25) that state percentages of climb gradients required for two-, three-, or four-engine transports during climb, approach (descent), and landing usually governed by specifying a minimum performance requirement with one engine inoperative.

For slow acceleration rates, the rate of climb can be obtained in terms of Mach number by rewriting Eq. (3.25) as

$$\frac{dh}{dt} = a_0 M \sqrt{\theta}\left(\frac{T/\delta - D/\delta}{W/\delta}\right) \tag{3.27}$$

where a_0 is the speed of sound at standard condition and $\theta = T/T_0$ the ratio of temperature to standard temperature.

Maximum rate of climb for a given weight and altitude will occur at the speed where $(T - D)/\delta$ multiplied by the Mach number is greatest. This will be at a speed slightly higher than that for $[(T - D)/\delta]_{max}$. The true airspeed at the maximum rate of climb increases with altitude; therefore, the airplane must accelerate along the flight path to maintain the maximum rate of climb. If the thrust rating cannot be increased, the rate of climb will be reduced to account for this lack of acceleration, as seen from Eq. (3.27).

Figure 3.15 shows that the sine of the climb angle is equal to rate of climb divided by the climb velocity V. The climb gradient is defined as the tangent of the climb angle. For small angles, the sine of an angle is approximately equal to the tangent of the angle; therefore,

$$\text{Climb gradient} = \tan\phi \cong \sin\phi = \frac{dh/dt}{V} \qquad (3.28)$$

Substituting Eq. (3.26) into Eq. (3.28) and noting that

$$\frac{D}{W} = \frac{C_D}{C_L}$$

results in the climb gradient for unaccelerated climb,

$$(\tan\phi)_{V=\text{const}} = \frac{T}{W} - \frac{C_D}{C_L} \qquad (3.29)$$

Substituting Eq. (3.25) into Eq. (3.28) results in the climb gradient for accelerated climb,

$$\tan\phi = \frac{(T/W) - (C_D/C_L)}{1 + (V/g)(dV/dh)} \qquad (3.30)$$

Climb gradients are often expressed in percent.

For level accelerated flight the equation of motion becomes

$$T - D = \frac{W}{g} a \qquad (3.31)$$

or

$$\frac{T}{W} - \frac{C_D}{C_L} = \frac{a}{g} \qquad (3.32)$$

where a/g is the acceleration gradient for level flight.

Equations (3.29), (3.30), and (3.32) show that the quantity

$$T/W - C_D/C_L$$

can be used to measure climb or acceleration performance.

If the horizontal velocity V_H is plotted against the vertical velocity V_V a hodograph plot results. The radius vector from the origin to any point on the curve is proportional to the flight path speed and the angle from the horizontal is equal to the climb angle ϕ. Figure 3.16 shows a hodograph typical of full power performance where the distinction between the maximum climb angle (ϕ_{max}) and maximum rate of climb are delineated. Figure 3.17 (which is not a hodograph) displays the speed for best climb angle and speed for best rate of climb.

Altitude affects both the velocity and the maximum rate of climb as illustrated on Fig. 3.18. (The assumption of constant weight is appropriate for long-range subsonic transports, which burn only about 10% of their total fuel during climb.) Figure 3.19 depicts the true airspeed variation with altitude necessary to fly at either constant indicated airspeed or at the speed for maximum rate of climb. Above 36,089 ft, jet aircraft obtain the maximum rate of climb at a constant true airspeed and, hence, constant Mach number.

The T/δ available for climb decreases with increasing temperature. On a hot day, the rate of climb decreases and the time to climb, range, and fuel flow increase.

Excess power decreases as altitude increases. The altitudes at which the unaccelerated rates of climb are 100 ft/min and zero are referred to as the service ceiling and absolute ceiling, respectively.

The time required to climb to desired altitude h_1 is

$$t = \int_0^{h_1} \frac{dh}{dh/dt} \tag{3.33}$$

where dh/dt is a function of h as shown in Fig. 3.18. Computer programs conveniently obtain the time as a summation of small increments using the

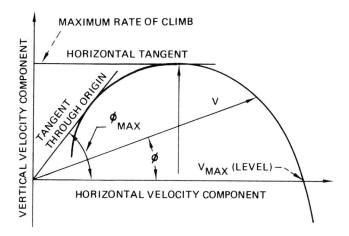

Fig. 3.16   Climb hodograph showing locations of maximum climb angle and rate.

average rate of climb within the increment,

$$\Delta t = t_2 - t_1 = \frac{h_2 - h_1}{(dh/dt)_{av}}$$ (3.34)

For precise computations, care must be taken to distinguish between pressure altitude and actual altitude on nonstandard days.

The range during climb is the summation of the incremental values of $V \cos\phi$ multiplied by time.

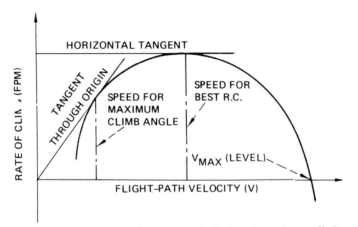

Fig. 3.17   Speeds for maximum rate of climb and maximum climb angle.

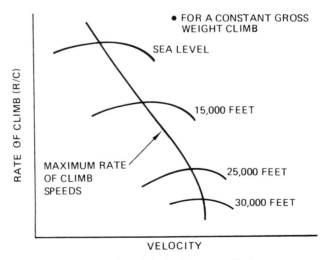

Fig. 3.18   Altitude effects on climb.

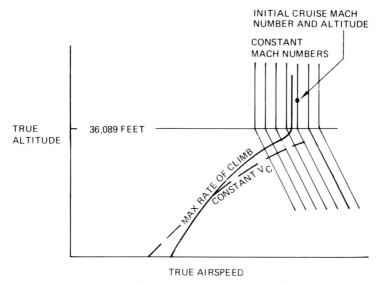

Fig. 3.19   True airspeed vs altitude for climb.

During the climb segment of a mission, the engine is sized to produce or exceed the desired rate of climb at the engine power setting provided for climb.

An example is a mission analysis computer approximation for climb is shown in Fig. 3.20. Thrust, fuel flow, and drag are determined at the midpoint of each increment i and applied over the increment to approximate time, fuel weight W_F, and range R. Five to ten increments usually provide sufficient accuracy.

Descent calculation is identical to climb except the power required is less than the power required for level unaccelerated flight. To avoid acceleration during descent, low-power settings are used and the drag polar is often changed to represent the increase in drag and decrease in lift made possible from spoilers or dive brakes. If the descent angle is less than 15 deg, the rate of sink is approximated as

$$\text{Rate of sink} = \frac{V}{W}\left(T_{\text{av}} - \frac{\rho V^2 C_D S}{2\cos\alpha_T}\right) \qquad (3.35)$$

For designing commercial aircraft, the descent rate is specified to provide reasonable rates of change for cabin pressure. In general, descent analysis can be computed using the technique of Fig. 3.20, but with the altitude and Mach number path calculated in descending order. Because the power setting must be very low, in cases where descent is either a small portion of the total range or the descent is flown in a spiral, preliminary mission analysis often credits zero range and zero fuel expenditure to the descent segment.

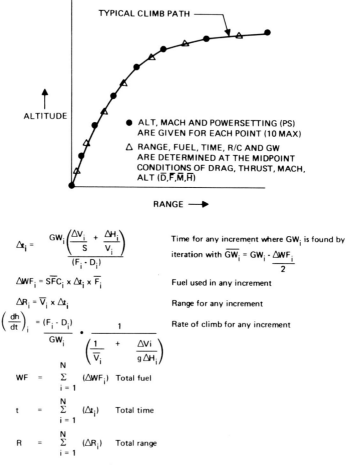

$$\Delta t_i = \frac{GW_i \left(\frac{\Delta V_i}{S} + \frac{\Delta H_i}{V_i} \right)}{(F_i - D_i)}$$

Time for any increment where GW_i is found by iteration with $\overline{GW}_i = GW_i - \frac{\Delta WF_i}{2}$

$$\Delta WF_i = \overline{SFC}_i \times \Delta t_i \times \overline{F}_i$$

Fuel used in any increment

$$\Delta R_i = \overline{V}_i \times \Delta t_i$$

Range for any increment

$$\left(\frac{dh}{dt} \right)_i = \frac{(F_i - D_i)}{GW_i} \cdot \frac{1}{\left(\frac{1}{\overline{V}_i} + \frac{\Delta V_i}{g \Delta H_i} \right)}$$

Rate of climb for any increment

$$WF = \sum_{i=1}^{N} (\Delta WF_i) \quad \text{Total fuel}$$

$$t = \sum_{i=1}^{N} (\Delta t_i) \quad \text{Total time}$$

$$R = \sum_{i=1}^{N} (\Delta R_i) \quad \text{Total range}$$

• Descent is calculated with the same procedure except that the altitude and mach number variation are decreasing.

Fig. 3.20 Method of climb (and descent): fuel, range, and time calculation.

Deceleration at constant altitude can be represented with a descent segment. This can be done by specifying a minimal descent of, say, 1 ft/increment. When using an incremental technique for any segment, it is safest to require monotonic increases or decreases in altitude/Mach number to prevent division by zero.

Figure 3.21 illustrates the use of increments to calculate the time, range, and fuel consumed during acceleration in level flight. To accurately estimate acceleration, small increments of Mach number (0.05–0.10) should be used. In the transonic region, where the aircraft drag changes rapidly, very small increments should be used.

An acceleration segment can be used at any point in the mission to meet constraints that may size the engine or airframe, but are not necessarily a part of the design mission. If the aircraft, as defined when it enters the acceleration constraint segment, is incapable of fulfilling the constraint, the program will automatically resize the engine and initiate computation from the original mission start condition. When the constraint is satisfied, the analysis proceeds with the aircraft in a new configuration. Because the design mission does not include the requirement that the aircraft actually fly this segment, but rather only be capable of flying it if the need arises, the total mission is credited with zero distance and zero time for the constraint mission. (It is assumed that should the constraint segment actually be flown by the aircraft, the aircraft will not proceed along the design mission because of fuel constraint.) For example, an airplane may be required to accelerate to a high Mach number if an intercept is necessary. If the design mission for the aircraft is to penetrate enemy territory without making an intercept, the intercept capability may be the most stringent engine sizing criteria. To insure that the intercept could be made, if necessary, an acceleration to the high Mach number, intercept, and return to base must be verified, but will not contribute to flying the full mission requirement.

Cruise.   The fuel consumption is usually stipulated in terms of fuel flow per unit thrust, commonly called specific fuel consumption (SFC). The SFC of an engine or engines at constant flight Mach number decreases with altitude, up to a particular altitude (usually between 30,000 and 40,000 ft depending on design) and then begins to increase with altitude. The best cruise condition produces the most range for the least fuel consumption. Over a small increment of range ΔR, an airplane consumes ΔW pounds of fuel. The mileage per pound of fuel S is

$$S = -\Delta R / \Delta W \qquad (3.36)$$

(Note ΔW is negative.) In the limit, as the increment becomes small,

$$dR = -S \cdot dW = -\left(W \cdot S \cdot \frac{dW}{W} \right) \qquad (3.37)$$

where the quantity $W \cdot S$ is referred to as the range factor.

If the range factor for any increment of a cruise segment can be considered constant, Eq. (3.37) can be integrated and evaluated between the initial weight W_1 and the final weight W_2, yielding

$$R = W \cdot S \cdot \ell n \frac{W_1}{W_2} \qquad (3.38)$$

By noting that

$$S \equiv \frac{\text{miles}}{\text{pound}} = \frac{\text{miles}}{\text{hour}} \cdot \frac{\text{hours}}{\text{pound}}$$

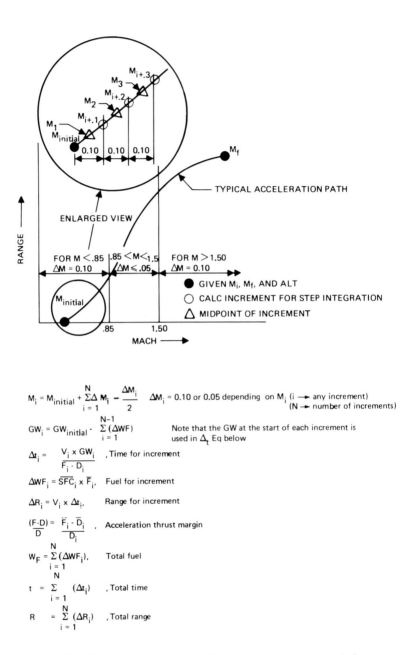

$$M_i = M_{initial} + \sum_{i=1}^{N} \Delta M_i - \frac{\Delta M_i}{2} \qquad \Delta M_i = 0.10 \text{ or } 0.05 \text{ depending on } M_i \; (i \rightarrow \text{any increment})$$
$$(N \rightarrow \text{number of increments})$$

$$GW_i = GW_{initial} - \sum_{i=1}^{N-1} (\Delta WF) \qquad \text{Note that the GW at the start of each increment is}$$
$$\text{used in } \Delta_t \text{ Eq below}$$

$$\Delta t_i = \frac{V_i \times GW_i}{\bar{F}_i \cdot D_i} \quad , \text{Time for increment}$$

$$\Delta WF_i = \overline{SFC}_i \times \bar{F}_i, \quad \text{Fuel for increment}$$

$$\Delta R_i = V_i \times \Delta t_i, \quad \text{Range for increment}$$

$$\frac{(F-D)}{D} = \frac{\bar{F}_i \cdot \bar{D}_i}{D_i} \quad , \text{Acceleration thrust margin}$$

$$W_F = \sum_{i=1}^{N} (\Delta WF_i), \quad \text{Total fuel}$$

$$t = \sum_{i=1}^{N} (\Delta t_i) \quad , \text{Total time}$$

$$R = \sum_{i=1}^{N} (\Delta R_i) \quad , \text{Total range}$$

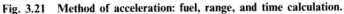

Fig. 3.21   Method of acceleration: fuel, range, and time calculation.

the range factor can be rewritten in the more useful form as

$$W \cdot S = W \cdot V \cdot \frac{1}{T \cdot \text{SFC}} \tag{3.39}$$

For level unaccelerated flight, the net thrust is equal to the total airplane drag and the thrust term T in Eq. (3.39) may be replaced by the drag D; the numerator may be multiplied by the airplane lift L and the denominator by the airplane weight W, since the lift and weight are equal in this flight condition. Thus, an alternate form for the range factor is obtained,

$$W \cdot S = W \cdot \frac{V}{\text{SFC}} \cdot \frac{L}{D} \cdot \frac{1}{W} = \frac{V}{\text{SFC}} \cdot \frac{L}{D} = \frac{V}{\text{SFC}} \cdot \frac{C_L}{C_D} \tag{3.40}$$

Substituting Eq. (3.40) into Eq. (3.38) produces the Breguet range equation,

$$R = \frac{V}{\text{SFC}} \cdot \frac{L}{D} \cdot \ell n \frac{W_1}{W_2} \tag{3.41}$$

If the SFC is nearly constant, then the ariplane will achieve the greatest level unaccelerated range if $V \cdot L/D$ (or $M \cdot L/D$) is maximized.

Figure 3.22 shows a C_L vs C_D polar on which lines have been drawn from the origin tangent to the polars at various Mach numbers. The points of tangency define the maximum L/D for that Mach number. By plotting the identified $M(L/D)$ vs Mach number (Fig. 3.22), the Mach number producing the maximum $M(L/D)$ can be found. The maximum $M(L/D)$ will occur at slightly higher Mach number than that associated with the highest possible L/D. For subsonic cruise, it is found that maximum range occurs at conditions where compressibility phenomena are important, that is, where the airplane experiences transonic drag rise.

The airspeed for optimum range occurs at the airspeed corresponding to maximum $\sqrt{C_L}/C_D$. Because maximum T/δ is independent of altitude (at high altitude and fixed engine setting), the range factor will remain constant if the flight Mach number and W/δ are held constant. As the weight is reduced by fuel burnoff, the altitude must increase to keep W/δ constant. Thus, to maximize range, as the weight decreases due to fuel consumption, the altitude must increase to maintain a constant value of W/δ. Air traffic control requirements prevent gradual change of altitude (except on some very low-density routes), so in practice the desired gradual altitude increase is approximated by changing altitude in discrete climbing steps. This nonoptimal procedure can cause a fuel consumption penalty on the order of 3% in total fuel consumption for a long-range subsonic jet transport.

Examination of Eq. (3.41) shows that supersonic cruise airplanes can be designed for long-range operation if the high velocity offsets the high SFC and drag.

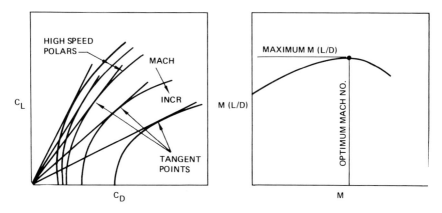

Fig. 3.22   Determination of maximum $M(L/D)$.

Figure 3.23 illustrates methods of representing the cruise segment for computer mission analysis. If the range is specified, the calculations give time and fuel consumption. Alternatively, if amount of fuel burned is specified, the corresponding cruise distance and time can be computed.

Endurance or loiter (for military missions).   Maximum endurance implies staying in the air for the maximum amount of time using the specified amount of fuel. Maximum endurance occurs at the flight conditions corresponding to the best C_L/C_D (i.e., the tangent to the drag polar that passes through the origin). Maximum endurance is used for hold or loiter (for example, surveillance). The endurance portion of a mission is not considered to contribute to the total mission range.

Combat (for military missions).   A combat segment is often pictured as consisting of a series of turning maneuvers. The assumed battle scenario leads to a specification of combat Mach number, altitude, power setting, and time duration. No range credit is taken. The mission analysis computes the fuel consumed over the time duration, taking into account changes in weight. Since the power setting is specified, the fuel consumption can be calculated directly for the time of combat. The engine may be sized by the combat condition if a specific excess power P_S is specified.

Refuel or munitions drop (for military missions).   Refueling, weapon expenditures, and fuel tank release are treated as weight changes that sometimes affect the airplane drag polar.

Landing.   The problem of determining landing performance is, in most respects, similar to the takeoff calculation, varying only in the treatment of the approach and flare and in the consideration of auxiliary stopping devices such as the speed brakes or thrust reversers. Detailed analysis and derivation of equations are omitted because of the similarity of the calcula-

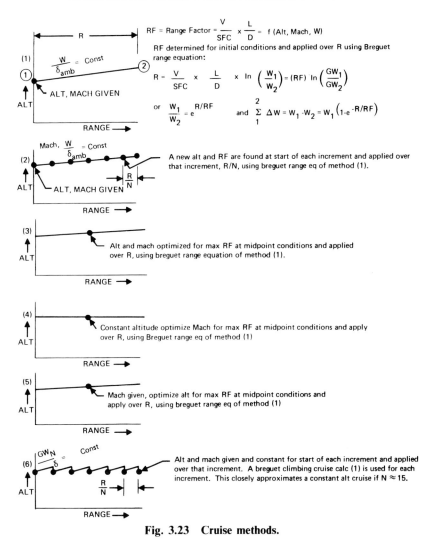

$$RF = \text{Range Factor} = \frac{V}{SFC} \times \frac{L}{D} = f(\text{Alt, Mach, W})$$

RF determined for initial conditions and applied over R using Breguet range equation:

$$R = \frac{V}{SFC} \times \frac{L}{D} \times \ln\left(\frac{W_1}{W_2}\right) = (RF)\ln\left(\frac{GW_1}{GW_2}\right)$$

$$\text{or}\quad \frac{W_1}{W_2} = e^{R/RF} \qquad \text{and}\quad \sum_{1}^{2}\Delta W = W_1 - W_2 = W_1\left(1-e^{-R/RF}\right)$$

A new alt and RF are found at start of each increment and applied over that increment, R/N, using breguet range eq of method (1).

Alt and mach optimized for max RF at midpoint conditions and applied over R, using breguet range equation of method (1).

Constant altitude optimize Mach for max RF at midpoint conditions and apply over R, using Breguet range eq of method (1)

Mach given, optimize alt for max RF at midpoint conditions and apply over R, using breguet range eq of method (1)

Alt and mach given and constant for start of each increment and applied over that increment. A breguet climbing cruise calc (1) is used for each increment. This closely approximates a constant alt cruise if N ≈ 15.

Fig. 3.23   Cruise methods.

tion procedure for the landing segment to the procedure for the takeoff segment already discussed. For commercial transports, FAR regulations specify landing requirements.

Reserves.   Reserves are simply treated as an additional increment of takeoff gross weight or as a fraction of the full fuel load that is not expendable in the normal mission analysis.

Engine Performance Representation

Engine performance is usually specified in terms of thrust, fuel consumption, and airflow rate. Each of these variables may be used in a variety of

Table 3.1   Variables Used for Engine Performance Representation

Thrust	F_N, $\boxed{F_N/\delta_2}$, F_N/δ_{amb}, $(F_N/\delta_{amb})/(F_N/\delta_{amb})_{ref}$
Fuel flow rate	W_F, $\boxed{W_F/(\sqrt{\theta_2}/\delta_2)}$, $SFC/\sqrt{\theta_{amb}}$, $W_F/(\sqrt{\theta}/\delta_{amb})$
Air flow rate	W_0, $\boxed{W_0\sqrt{\theta_2}/\delta_2}$, $W_0\sqrt{\theta_{amb}}/\delta_{amb}$

corrected forms (Table 3.1). δ_2 and θ_2 are compressor face total pressure and temperature divided by reference values (2116 lb/ft and 519.7°R, respectively); while δ_{amb} and θ_{amb} are ambient static pressure and temperature at altitude, divided by the reference sea-level standard values.

Of these choices, the quantities boxed in Table 3.1 give the most nearly linear relationship with Mach number and altitude. Thus, their use is desirable because it allows more accurate interpolation or extrapolation between specified conditions.

The calculation of engine performance at representative flight conditions independent of the mission is shown schematically in Fig. 3.24. At each of the selected flight conditions, a complete power line is obtained for an engine with a reference design airflow size. For this engine, the corrected fuel and airflow can then be computed for any desired corrected thrust. Using

$$W_0 = \frac{F_N}{F_{N_{ref}}} \cdot W_{0_{ref}} \tag{3.42}$$

the required engine airflow size can be scaled from the reference airflow value so that the required thrust at the most demanding flight condition equals the maximum thrust available. At this new airflow size, the power lines of all other flight conditions can be scaled by the same amount and the part power fuel flow rate necessary to produce the required thrust at any flight condition can be calculated.

Linear interpolation is used whenever the flight condition being evaluated falls between two or four given points. For example, assume the increment of a mission segment being considered is at conditions represented by the point g on the Mach number and altitude map shown in Fig. 3.24. The type of mission segment being flown combined with the airplane drag polar determines the required net thrust for the flight condition. The fuel flow at point g is found from interpolating the fuel flow required to produce the same thrust at points e and f, where the altitude is different from point g and the Mach number is the same. Similarly performance at points e and f are obtained from the fuel flow rate at the required thrust interpolated between given power lines at flight conditions a and b and c and d, respectively.

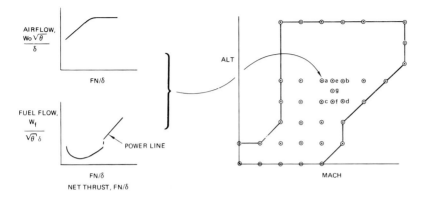

Fig. 3.24   Engine performance representation.

Propulsion System Installation

Engine/airplane matching requires an accurate evaluation of the interactions between the propulsion system and the airframe. The evaluation is traditionally broken into inlet effects (Chapter 4) and nozzle/aftbody effect (Chapter 5). Depending upon the accuracy desired from the mission analysis, different techniques are used to evaluate the installation effects. For conceptual and preliminary design evaluations, estimates are usually based on past experience and limited computations.

As the complexity of the analysis and the need for accuracy increase, wind-tunnel tests may be required. Typically, a complete airplane model is tested and the drag information is used as a baseline or reference value. The inlet (and forward portion of the aircraft, if necessary) is then tested in detail and changes in drag that result from changes in the configuration or engine power setting are tabulated as drag increments relative to the reference configuration. For fighters, detailed models of the aftbody (usually from the maximum diameter aft) are used to evaluate incremental changes in thrust minus drag due to a change in configuration or power setting.

Although much of this chapter discusses computer simulations, the importance of experimental evaluation of the interaction effects on the engine/airplane matching process cannot be overemphasized. Wind-tunnel and/or scale-model remotely piloted free flight testing is a very expensive portion of the aircraft system development process. Useful results depend on comprehensive planning and careful and accurate execution of tests. As the engine/airplane matching process continues toward the identification of the optimum aircraft, the validity of the estimation of the interaction effects must be constantly re-evaluated and updated.

Results of Mission Analysis

The mission analysis requirements, discussed in the preceding sections, form the starting point for a computer program with the logic shown in

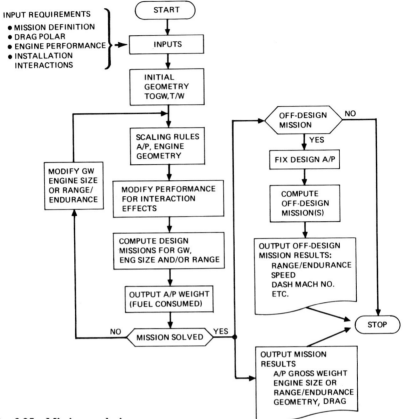

Fig. 3.25   Mission analysis process.

Fig. 3.25. As indicated, the iterative process uses the specified drag polar, engine cycle, and scaling rules to define an engine/airplane match that satisfies the mission. This integrated engine/airplane system represents the combination that simultaneously meets the conditions of appropriate engine airflow size and airplane gross takeoff weight or maximum range (or endurance).

The engine is sized by the airflow required to produce the net thrust for the most demanding point of the mission in the following way. The engine performance relates the thrust per unit (or reference) airflow size for each Mach, altitude, and temperature used in the mission. The inlet and nozzle losses and the installation-interaction effects account for the difference between installed and uninstalled thrust. The maximum thrust available from an engine varies with the flight segment. The highest power setting is usually takeoff power, which can be used for 2–5 min. Slightly lower power settings are allowed for continuous operation and a continuum of part-power settings exist down to an idle setting usually determined by minimum airflow, fuel flow, or engine rpm acceleration requirements. Thus, when combined with the scaling rules, the net thrust required at the most demanding condition in the mission defines the required airflow size that, in turn, fixes the engine weight, dimensions, and full-power fuel consumption. With the engine size fixed to meet this condition, all other points in the mission are calculated with the engine operating at part power. The fuel consumption is determined from the specific fuel consumption at the appropriate part-power setting. When afterburning or variable cycle engines are used, the iterative process must account for the complexity of additional parameters in the thrust/fuel flow relationship.

The computation of airplane takeoff gross weight results from a fixed range mission analysis. Simulation of a specified mission with a given engine cycle and size requires a specific fuel quantity for a given airplane size. Since the weight of the airplane is the sum of all the component weights, payload, and fuel, there exists a unique size of scaled candidate airframe that will carry enough fuel to be able to fly the required mission range.

As the airplane size changes, so does the lift and drag, which implies a changing reference area (if the polars are similar). The change in drag results in a change in thrust that must be supplied and, hence, a different engine size and fuel consumption. The change in fuel consumption requires a change in airframe size and the iterative loop required to find the unique size of the candidate engine and airframe continues until a match is found to the desired degree of accuracy.

If the final airplane is substantially different in size from the first guess, the scaling laws should be re-examined. The installation of the engine should be re-examined to see that the engine size has not so altered the airframe as to change the drag polar. Finally, the interaction effects must be re-examined.

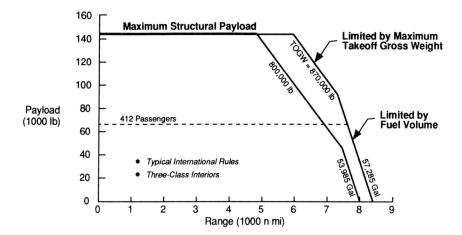

Fig. 3.26 Payload range comparison for 747.

Once the appropriate airplane weight and engine size have been established, the characteristics of the fixed configuration can be evaluated on alternative (off-design) missions. These missions typically leave the distance or time of one leg unspecified and the range or endurance of the fixed aircraft is computed for the alternate operating conditions. Off-design missions can also be used to evaluate configuration changes, such as external munitions, in which case the drag polar should be altered.

The end result of a design mission analysis is the airplane weight and engine size. Off-design computation provides additional information about range, endurance, dash capability, etc. These results can be used to re-evaluate the inputs, but the final result is one engine and airplane combination that can fly the specified mission. The analysis indicates nothing of the quality of the choice of engine cycle and airframe geometry. Optimizing the choice of cycle and airplane parameters is the principal task of the preliminary design process discussed in the next section. In preliminary design, mission analysis is used repeatedly in order to identify the region of optimum cycle and geometry characteristics. Following the identification of these characteristics, mission analysis is again used during system development. In the latter application, the output is used to refine the airplane and engine characteristics to the next level of development that carries requirements for increased accuracy.

For specified mission, the resulting capability of an airplane is often summarized on a payload range chart as shown in Fig. 3.26.

3.3   OPTIMIZATION OF ENGINE/AIRPLANE MATCH

Having discussed the process of sizing a candidate engine cycle and "weighing" a given airplane geometry to meet the requirements of a specified mission, we are now in a position to ask if this was a good choice or if a different engine cycle and airplane geometry could do the mission better. This section describes a process for obtaining the best engine/airplane match for a specified mission. Before the optimum can be chosen, it is necessary to decide by what criteria "best" can be defined.

Figures of Merit

While it is obvious that the best airplane can be chosen only after criteria for its "goodness" or viability have been established, the commonly used selection criteria, or figures of merit, are not all that obvious and are often difficult to define quantitatively during a preliminary design study.

Airplane selection procedures seldom seek the aircraft that flies the mission best without any regard for cost. The desired mission performance and constraints must be weighted against the cost of ownership. The total cost of purchasing, operating, and maintaining the aircraft during the entire period of ownership, less the final resale value, is referred to as the life cycle cost (LCC) or life cycle operating costs (LOC). Equally obvious a low-cost system may lack mission effectiveness—that is, it might take too long to fly a mission (low productivity) or it may lack flexibility to fly off-design missions. This has caused the U.S. Department of Defense to choose an optimization criterion called cost effectiveness, a blend of cost and mission effectiveness, that must be formulated for each system, often using judgmental factors. Defense budget constraints have forced the military to balance the aircraft system effectiveness against estimated life cycle costs. Due to the requirement for "crystal ball" estimates needed for accurate life cycle costing, it is more appropriate to say that life cycle cost as a figure of merit is a goal toward which evaluation techniques are evolving.

Aircraft utilization has a significant impact on the relative cost of ownership vs cash operating costs. A typical commercial aircraft will have an annual utilization of 3500–4000 h, while a military fighter may fly less than 350 h/yr.

The figures of merit for commercial airplanes are driven by the marketplace, i.e., worldwide airline requirements. Predictions are made for traffic growth and replacement airplane needs of airlines for the future. These needs are categorized by capacity (number of seats) and range requirements. The market requirement for passenger comfort and cargo hauling capability must be anticipated. Typically, as passenger comfort and cargo capability increases, so does the cost to build the airplane and the cash operating costs; so the market segment must be analyzed to see that the demand will produce the revenue to cover costs. Constraints such as noise need to be considered, as well as airline-perceived performance requirements such as field length, approach speeds, initial cruise altitude, and block time per trip.

The life cycle direct operating costs (DOC)—i.e., net present value of ownership cost plus cash operating costs per mile and per seat-mile—can be used as a figure of merit for competing airplanes. Typically, cash operating costs include fuel, flight crew, maintenance, and insurance. The market-based price for new or derivative airplanes can be calculated to match or improve on the DOC of competitive airplanes within the market segment. The associated cost to build an airplane with this market-based price is referred to as the "must cost." The estimated cost per airplane to actually build the airplane is referred to as the "will cost." The "will cost" is the total nonrecurring cost of investment by the manufacturer to design, engineer, tool-up for, and certify the type of aircraft, divided by the number of these airplanes the market place is predicted to buy, plus the recurring cost to produce the airplane.

A manufacturer will likely offer an airplane when its management is comfortable that the "will cost" is less than the "must cost" of the candidate airplane. An airplane meeting this condition will sell enough units in a competitive marketplace at a net price that will pay off the nonrecurring and recurring costs and provide the manufacturer with a reasonable rate of return on the investment for taking the risk and providing the product. Meeting this condition is referred to as "satisfying the cost-price-market loop." It is usually the responsibility of the marketing and finance departments and will not be pursued further here.

Even though LCC or LOC are important figures of merit, they do not easily lend themselves to engineering analysis. Therefore, engine/airplane systems are usually optimized to some performance figure of merit and the cost optimization is inferred indirectly. For example, takeoff gross weight is used as a performance figure of merit because a smaller airplane should cost less to build and operate. This perceived, or inferred, minimum cost is valid only if the optimization procedure uses consistent technology levels throughout. Traditionally, minimum engine takeoff thrust-to-weight ratio was assumed to be a measure of goodness. Recent analyses have shown, however, that in some cases overall system performance and cost need not meet this requirement. The high-throttle-ratio engine concept, for example, sacrifices takeoff thrust in favor of optimizing cycle characteristics in supersonic cruise, as will be demonstrated by example later.

The propulsion system has both direct and indirect impacts on system costs. The direct cost is due to engine development, procurement, and support (maintenance) costs. Indirect costs include the performance and installation characteristics that affect the airplane size, takeoff gross weight, and mission fuel. The operating cost benefit of a better matched, newer technology engine must be weighed against the cost of development for a new engine. The decision to develop a new engine or to use an existing one for a new or derivative airplane requires that the direct and indirect operating costs be considered along with the system effectiveness.

The date of initial operational capability and mission requirements define the level of technology and thus the component efficiencies and weights available to reduce the cost of ownership and improve system effectiveness. Occasionally, engine/airplane matching has been used to identify and prioritize the advanced technology efforts.

Other figures of merit used in the military arena are survivability/vulnerability and operationally ready rates. Airplane systems evaluations often optimize to a combination of factors over the most probable mission mix. Commercial airplane buyers also consider the value of reliability and commonality.

The following section shows a procedure for optimizing the engine/airplane match based on performance figures of merit that can be evaluated by computer. These figures of merit become dependent variables in the computer solution. The more subjective figures of merit are not treated further.

Methods of Optimization

Historically, the selection of an optimum engine/airplane combination was made on the basis of experience. Mission analysis was conducted on selected engine cycles and aircraft geometries and the best airplane selected as the combination that best satisfies the figure of merit. For missions that are similar to ones already being flown, it is possible for experienced designers to consider good candidate configurations. Historically, airplanes have been designed in this manner. Two weaknesses of this approach are the reliance on the insight required by the designers and the tendency to perpetuate design philosophies without due consideration for alternatives that new technology may have made available.

The advent of computer assistance to the designer has made it possible to consider many more design alternatives and has created a preliminary design approach throughout the aerospace industry. To provide insight into

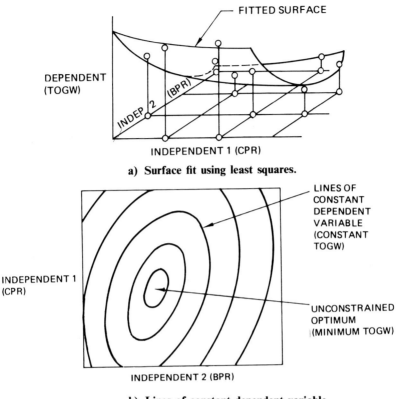

a) **Surface fit using least squares.**

b) **Lines of constant dependent variable.**

Fig. 3.27   **Representation of a surface.**

the nature of computer optimization, the philosophy is presented for one
method currently used in industry as an operational design tool.

The procedure applies statistics to independent cycle and airframe variables to determine a minimum number of mission analysis cases that must
be run to create parametric equations for the solutions for all variables and
constraints under consideration. These equations can be considered to form
multivariant surfaces as a function of the independent variables. An
optimum can then be obtained on any solution surface by standard
methods, although the solution of a nonlinear equation may be required.

For any two independent variables, for example, compressor pressure
ratio (CPR) and bypass ratio (BPR), the performance of a dependent
variable, say takeoff gross weight (TOGW), can be determined. By computing TOGW for many combinations of CPR and BPR, an equation can be
fitted to the points to represent TOGW as a continuous function of BPR
and CPR; the equation so obtained can be represented as a surface (Fig.
3.27). The form of the equation is found by regression techniques using the
computed points to obtain coefficients for polynominal equations representing the surface. A second-order equation for this example would have

the form

$$TOGW = C_0 + C_1(BPR) + C_2(CPR) + C_3(BPR)^2$$
$$+ C_4(BPR)(CPR) + C_5(CPR)^2 \qquad (3.43)$$

where $C_0, C_1, ..., C_5$ are constants to be determined.

A regression analysis could be used to compute the coefficients C_i using a least squares fit to the data points; goodness-of-fit statistics are also calculated by which the suitability of the form of the regression equation can be judged.

Once the surface is established, the surface fit polynomials are interrogated by nonlinear optimization methods that use gradient vectors to search for maximum or minimum dependent variable (TOGW) values within the specified boundaries of each independent variable. Conceptually, this amounts to plotting contour lines on the surface of Fig. 3.27a as shown in Fig. 3.27b and identifying the extreme values. In this case, the resulting minimum TOGW is referred to as an unconstrained optimum or the optimum combination of BPR and CPR for the desired mission. Constraints such as minimum specific excess power P_S or thrust margin (TM) can be superimposed on the contour maps after the unconstrained optimum is obtained. Figure 3.28 shows the choices of CPR and BPR that provide optimum TOGW when the design constraints include $P_S > 75$ and $TM > 2.8$. An advantage of this technique is that, since the constraints are placed on the system after the regression analysis is completed, the constraints can be changed without recomputing the optimization. This simple example, drawn from Ref. 6 shows the optimization process for two independent variables. The actual optimization program uses the same procedure, solving for the optimum as a function of a larger number (10–20) of independent variables. Conceptually, this amounts to a process of determining a quadratic surface fit and contour mapping in multidimen-

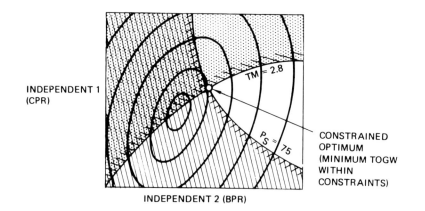

Fig. 3.28   Constrained optimum.

sional space. The solution is not difficult on a large modern computer, provided the number of input values for each of the independent variables can be restricted.

Dependent and Independent Variables

Independent variables of most significance for engines/airplane matching are shown in Table 3.2. The engine variables identified as secondary are not commonly used except in exercises concentrating on engine size and weight. Table 3.3 displays dependent variables commonly used as the performance figures of merit to be optimized. In a given study, a surface fit and contour mapping are computed for each dependent variable. The contour line values of one variable can be constrained to a specific value and cross plotted. For example, the line of $P_S = 75$ in Fig. 3.28 was obtained from a contour map of P_S as a function of CPR and BPR (Fig. 3.29). By selecting the appropriate contour line on the P_S map, any value of P_S can be used as a constraint on TOGW.

Minimizing Required Combinations of Independent Variable Values

The number of mission analyses necessary to define an equation grows very rapidly with the number of independent variables. If each of 10 independent variables is allowed to take on 4 values to define the surface shape, a total of 1,048,580 missions would have to be calculated! Fortunately, statistical methods can be used to reduce this number of combinations without jeopardizing the validity of the results.

One such technique is known as orthogonal Latin squares (OLS). OLS is used to select the values of each independent variable to provide a sparse

Table 3.2   Typical Independent Variables

	Engine	Airframe
	Fan pressure ratio	Takeoff gross weight
	Bypass ratio	Thrust/weight
	Overall pressure ratio	Wing loading
	T_4	Aspect ratio
	Turbine nozzle area variation	Sweep angle
	Throttle ratio	Wing thickness ratio
	Exhaust nozzle area ratio	Wing taper ratio
	⎡ Number of spools	Operating weight increment
	⎢ Number of stages	Mission radius
Secondary variables	⎨ Afterburning	
	⎢ Stage loading	
	⎣ Hub/tip ratio	

Table 3.3   Typical Dependent Variables

Takeoff gross weight	Dash time required
Cruise range factor	Specific excess power
Loiter factor	Thurst margin
Fuel weight	Takeoff roll distance
Rate of climb or climb gradient	Service ceiling
Cruise Mach/altitude engine-out performance	Spotting factor (airplane size for carrier handling)

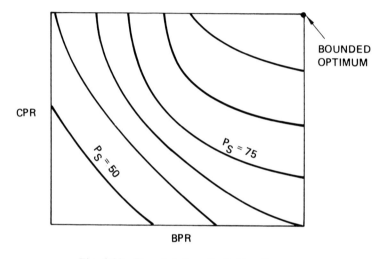

Fig. 3.29   Bounded (box-limited) optimum.

but uniformly distributed set of data points to which the curve fitting routines can be applied. Reference 6 discusses the theory of OLS in some detail. Here, it is sufficient to note that by applying OLS, the number of missions that must be analyzed is reduced to N^2, where N is the smallest prime number (or power of a prime number) larger than the number of independent variables being considered. For example, when 10 independent variables are being analyzed, 121 (11^2) missions must be calculated. Regression methods produce quadratic equations for each dependent variable as a function of the independent variables. For 10 independent variables, each equation requires 66 coefficients. Experience has shown that some dependent variables are not a strong function of all independent variables and the corresponding coefficients may be insignificant. A least squares fit of the data to 10 independent variables can usually be done adequately with less than 25 coefficients and still define the proper optimum engine/airplane combination. The logic flow for finding the optimum is reviewed in Fig. 3.30.

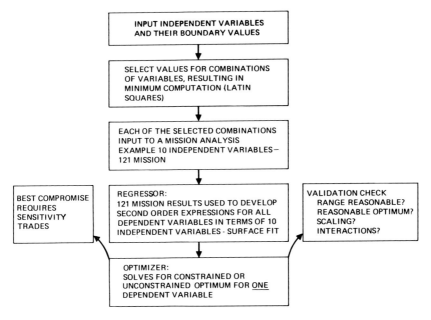

Fig. 3.30   Optimization logic.

Boundary Values for Independent Variables

The computer optimization technique selects the combination of values for the independent variables representing the best engine/airplane system combination measured in terms of a specified figure of merit for the mission. If the optimum of any independent variable equals its boundary value, this value can be changed and the optimization rerun to see the relevance of this possibly arbitrary constraint. Suppose, for example, fan pressure ratio (FPR) is an independent variable and the range of values selected was FPR = 2.5–4.5. If the optimization program selected a "best" configuration with FPR = 2.5, it is not possible to ascertain if this FPR is really optimum or if FPR is constrained by the boundary value initially specified. New limits for FPR of say 1.5–3.0 can be specified and after the optimization program is rerun, the true optimum FPR may turn out to be 2.15. This feature of the technique leads the engineer to an optimum even if he is not experienced enough to guess the limits for each variable appropriately. Whenever the limits are changed, the mission analyses program should be re-examined to insure that the geometry, scaling, and interaction relationships remain valid.

Thus, the boundary conditions on each independent variable are self-correcting, but the designer does not know if he has selected the most important independent variables!

Re-examination and Validation

The validity of the airplane configuration identified by the computer optimization must be established since the real optimum configuration rarely corresponds to one of the 121 configurations computed through actual mission analysis. Therefore, the values of each of the independent variables specified by the optimum are used as input values to a mission analysis. If the value of the figure of merit for single-mission analysis agrees with that obtained from optimization, then some confidence in the optimum is justified. Because the surface (equation) is a least squares quadratic fit, the values of the single-mission analysis may not be on the surface, particularly if the true surface is a higher-order polynomial or is discontinuous. Fortunately, most configurations analyzed have matched reasonably well near the optimum.

Computer optimization is a preliminary design tool whose main value is to quickly obtain an unbiased estimate of the "neighborhood" in which the real optimum configuration can be expected. The solution is unbiased only to the extent that the engine/airplane combinations in the mission analysis are realistic. Thus, the mission analysis should be re-executed to insure that the scaling, interactions, and geometries are appropriate and that the optimum is not artificially constrained by the range of one or more independent variables. Finally, it should be emphasized that the optimum can be defined for only one independent performance figure of merit. That is, the optimum engine/airplane combination can be found using either takeoff gross weight or specific excess power as the unconstrained figure of merit. Alternately, a constrained optimum can be obtained for either figure of merit with the other constrained, but it is not possible to use this

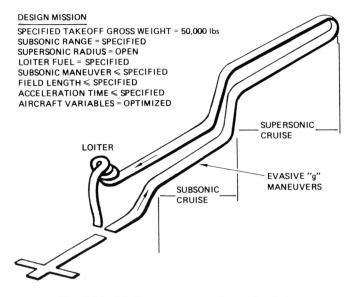

DESIGN MISSION

SPECIFIED TAKEOFF GROSS WEIGHT = 50,000 lbs
SUBSONIC RANGE = SPECIFIED
SUPERSONIC RADIUS = OPEN
LOITER FUEL = SPECIFIED
SUBSONIC MANEUVER ≤ SPECIFIED
FIELD LENGTH ≤ SPECIFIED
ACCELERATION TIME ≤ SPECIFIED
AIRCRAFT VARIABLES = OPTIMIZED

SUPERSONIC CRUISE

LOITER

EVASIVE "g" MANEUVERS

SUBSONIC CRUISE

Fig. 3.31   Medium-range ground attack mission.

technique to find the configuration that simultaneously optimizes both figures of merit.

Example Mission

To illustrate the results of mission analysis, engine/airplane optimization, and the impact of mission constraints, an example is presented for a military airplane designed to perform a medium-range ground attack mission.* The design mission is shown in Fig. 3.31 and consists of the following maneuvers: (1) takeoff from a fixed maximum field length, (2) climb to altitude for maximum subsonic cruise range (during the fixed range subsonic cruise leg, the aircraft is required to perform evasive maneuvers of substantial g force), (3) using afterburning (A/B), accelerate and climb to supersonic cruise altitude in minimum time, (4) level off at the supersonic cruise design Mach number, cruise into hostile territory (with or without A/B), and perform the radius mission, (5) returning from hostile territory, decelerate to subsonic cruise, (6) loiter for a fixed period of time, and (7) land.

Five distinct engine cycle concepts were analyzed to show how they affect the geometry of a fixed-weight airplane and how the optimum aircraft compared (with and without various mission constraints). The mission includes a relatively long subsonic leg of fixed distance, but does *not* fix the length of the supersonic range (twice the supersonic radius). Supersonic range was the figure of merit to be optimized for a fixed-gross-weight airplane. For each engine type, an optimization was conducted using airplane polars and configurations that properly accounted for scale and installation effects. (If flight to a specific target at a known range had been desired, takeoff gross weight would have been the optimization variable.)

For the five engine types under investigation, the design bypass ratio, pressure ratio, and cruise throttle ratios are presented in Table 3.4; a complete study would include these as independent variables. The technology level, combustor exit temperature, and cooling air requirements for all engines are held constant. The fixed values shown in Table 3.4 are representative for each engine type. The lapse in thrust as a function of Mach number for the chosen engines is shown in Fig. 3.32. Figure 3.33 shows the relative dimensions of the engines producing the same takeoff thrust.

These five fixed cycles affect the optimum geometry of the airplane designed for maximum supersonic cruise. For each independent engine cycle study, five airplane variables were analyzed to give the optimum supersonic range. The independent variables and their boundary values are shown on Table 3.5. As described in the previous section on minimizing independent variable values, 49 mission analyses (the square of the prime number greater than the number of independent variables) were run for each engine. Each of the 49 had a different combination of values for the

---

*Specific numbers are not shown in the example. The results are from an actual mission analysis and all characteristics and constraints are representative of military airplane design.[7]

Table 3.4 Selected Engine Cycle Definition

Engine	Bypass ratio	Overall pressure ratio	Maximum combustor exit temp (CET)	Throttle ratio[a]	Augmenter type
Low-throttle-ratio (LTR)[a] turbojet	0.20	15		1.05	Afterburner
High-throttle-ratio (HTR) turbojet	0.20	15		1.31	Afterburner
Variable-geometry - turbine (VGT) turbojet	—	15	Constant technology	1.00	Afterburner
Mixed-flow turbofan (AB fan)	1.30	20		1.15	Afterburner
Separate flow turbofan (DB fan) with fan duct burner	1.30	20		1.17	Duct burner

[a]Throttle ratio = $CET_{cruise}/CET_{takeoff}$.

Table 3.5   Airframe Variables

Independent variables	Range of interest
Wing aspect ratio	1.3–4.3
Wing loading	50–150 lb/ft²
Wing thickness-to-chord ratio	0.030–0.060
Wing leading-edge sweep	34–74 deg
Airplane thrust-to-weight ratio	0.70–1.40
Dependent variable to be optimized	Supersonic range

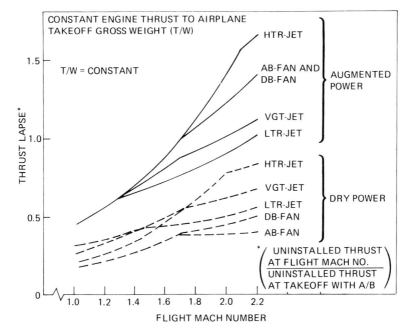

Fig. 3.32   Comparative thrust lapse.

independent variables as prescribed by the Latin square technique. A surface fit (regression analysis) then provided the unconstrained optimum range and "contour lines" of configurations producing constant values of range. These results can be used to evaluate the effects of changing mission constraints without rerunning the mission analysis.

Table 3.6 shows the resulting optimum airplane for each engine type. The optimum is constrained only by a constant thrust-to-weight ratio. The optima were not constrained by field length or acceleration requirements. Notice that each engine cycle requires a substantially different optimum airframe geometry. The wing loading range is 115–135 lb/ft² and the wing

Table 3.6 Airplane Optimization Results
(Maximum supersonic radius for 45,000 lb thrust engines
with no performance constraints)

	LTR jet	HTR jet	VGT jet	DB fan	AB fan
Airplane					
AR	4.3	4.2	4.3	4.3	4.3
T/C	0.030	0.030	0.030	0.030	0.030
W/S	120	120	135	115	120
Sweep, deg	40	45	52	34	43
Supersonic					
P_S	Aug.	Dry	Dry	Aug.	Aug.
SFC	1.87	1.37	1.49	1.44	2.08
L/D	3.7	4.9	4.9	4.6	4.8
Subsonic					
P_S	0.42	0.62	0.44	0.70	0.71
SFC	1.25	1.15	1.49	0.94	1.04
L/D	11.1	10.2	9.9	10.5	10.7

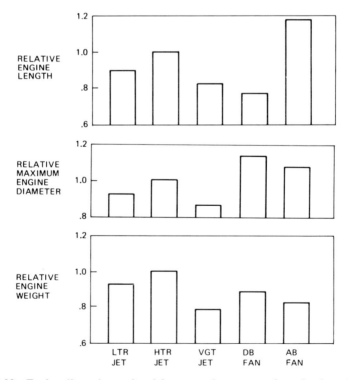

Fig. 3.33 Engine dimension and weight comparison at equal sea-level static maximum thrust (A/B).

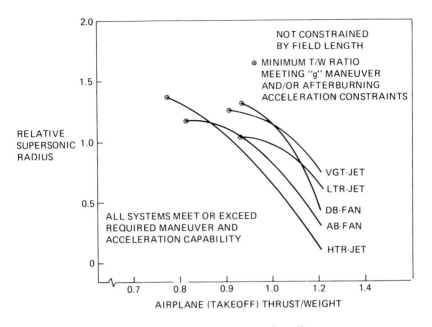

Fig. 3.34   Maximum supersonic radius.

sweep range 34–52 deg. Thus, it is evident that it is *not* appropriate to evaluate the merits of various engine concepts installed in an airplane of fixed geometry.

Figure 3.34 shows the relative supersonic range as a function of airplane takeoff thrust-to-weight ratio T/W. At each T/W, the airplane variables are optimized to produce curves that represent the loci of constrained optima (constrained by T/W). For each curve, a minimum T/W is obtained where the constrained optimum engine/airframe match can no longer meet the selected g maneuver or A/B acceleration time (indicated by the dots on Fig. 3.34). Figure 3.35 is a comparison of configurations represented by the constrained optima (dots).

From Fig. 3.35 the optimum aircraft appears to be a high-throttle-ratio (HTR) turbojet yielding the lowest T/W and longest range while meeting the acceleration and maneuvering requirements. In this case, the field length is not restricted.

If short field lengths are important, higher takeoff T/W ratios are necessary. Airplanes with $T/W = 0.9$ produce takeoff field lengths representative of medium-range attack aircraft. As shown in Fig. 3.36, the required T/W ratios, necessary to produce reasonable takeoff distance, result in a different engine providing the best supersonic radius. The HTR concept has moved from the best choice to worst choice, because of the field length restriction. This result is to be expected because the concept of HTR engines is to design for supersonic cruise using a constant corrected engine weight flow. The resulting design produces the required cruise thrust from an engine that is smaller in length, diameter, and airflow rate at a low Mach

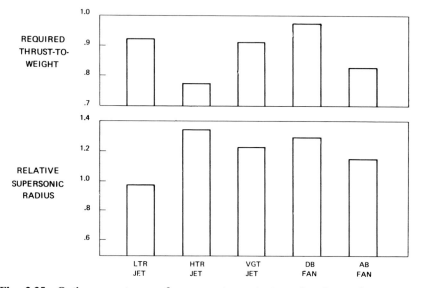

Fig. 3.35 Optimum system performance at constant acceleration and maneuver requirement.

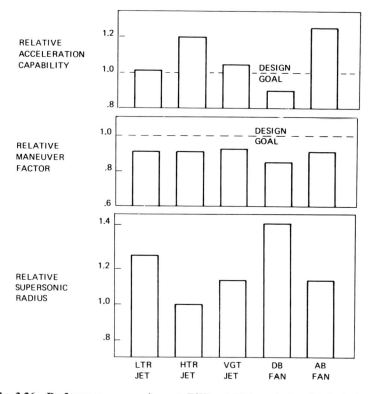

Fig. 3.36 Performance comparison at $T/W = 0.90$ (constraint: fixed airplane design and equal, reasonable field length).

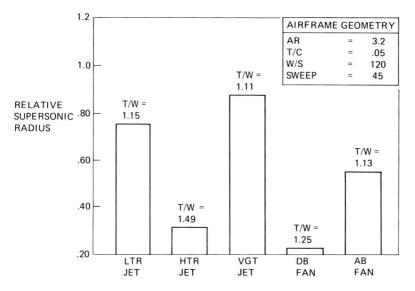

Fig. 3.37 Supersonic radius with required maneuver and acceleration capability (fixed airplane design).

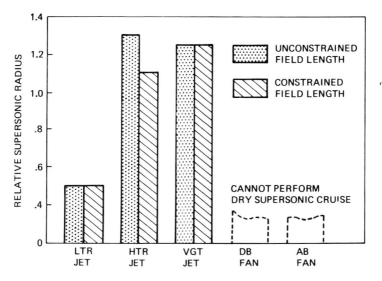

Fig. 3.38 Optimum systems for dry supersonic cruise.

number. As a result, the takeoff thrust is relatively low. This deficiency can be somewhat offset by incorporating variable turbine geometry or variable burner bypass, which was not included in this study.

If a fixed airplane geometry is chosen and the engines are sized to meet given maneuver and acceleration time requirements, the relative engine performance shifts and the duct burning turbofan that was best for supersonic range becomes the worst (Fig. 3.37).

In the considerations thus far, the amount of afterburning has been allowed to vary. Flying supersonic missions and turbofan cycles requires more A/B than do turbojets. If the infrared (IR) signature of the aircraft over hostile territory is considered, the engines may be constrained to dry (non-A/B) operations. Such a requirement applied to the aircraft discussed in Fig. 3.37 results in the characteristics shown in Fig. 3.38. The fan engines cannot perform even the desired mission under this constraint.

This example has illustrated that the engine designer cannot work alone. Optimum engine/airplane matching cannot be accomplished unless an array of both engine and airframe variables are considered together with mission rules and constraints. The next section will consider ways to evaluate the sensitivity of the mission and aircraft to changes in the variables and constraints.

3.4 SENSITIVITY AND INFLUENCE COEFFICIENTS

Sensitivity

Sensitivity studies use influence coefficients to measure the change of a dependent variable y due to the change of a single variable x of several independent variables. Therefore, influence coefficients involve the use of partial derivatives,

$$I_{y,x} = \frac{x}{y}\frac{\partial y}{\partial x} = \frac{\partial \ell n y}{\partial \ell n x} \quad \text{or} \quad I_{y,x} = \frac{\partial y}{\partial x} \tag{3.44}$$

Dependent variables may be performance or cost figures of merit, while independent variables are usually the design parameters. Examples of sensitivity studies commonly conducted during engine/airplane matching include:

(1) The sensitivity of a configuration to the proximity of the optimum.
(2) The sensitivity to relaxation of a mission constraint.
(3) The influence of changing the level of technology.

Item 1 is a logical extension of the optimization program. Using the optimum value of each of the independent variables as a central configuration, a careful analysis is usually conducted on designs lying in a neighborhood near the optimum. Mission analyses are conducted on each combination of independent variables in order to approximate more closely the surface fit polynomials near the optimum. Once the polynomials fit the mission analysis results with the desired certainty, a comprehensive sensitivity study can be done by partial differentiation of the polynomial. Using the simple example introduced in the previous section, the influence on takeoff gross weight due to changing the bypass ratio or compressor pressure ratio can be evaluated, i.e.,

$$\frac{\partial(\text{TOGW})}{\partial(\text{BPR})}\bigg|_{\text{CPR = const}} \quad \text{or} \quad \frac{\partial(\text{TOGW})}{\partial(\text{CPR})}\bigg|_{\text{BPR = const}}$$

The resulting values may be nondimensionalized and ranked according to their importance (influence) on performance or cost. As a result of ranking the independent variables and establishing the cost of changing any of them, the engine and airframe companies identify the areas where improvements will be most beneficial. The customer or systems integrator can also weigh cost vs performance to see if moving away from the specified optimum will decrease cost sufficiently to justify reduced performance.

If the optimum engine/airframe combination is constrained by a customer, the penalties associated with the constraint may be identified. In preliminary design, the sensitivity of the aircraft system to mission constraints is of paramount importance in the interaction between the engine and airframe designers and the customer. For military applications, the results of sensitivity studies provide a quantitative guide allowing the customer to establish the tradeoffs between the mission requirements he foresees as important and the benefits of relaxing one or more constraints.

The other important use of sensitivity is to establish the priorities in technology development. It can be a very costly decision to increase the performance of an engine component. Similar commitments are necessary from structures, aerodynamics, and production groups in the airframe companies. Budget constraints necessitate that the performance of only some of the components may be advanced at the same rate. Priorities are established within industry and government in an effort to find those components whose development will produce the largest benefit to the system in the given time period with the available funding. Such decisions are best made when a priority of the independent variables is established and the performance level available for a given expenditure has been estimated.

Importance of the Choice of Independent Variables

Section 3.3 demonstrated that the optimum engine/airplane combination can be selected for a given set of independent variables and that the boundary values selected for the variables are self-correcting. The previous subsection indicated that required levels of component technology and tradeoffs between performance and cost can be established by analysis of the surface fit polynomial that is obtained as a function of the chosen independent variables. As a result, independent variables can be ranked so that those having the most influence can be emphasized. The choice of independent variables—based on intuition, new evidence, or past experience—must be made prior to optimization and sensitivity studies. Unfortunately, there is no method of guaranteeing that an important independent variable has not been overlooked. The most important variables, identified through experience, were shown on Table 3.2. The occasional importance of other variables is usually realized during component design or mission analysis. If a very strong variable has been ignored, its influence will be obvious if that variable is changed for the "optimum" aircraft design. Sometimes, variables can significantly influence the true optimum aircraft design even though they do not seem important when the optimum (chosen

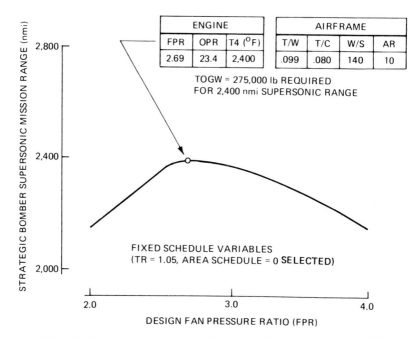

Fig. 3.39   Optimum bomber configuration using eight independent variables.

using fixed values for these elements) is perturbed. Consider the following example.

For a specified mission, the optimum configuration of a variable-sweep wing strategic bomber was desired. The mission included both subsonic and supersonic segments. During the subsonic flight the aircraft used a low-altitude, terrain-following path that imposed a q loading constraint; a 2400 mile supersonic range and a fixed payload were specified. Three engine variables (fan pressure ratio, overall pressure ratio, and turbine inlet temperature) and five airframe variables (takeoff gross weight, takeoff thrust/takeoff gross weight, wing loading, wing thickness-to-chord ratio, and aspect ratio) were chosen. The selected figure of merit was the takeoff gross weight that the analysis showed to be minimum at 275,000 lb TOGW. The table in Fig. 3.39 shows the values of the independent variables at that weight.

Suppose one would like to know the sensitivity of the supersonic range to changes in the design or fan pressure ratio. Recall from the mission analysis that either the TOGW is evaluated for a given range or the range is evaluated for a fixed TOGW. Therefore, for the sensitivity study, TOGW was fixed to the optimum value, the design FPR was calculated away from the optimum, while the other six independent variables were allowed to reoptimize. The resulting variation of range for the constrained optimum airplane is plotted as a function of design fan pressure ratio in Fig. 3.39. For each value of FPR, the range shown is the best that can be obtained from a 275,000 lb airplane and the values of each of the other six independent variables may vary accordingly. Therefore, a different aircraft configuration is implied by each point on the curve. Each point is just the constrained optimum range value (found by the optimization process of Sec. 3.3) for constraints of TOGW and FPR. Looking at the curve of Fig. 3.39, one would conclude that range is sensitive to fan pressure ratio and that the influence becomes strongest as the FPR moves away from optimum toward a lower fan pressure ratio.

Now consider the importance of the nozzle throat area schedule or the throttle ratio. Since neither of these was considered an independent variable, the sensitivity of range to either of these must be evaluated externally to the optimization process. This is done by fixing the optimum aircraft (from the table in Fig. 3.39) as a reference configuration; then, using mission analysis on the fixed aircraft, establish the effect of variations in nozzle area schedule and throttle ratio on range. Figure 3.40 shows the results. The throttle ratio extremes were chosen to represent the available range consistent with expected engine technology levels. The nozzle schedule represents the maximum dry nozzle throat area increase beyond that used in the original optimum (nondimensionalized to fall between -1 and $+1$). The values of the nozzle schedule are not important, per se, but the results show that changing the nozzle throat area can produce slightly more range (0.5–0.8%) and that the throttle ratio has less effect than the area schedule.

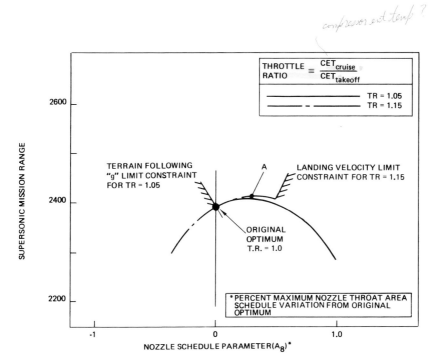

Fig. 3.40 Effect of additional independent variables.

Three questions arise at this point:

(1) Is the airplane specified by the table in Fig. 3.39 but with an area scheduled of 0.22 and throttle ratio of 1.15 (point A in Fig. 3.40) the true optimum airplane?

(2) Is the sensitivity of the range-to-throttle ratio and nozzle area schedule sufficient to imply the two variables have a strong enough influence that they should have been included as independent variables?

(3) If they had been included as independent variables, what would have been the configuration of the optimum aircraft?

When the best airplane is selected by the optimization technique, all independent variables must be optimized at the same time. The airplane described in question 1 is therefore not necessarily the best because eight of the variables were fixed before the other two were introduced. Only a new optimization using 10 variables will establish how the other variables change.

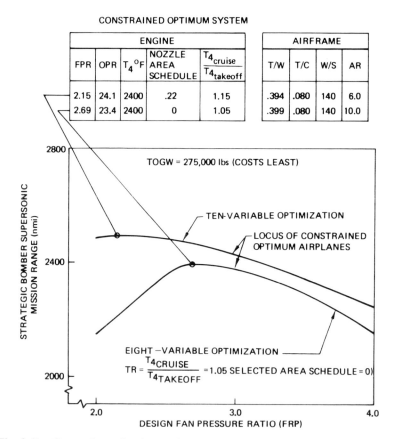

Fig. 3.41   Comparison of optimum aircraft resulting from 8 and 10 variable optimization.

Based on the results shown in Fig. 3.40, the 0.8% increase in range is not usually sufficient to justify the time and expense of running a new optimization with 10 variables. However, if the 10 variable optimization is run, the following results are obtained. As can be seen on Fig. 3.40, the aircraft with a throttle ratio of 1.05 (original optimum) was very near the limit imposed by subsonic g loading for terrain following, while the throttle ratio of 1.15 is not affected by this limit. The results of the 10 variable analysis are shown on Fig. 3.41 with the results of the original 8 variable optimization repeated for comparison. If the FPR and TOGW are constrained to the same values as for Fig. 3.39, a new set of constrained optimum airplanes is produced that have more range than those with fixed throttle ratio and nozzle area schedules over the entire range of fan pressure ratios. The new optimum airplane has 4% more range, but more significantly, the engine and airframe changes substantially. Note the change in the fan pressure ratio and wing aspect ratio.

The advantage of using an optimization process rather than merely comparing mission analysis results can be appreciated when on considers the experience necessary to realize that g loading and nozzle area schedule profoundly affect the optimum values for wing aspect ratio and fan pressure ratio. For these reasons, it is recommended that the maximum numbers of independent variables allowed by time, cost, and simulation technique be used whenever a preliminary design study for a new mission is being conducted.

A final point this example illustrates is the impact of technology level. The turbine inlet temperature for both optimizations was 2400°F, which was the maximum value felt to be reasonable. One might conclude from a sensitivity study on turbine temperature that increasing T_4 is very worthwhile. If the metal temperature in the turbine is as high as possible for the method of cooling chosen, then the only way to increase turbine temperature is to increase the amount of compressor bleed air used for cooling. The requirements for additional bleed air that affect engine size, weight, and performance may well mean that increasing T_4 is not beneficial. This tradeoff between turbine temperature and the bleed air requirement to maintain metal temperature limits is critical for missions requiring long endurance at part power. These missions frequently optimize at relatively low turbine temperature for these reasons. (This problem could be alleviated by using a cooling system with throttling capability.)

3.5 COMPUTER SIMULATION OF GAS TURBINE ENGINES

Engine/airplane matching has been shown to require inputs from many disciplines, as shown schematically in Fig. 3.42. Figure 3.43 displays a breakdown for the performance and weight contributions to the matching process. The figure indicates that mission analysis and optimization programs require inputs from aerodynamics and structures, as well as propulsion. The propulsion contribution includes performance and weight analysis for the gas generator, inlet, exhaust, and their interactions. The

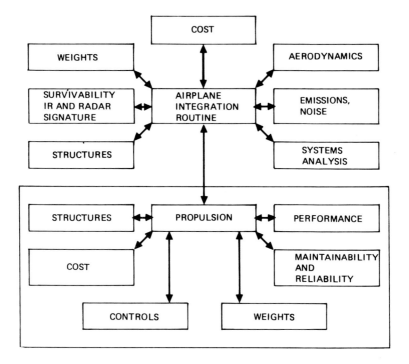

Fig. 3.42   Engine/airplane matching disciplines.

previous chapters have discussed the aerothermodynamics of engines in general. The chapters that follow will develop the details of individual components. This section demonstrates the usefulness of the computer in combining the results of detailed component design with basic cycle analysis to simulate the performance of gas turbine engines accurately.

The computer simulation to be described assumes a quasi-steady-state situation where the rates of change of Mach numbers, altitude, engine rpm, and power setting are assumed small. While this assumption is sufficient for mission analysis, where the flight profile is broken into a series of quasi-steady conditions, it must be acknowledged that the performance of the final aircraft will also depend on the engine dynamics and controls, which are beyond the scope of the present text.

Method of Performance Simulation

The computer simulates gas turbine engine performance by an appropriate matching of components. Two philosophies are common in engine simulation programs. One incorporates a "generalized" engine, say a three-spool duct burning afterburning turbofan engine, and provides a technique for elimination of unneeded components. The other philosophy, used by industry and government for more than two decades, allows the interconnection of components in a building block manner to simulate any

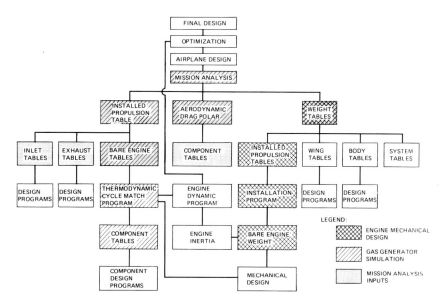

Fig. 3.43   Engine/airplane matching (summary of considerations for engine performance and weight).

conceivable combination of engine components. The latter technique is considered in detail at the end of this chapter. The generalized engine will be shown to be a particular preprogrammed version of the building block technique that simplifies the work required by the user in order to simulate most engines.

For any arbitrary choice of engine components, power and airflow paths through the engine can be drawn connecting the output of one component to the input of the next component. The output/input connection points are referred to as nodes. The nodes do not have to connect physically adjoining elements. For example, the airflow path of a simple turbojet has a node connecting the compressor and burner and another connecting the burner and turbine, while the power path of the same engine has a single node connecting the compressor and turbine. Either design or off-design engine performance calculations are performed on each component based on flight conditions input to the node in front of the engine and information given the input node to each component. As each component calculation is completed, all output flow properties and other characteristics of that component are transmitted to the next node. Connecting components are therefore related to one another as functions of the program inputs, system variables, and specified component characteristics (equations and component maps). The component calculations produce a system of coupled algebraic transcendental and partial differential equations to be solved.

One method is illustrated by calculating design and off-design performance of a single-spool turbojet engine.

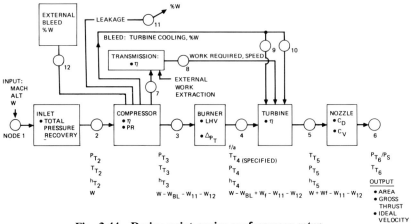

Fig. 3.44   Design point engine performance setup.

Design Point Simulation

The simulation of engine design point performance is a direct computation requiring that the efficiency of each component be known. The design flight conditions (Mach number, altitude, and temperature) and inlet mass flow are specified at a node in front of the engine. The schematic for a single-spool turbojet is shown in Fig. 3.44. The primary airflow path is through nodes 1-2-3-4-5-6. The power demand path (such as turboprop, turboshaft, or geared propfan) connects the compressor to the turbine through nodes 7 and 8. The transmission represents either gears or a shaft. Compressor bleed is supplied to the turbine through node 9 if the air becomes working fluid in the turbine and through node 10 if bleed re-enters the airstream from the trailing edge of the last turbine stage. More complex cooling bleed paths can be specified if a detailed stage-by-stage study is being performed. Bleed air can be taken from intermediate compressor stages and reintroduced through the leading and trailing edges of turbine vanes and rotors, as desired. Node 11 can be used to account for leakage through seals (similar nodes could be established to account for leakage from any other component). In the example, the leakage is assumed to be lost overboard without producing a thrust or drag increment.

Bleed for air conditioning, anti-icing, avionics cooling, etc., is removed from the engine through node 12 as a percent of the inlet weight flow. Power extraction from the engine to operate pumps and generators can be treated as a horsepower addition to the work demanded by the compressor from the turbine. At times, it is convenient to specify the external work requirement as a constant percent of the turbine work available by subtracting it from the transmission efficiency.

The input node values and inlet total pressure recovery are sufficient to establish total pressure P_T, total temperature T_T, total enthalpy h_T, and weight flow W at node 2. These properties at node 2, plus the compressor pressure ratio (PR), adiabatic efficiency, and the bleed fraction,

$$\text{Bleed fraction} = \frac{W_9 + W_{10} + W_{11} + W_{12}}{W_1}$$

are sufficient to determine the compressor outlet properties (P_T, T_T, h_T, and $W_{\text{air remaining}}$) at node 3.

Burner outlet properties at node 4 are determined from specified values of combustor exit temperature T_4, fuel-to-air ratio (f/a), fuel lower heating value (LHV), and the total pressure drop ΔP_T. Alternatively, the rotor inlet temperature (RIT) may be specified and a short iteration loop set up to establish the T_4 that together with the turbine nozzle cooling air will produce the desired RIT.

The turbine exit properties at node 5 are calculated using the properties of node 4, the cooling air from nodes 9 and 10, the work to supply the compressor and extraneous work requirements, and the turbine adiabatic efficiency. The conditions at node 5 plus the nozzle discharge coefficient C_D and nozzle pressure ratio P_{T_5}/P_∞ are used to size the nozzle throat area. Following the calculation of the nozzle throat conditions, the gross thrust, ideal exit velocity, and nozzle exit area are computed. The design point solution for the engine cycle performance is thus complete and requires no iteration. The design point establishes the thermodynamic cycle and sizes the flow passages through the engine.

Off-Design

For off-design cycle calculations, the nozzle throat area computed from the design case is assumed unless a new value (simulating variable nozzle area) is specified.[8] The turbine temperature T_4 must be supplied along with flight conditions. Off-design cycle calculations are iterative processes requiring several guesses before all component properties can be established.

Using the example turbojet (Fig. 3.44), the flight conditions, inlet recovery, and an initial guess of inlet mass flow W_{inl} are necessary to establish the properties at node 2. Compressor data ($W\sqrt{T}/P$, PR, and η) can be determined from the compressor map (illustrated in Fig. 3.45) using values from node 2 and a guess at speed N and compressor grid line R. The corrected mass flow ($W\sqrt{T}/P$) is obtained at the intersection of the R grid line and the guessed value of corrected speed line ($N/\sqrt{\theta}$). The mass flow W_{map} is calculated using the corrected mass flow from the map and T_t and P_t at the compressor inlet (node 2). Discrepancies in mass flow rate are merely noted at this point for later iterative correction. Map properties η and PR are used to calculate the compressor outlet properties at node 3.

The inputs to the burner (LHV, ΔP_T, f/a), the T_4 estimate and the properties from node 3 are sufficient to define the fuel flow required and all other burner outlet properties at node 4.

The turbine speed is known (from the guess of N) and the turbine work requirement ΔH is obtained from a power balance between the turbine and compressor and external power requirements. The turbine pressure ratio

(PR) is found by an iterative method. A first guess is made for the turbine isentropic discharge temperature T_{ti} to determine the corresponding PR from T_{inl} to T_{ti}. The PR from this calculation is used with the turbine map (lower part of Fig. 3.46) to find the turbine efficiency η. A calculated value for h_{ti} (exit) is found by the equation: $h_{ti} = (h_{tinl} - \Delta H/\eta)$. This calculated h_{ti} is converted to T_{ti} (function of the f/a and h_{ti}) compared with the guess for T_{ti}. When the guessed and calculated T_{ti} values agree, the PR value from the last iteration is used with Fig. 3.46 (upper plot) to obtain the corrected mass flow ($W\sqrt{T}/P$) at the turbine inlet (at the intersection of PR and $N/\sqrt{\theta}$). The turbine inlet mass flow (W_{map}) is determined from the corrected mass flow from the map and T_t and P_t at the turbine inlet (node 4). The mass flow residual at the turbine inlet is: $\Delta W_T = (W_{map} - W_{node\ 4})$. The flow properties at node 4, the turbine work requirement ΔH, and the turbine pressure drop define the turbine outlet properties at node 5.

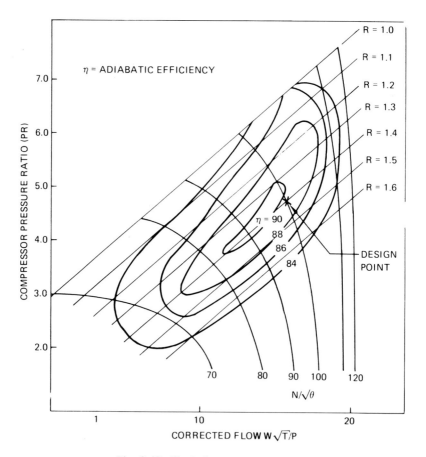

Fig. 3.45 Typical compressor map.

The throat area and discharge coefficient are specified for the nozzle. The pressure ratio across the nozzle is determined from $\Delta W_{noz} = (W_{cal} - W_{node\ 5})$, where W_{cal} is determined at the throat. The first pass through the engine is thus complete. The quality of off-design cycle simulation computer programs is often determined by the efficiency with which they converge to a solution. The Newton-Raphson technique described in the following subsection is an example of a convergence method. In practice, highly sophisticated iteration techniques have been developed and are described in the literature on numerical analysis.

Newton-Raphson Iteration

The off-design simulation of gas turbine engine performance required an initial guess for the inlet weight flow W_{inl}, rpm, N (if a geared transmission is used the gear ratio must be specified a priori), and compressor grid line R. These guesses must next be improved through iteration to converge to their correct values. The method of iterative solution illustrated is the matrix residuals technique of Newton-Raphson. The guessed variables W_{inl}, N, and R and the mass flow residuals ΔW_C for the compressor, ΔW_T for the turbine, and ΔW_{noz} for the nozzle are used to form a set of equations of the following form. [The mass residual is the difference between the calculated outlet mass flow (from maps and equations) and the inlet mass flow (arriving from an upstream component).]

$$\Delta(\Delta W_C) = \frac{\partial(\Delta W_C)}{\partial W_{inl}} \Delta W_{inl} + \frac{\partial(\Delta W_C)}{\partial N} \Delta N + \frac{\partial(\Delta W_C)}{\partial R} \Delta R$$

$$\Delta(\Delta W_T) = \frac{\partial(\Delta W_T)}{\partial W_{inl}} \Delta W_{inl} + \frac{\partial(\Delta W_T)}{\partial N} \Delta N + \frac{\partial(\Delta W_T)}{\partial R} \Delta R$$

$$\Delta(\Delta W_N) = \frac{\partial(\Delta W_N)}{\partial W_{inl}} \Delta W_{inl} + \frac{\partial(\Delta W_N)}{\partial N} \Delta N - \frac{\partial(\Delta W_N)}{\partial R} \Delta R$$

or in matrix form

$$\begin{vmatrix} \Delta(\Delta W_C) \\ \Delta(\Delta W_T) \\ \Delta(\Delta W_N) \end{vmatrix} = \begin{vmatrix} \frac{\partial(\Delta W_C)}{\partial W_{inl}} & \frac{\partial(\Delta W_C)}{\partial N} & \frac{\partial(W_C)}{\partial R} \\ \frac{\partial(\Delta W_T)}{\partial W_{inl}} & \frac{\partial(\Delta W_T)}{\partial N} & \frac{\partial(\Delta W_T)}{\partial R} \\ \frac{\partial(\Delta W_N)}{\partial W_{inl}} & \frac{\partial(\Delta W_N)}{\partial N} & \frac{\partial(\Delta W_N)}{\partial R} \end{vmatrix} \begin{vmatrix} \Delta W_{inl} \\ \Delta N \\ \Delta R \end{vmatrix}$$

The matrix partial derivatives

$$\frac{\partial(\ )}{\partial W_{inl}}, \quad \frac{\partial(\ )}{\partial N}, \quad \text{and} \quad \frac{\partial(\ )}{\partial R}$$

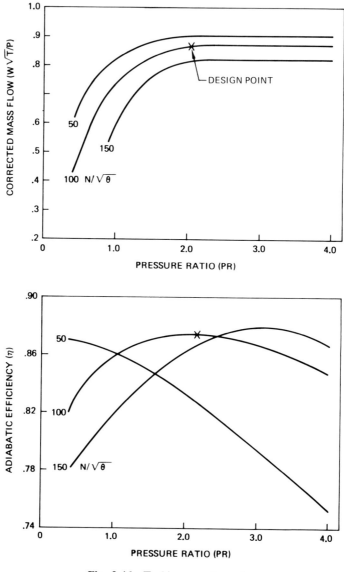

Fig. 3.46 Turbine map format.

represent the first-term approximation of a Taylor's series expansion and
are determined numerically in successive passes through the engine cycle
off-design equations. The first pass through the cycle is used as a base case
(the off-design case just discussed). These initial guesses can literally be
guesses or the values from the design case can be used as a first guess.
Succeeding passes (three are needed for this matrix) are used to determine
the mass flow residual changes associated with small changes in each of the

guessed variables. For example, the quantities

$$\frac{\partial(\Delta W_C)}{\partial N}, \quad \frac{\partial(\Delta W_T)}{\partial N}, \quad \text{and} \quad \frac{\partial(\Delta W_N)}{\partial N}$$

(in the second column of the matrix) are obtained by perturbing N by a small amount ΔN to give

$$\frac{\partial \Delta W_C}{\partial N} = \frac{(\Delta W_C)_2 - (\Delta W_C)_1}{\Delta N}$$

$$\frac{\partial \Delta W_T}{\partial N} = \frac{(\Delta W_T)_2 - (\Delta W_T)_1}{\Delta N}$$

$$\frac{\partial \Delta W_N}{\partial N} = \frac{(\Delta W_N)_2 - (\Delta W_N)_1}{\Delta N}$$

calculated at the compressor, turbine, and nozzle, respectively. Other partial derivatives are similarly developed with respect to the variables W_{inl} and compressor grid line R. The values $\Delta(\Delta W_C)$, $\Delta(\Delta W_T)$, and $\Delta(\Delta W_N)$ are obtained from the base case. The matrix is then solved for ΔW_{inl}, ΔN, and ΔR and the initial guesses for W_{inl}, N, and R are revised. These revised values are used in the next iteration until the residuals (ΔW_C, ΔW_T, ΔW_N) satisfy the convergence criteria.

As demonstrated, the number of guessed variables must equal the number of residuals in an off-design calculation.

In general, the variables in the calculation are those that must be estimated (guessed) to initiate calculation. Common examples are inlet mass flow, rotational speed, bypass ratio, burner outlet temperature, work split relationships, etc. Once input variable values are guessed, residuals are calculated at those components that must satisfy a mass balance (compressors, turbines, and nozzles) or speed balance (transmission).

Returning to the original off-design problem, the off-design thrust can be calculated, compared to the required value, and an iteration on T_4 commenced.

Generalized Engine and Mechanical Design Simulation

Any conceivable combination of components can be simulated in a building block fashion using nodes to connect airflow and power transmission paths. Given the equations governing their behavior, uncommon components (regenerators, air inverter valves, etc.) can be placed at any location within the engine. While this total flexibility is desirable in some situations, a great deal of time is consumed if the geometry has to be generated anew each time an engine simulation is desired. The vast majority of aircraft engines are relatively similar and a single geometry can represent them.

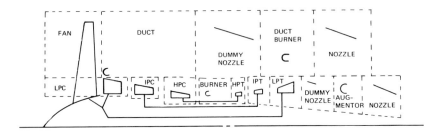

Fig. 3.47   Generalized thermodynamic engine.

Consider the "generalized" engine shown in Fig. 3.47. The figure represents a three-spool, duct burning, afterburning turbofan, which incorporates features common to a wide variety of engines. By creating a computer program capable of describing such a "general engine," but including the capability of excluding undesired components, considerable savings in setup time can be realized. The effective removal of unwanted compressor or turbine components can be accomplished simply by setting their pressure ratio and efficiency equal to 1. The entire fan stream can be ignored by specifying a bypass ratio of zero. Ducts have no effect if they have zero length and zero pressure loss. Augmenters and duct burners can be eliminated by specifying no fuel flow and zero pressure drop. Often, a program uses default values so that the output temperature is set equal to the input temperature for burners and augmenters when the temperature of the downstream node is specified as a lower number than the incoming temperature.

Thus, the simple turbojet of Fig. 3.44 can be simulated by the generalized engine by setting the following conditions: bypass ratio $= 0$; pressure ratio and efficiency $= 1$ for LPC, IPC, IPT, and LPT; dummy nozzle length $= 0$ and area ratio $= 1$; and augmenter exit temperature $\leq T_4$.

The concept of generalized engines can provide further savings in the time required to set up an engine simulation. For preliminary design studies, nondimensionalized maps can be prespecified for each component to represent a given technology level. For the representative state-of-the-art, default values can also be preprogrammed for component efficiencies and cooling bleed requirements. As the desired precision increases from preliminary design to the simulation of actual hardware, the default component maps, design efficiencies, bleed requirements, and leakages can be replaced with values from test results without having to alter the simulation program.

In addition to the thermodynamic analysis, engine evaluation requires consideration of the dimensions and weight. In one approach, the entire engine will be drawn. This "paper" engine can be measured and the weight can be obtained as the sum of the historical weights of components similar to those used for the design. Alternatively, a computer program can be used to "weigh" and "measure" an engine using the thermodynamic cycle results and a mix of empirical values, stress analysis, and material characteristics.

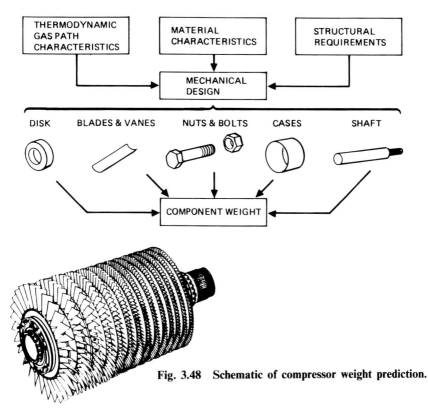

Fig. 3.48   Schematic of compressor weight prediction.

Figure 3.48 is a schematic of a weight prediction procedure for a compressor. Although the details of mechanical design programs are beyond the scope of this text, some general comments are in order.

The gas generator maximum diameter can be established if the maximum airflow, compressor (or fan) hub-to-tip ratio, and compressor face Mach number are known. Precise values of these vary from configuration to configuration. Approximate preliminary design values can be defaulted because each engine company tends to use hub/tip ratios and flow per unit area values that are consistent within the desired technology level and major types of application. Compressor face Mach numbers are typically near $M = 0.6$. With the addition of an historically derived case and flange thickness, the maximum diameter can be estimated.

The length of the engine depends on the engine type, technology level, number of stages in each component, ducts required to connect the flow paths between components, and burner length. The number of stages is determined by specifying the work per stage that can be done by the component within the specified state-of-the-art. The work per stage, which can use a default value for preliminary studies, is divided into the total work for the component (as computed by the thermodynamic cycle program) to arrive at the number of stages. The length of each stage is found

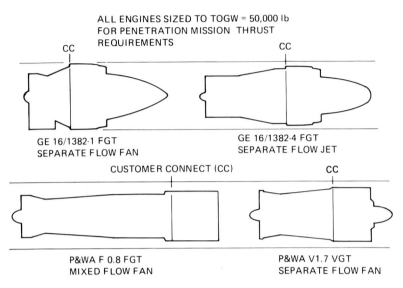

Fig. 3.49   Shape variation of some study engines.

from empirical correlations based on diameter, loading, temperature, and stresses.

Duct lengths are computed from empirical values for acceptable turning and diffusion rates and the radius changes necessary to connect the components. Burner length is again empirical but is governed primarily by burner type and the length necessary primarily by burner type and the length necessary to provide the needed residence time for flame front propagation.

Complete mechanical design requires many inputs, including: thermodynamic inputs, work/stage, hub/tip dimensions, blade pull stress, disk burst stress, blade aspect ratio, temperature gradients, load paths, etc. As a program incorporates more actual design and stress methods instead of empirical correlations, the program complexity and, hopefully, the simulation accuracy increase. While industry and government are continually working to increase the accuracy of the predictions, the combination of a large empirical data base and well-chosen correlation parameters provide useful data for preliminary design work.

Table 3.7   Change in Required TOGW Due to Changes in Engine Physical Characteristics

Physical characteristics (independent variable)	Change in engine characteristics, %	Resulting change in airplane TOGW required to fly mission, %
Engine diameter, max	10	5
Engine length	10	0.02
Engine weight	10	6

The computer simulation of engine aerothermodynamic performance discussed in previous sections specify nothing about the gas generator dimensions or weight. The engine/airplane matching process depends on the physical properties of the engine as well as the performance, as shown in Fig. 3.43. Airplane weight and drag are influenced by the engine dimensions, external shape, weight, and center of gravity location. Consider, as a final example, a Mach 2.7 fighter-type aircraft flying a penetration mission.[7] The shape differences between candidate engines producing the required thrust are shown in Fig. 3.49. The effect of engine weight and dimensions on the takeoff gross weight required to fly a mission of specified range is summarized in Table 3.7.

Acknowledgment

The author expresses appreciation to the Boeing Aerodynamics and Propulsion staffs, as well as the reviewers, including R. Decher, J. Ferrell, G. Oates, E. Tjonneland, A. Welliver, and R. Woodling.

References

[1]Perkins, C. D. and Em Hage, R., *Airplane Performance Stability and Control*, Wiley, New York, 1976.

[2]Hoerner, S., *Fluid Dynamic Drag*, P.O. 342, Bricktown, NJ, 1965.

[3]Dommasch, D. O, Sherby, S. S., and Connolly, T. F., *Airplane Aerodynamics*, Pitman Publishing Corp, New York, 1961.

[4]Oates, G. C., *Aerothermodynamics of Gas Turbine and Rocket Propulsion*, revised and enlarged edition, Education Series, AIAA, Washington, DC, 1988.

[5]Zavatkay, W. F., "Engine Cycle Selection Considerations in the Aircraft Design Process," presented at Aircraft Design Short Course, University of Dayton, July 1976.

[6]Healy, M. J., Kowalik, J. S., and Ramsay, J. W., "Airplane Engine Selection by Optimization on Surface Fit Approximations," *Journal of Aircraft*, Vol. 12, July 1975, pp. 593–599.

[7]Swan, W. C., Welliver, A. D., Klees, G. W., and Kyle, S. G., "Opportunities for Variable Geometry Engines in Military Aircraft," presented at 48th AGARD Propulsion and Energetics Panel, Paris, Sept. 1976.

[8]Yamagiwa, A. T. and Royal, S. G., Boeing Airplane Co., Renton, WA, Doc. D6-41915-1, Oct. 1969 (unpublished).

CHAPTER 4. INLETS AND INLET/ENGINE INTEGRATION

James L. Younghans and D. L. Paul
GE Aircraft Engines, Evendale, Ohio

4
INLETS AND INLET/ENGINE INTEGRATION

4.1 INTRODUCTION

The manner in which the inlet and engine designs are integrated plays a key role in determining the degree to which propulsion system and aircraft operational goals are ultimately realized in the production model. This is true for both subsonic transport applications and transonic and supersonic installations, although the latter's potential problems are generally more extensive. Perhaps the leading example of this situation occurred with the advent of turbofan engine installations in tactical aircraft during the 1960's, when inlet/engine operational difficulties well in excess of anticipated or tolerable levels were experienced. Resolution of this problem involved assessing the importance of time variant, or unsteady, pressure distortion produced by the inlet. This is a prime example of the multidimensional effects, mentioned in Chapter 6 of Ref. 1, associated with inlet operation that can degrade installed performance. In this particular case, the absence of one-dimensionality extends to both dimensions of space (variation of flow properties over the engine face at a given instant of time) and time (similar variation, except as a function of time). As the subject of measurement and evaluation of time-variant distortion is specifically developed in Chapter 6 of this volume, which deals with engine stability, it is sufficient to say here that this problem revolutionized integration of the aircraft and propulsion system, introducing a necessary complication that has continued to the present.

Lack of an effective program to integrate a fighter aircraft inlet and engine during their development phases can result in unacceptable aircraft performance, as reflected by one or more of the following figures of merit: range/payload, maneuverability, maximum Mach number, weapons delivery and gun firing, and engine life and/or maintenance costs.

Since the desirability of successfully blending the inlet and engine seems self-evident, its accomplishment may appear straightforward. In practice, however, various impediments exist during the development cycles. For example, component performance may be emphasized at the expense of obtaining the best overall integrated system performance in order to meet or exceed contract guarantees. This "component optimization" may result in "system suboptimization" in terms of aircraft capability. This somewhat natural tendency can be amplified by the multiorganizational structure that generally prevails during inlet and engine development—consisting of the customer, airframe manufacturer, and engine manufacturer. In addition, cost and time considerations may strongly shape the available development integration effort.

241

The ensuing discussion will develop the elements of a successful inlet/engine integration program. These elements include careful definition of the required inlet and engine operational requirements, consideration of the various ways in which the inlet and engine may interact to influence the other's design, several phases of developmental component and system testing, and frequent information exchange supplemented by formalized, periodic meetings between cognizant aircraft and engine design personnel. The basic objective—maximizing overall aircraft performance—must be kept firmly in mind during an ongoing development program. This may require modification of individual component performance goals/guarantees. This chapter should convince the reader that more than serendipity is required for achievement of a successful flight test demonstration. Rather, a successful flight test demonstration is characteristically the product of at least three to four years of concentrated, well-coordinated effort between aircraft and engine developers with ultimate customer cognizance.

4.2 ELEMENTS OF A SUCCESSFUL INLET/ENGINE INTEGRATION PROGRAM

The first step in embarking on the design integration of the inlet/engine combination, as with any design task, is establishment of the functional requirements to be placed upon the design. These are primarily dictated by the desired aircraft capability, toward which end all aspects of the propulsion system development must be responsive. Basic component performance levels must sometimes be compromised, when necessary and where feasible, to achieve this overriding objective. Once initiated, the individual inlet and engine design efforts must consider the interactions that will inherently exist, such that various pitfalls in achieving a harmonious installation are avoided. These factors can be increasingly investigated during the development phases, which involve various component and system tests. However, a point is generally reached, during this period but well before first flight, where the design configurations must of necessity be "frozen" to facilitate efficient fabrication. Thus, design modifications must be made under some time pressure, often without complete knowledge of all pertinent factors. Integration planning must consider this fact of life so that development activities are structured to provide a maximum of needed information to the designers at a time when configuration flexibility still exists. Provision for frequent transmittal of information and coordination between the aircraft, engine, and customer organizations, together with periodically scheduled audits to review progress and adjust future plans as required, are keystones of success in this endeavor.

The following remarks will elaborate upon these considerations. Due to the complexity of the subject and the wide range of configurations with singular operational requirements, the discussion is not intended to be all-inclusive. Rather, it should serve to illustrate the kinds of factors requiring attention to produce a successful inlet/engine interface.

4.3 DEFINITION OF SUBSONIC INLET/ENGINE OPERATIONAL REQUIREMENTS

Specification of operational requirements for the inlet and engine is largely driven by the intended aircraft capabilities. For example, a subsonic transport will spend a dominant proportion of its flight time at cruising speed and altitude, with the main objective being delivery of a specified payload over a certain distance or range within some allowable total fuel consumption. This is the classic example of a (primarily) single design point application. By contrast, fighter aircraft are generally designed for multi-mission capability. These roles may include high-speed/low-altitude penetration, transonic maneuvering, and supersonic dash to fulfill the vehicle's intended objectives. Such specifications strongly impact the propulsion system's design and operating conditions. Requirements typically facing subsonic installations are discussed next. It is understood that some of these are mandatory for the basic vehicle success, while others are less crucial and may be traded to some extent in the interests of overall capability. The prioritization involved is beyond the scope of this discussion, but it must obviously be addressed in an actual development program, via figures of merit or ranking criteria appropriate to the particular application.

Figure 4.1 shows a cutaway view of a current high-bypass-ratio turbofan engine installation. The key features of the inlet internal and external components are indicated. In developing a subsonic transport inlet/engine installation, the major considerations may be categorized and explained in the following discussion.

Flight Conditions Producing High Inlet Inflow Incidence

There are several common flight conditions that involve large incidence of the flow approaching the inlet lip relative to the inlet axis. Foremost among these are takeoff rotation and climb and landing approach. Other less common but design-significant points are takeoff during crosswind and takeoff climb with a failed engine. Such conditions must be examined because they typically result in a rapid acceleration and deceleration of the flow as it proceeds around the inlet lip, raising the possibility of flow separation during the flow recompression. Flow separation can degrade vehicle performance either through decreased inlet pressure recovery and/or increased flow distortion. The separation may result in compression system surge and possibly flameout in the case of internal flow separation or increased inlet drag for external flow separation.

The inlet mass flow ratio developed in Chapter 6 of Ref. 1 is the parameter that best describes the approaching flow pattern. For ease of computation the mass flow ratio may be expressed as

$$\frac{A_o}{A_i} = \frac{(\dot{m}_2 \sqrt{\theta_2}/\delta_2)\,(\delta_2/\delta_0)}{(\dot{m}\sqrt{\theta}/\delta A)_o\,(A_i)} = \frac{(\dot{m}\sqrt{\theta}/\delta)_2\,(\pi_d)}{(\dot{m}\sqrt{\theta}/\delta A)_o(A_i)} \tag{4.1}$$

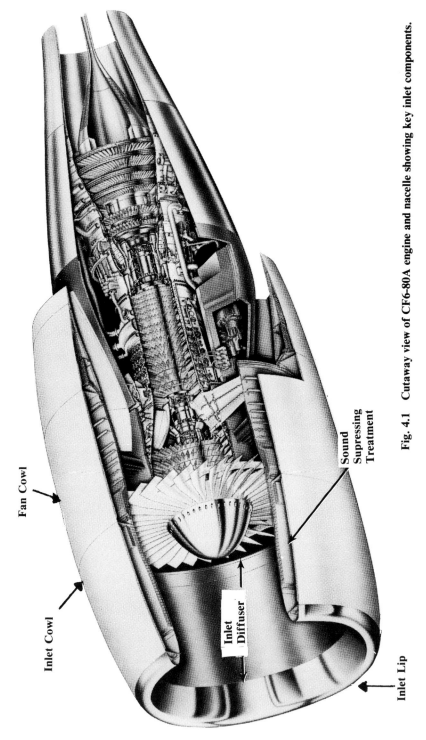

Fan Cowl

Inlet Cowl

Sound
Supressing
Treatment

Inlet
Diffuser

Inlet Lip

Fig. 4.1   Cutaway view of CF6-80A engine and nacelle showing key inlet components.

in which the inlet compression is adiabatic ($\theta_2 = \theta_0$) and typically no mass transfer occurs in a subsonic inlet ($\dot{m}_2 = \dot{m}_0$). The inlet pressure recovery (π_d) is generally within 1% of unity for most subsonic installations and may be neglected without major effect in most preliminary studies. The specific corrected flow term in the denominator of Eq. (4.1) may be expressed, utilizing continuity, the perfect gas equation, and the relationships for stagnation temperature and pressure in terms of Mach number, as

$$\frac{\dot{m}\sqrt{\theta}}{\delta A} = p_{STD} \sqrt{\frac{\gamma g}{RT_{STD}}} \frac{M}{\left(1 + \dfrac{\gamma - 1}{2} M^2\right)^{(\gamma + 1)/2(\gamma - 1)}} \tag{4.2}$$

where $\gamma = 1.4$ and $R = 53.34$ lbf-ft/lbm-°R; and inserting the standard temperature and pressure,

$$\frac{\dot{m}\sqrt{\theta}}{\delta A} = 85.386 \frac{M}{(1 + 0.2M^2)^3} \frac{\text{lbm/s}}{\text{ft}^2} \tag{4.3}$$

Thus, the specific corrected flow is a function only of Mach number for a given specific heat ratio. This relationship may be tabulated or plotted to facilitate computation. A plot is shown in Fig. 4.2, illustrating the familiar maximization at the sonic point.

For convenience, the subsonic inlet area A_i is generally defined as the highlight value, i.e., the projected area at the point where a normal to the inlet axis is tangent to the inlet nose, as shown in Fig. 4.3. This terminology eliminates the complexity of determining the stagnation point location for

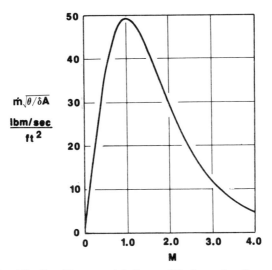

Fig. 4.2 Specific corrected flow vs Mach number for $\gamma = 1.4$.

a variety of operating conditions, which would be impractical. For subsonic inlets, the numerical value of A_o/A_i is a direct indication of the general incidence of the flow approaching the inlet. Referring to Fig. 4.3, a value of unity means that the inlet is capturing its projection in the freestream and the stagnation point will occur at the inlet highlight for level flight. A_o/A_i less than unity indicates flow is prediffusing in the freestream, such that an outward flow incidence occurs; this is generally the case for cruise flight speeds. Conversely, A_o/A_i will exceed unity at low flight speeds and moderate to high power settings, such that an inward flow incidence develops with the stagnation point on the outer portion of the lip.

The specific operating conditions mentioned previously are schematically represented in Fig. 4.4 in terms of the approaching flow incidence. These situations add the effect of a velocity component normal to the freestream to the basic mass flow ratio effect. The top three conditions (takeoff rotation and climb, landing approach, and crosswind takeoff) involve a mass flow ratio greater than one that is accentuated either by a nonzero pitch attitude or a prevailing ground wind normal to the inlet axis. This introduces the possibility of internal lip flow separation with an attendant performance degradation and reduction in engine stability margin that may lead to compressor surge, blade damage, or life reduction—and possibly engine flameout.

An example of typical low-speed inlet flow incidence requirements for a transport aircraft is given in Fig. 4.5. Lift augmentation concepts, such as a blown flap with an under-the-wing engine location, develop a different wing circulation and attendant upwash characteristics, such that greater inlet inflow angles are produced. However, the general procedure involved in defining the required inflow envelope for a high-incidence subsonic installation is similar.

A typical format of inlet/engine operational boundaries for a subsonic transport application at low-speed, angle-of-attack conditions is shown in

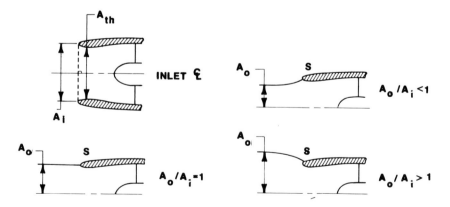

Fig. 4.3   Definition of inlet capture area A_i and characterization of flow incidence approaching inlet via mass flow ratio A_o/A_i.

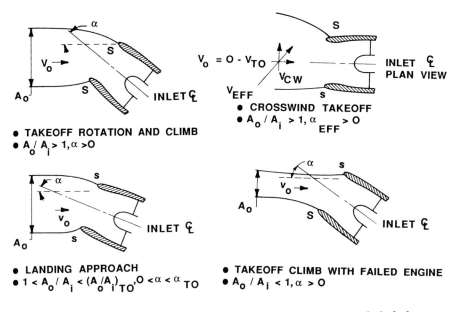

Fig. 4.4   Schematic illustration of key operating conditions involving relatively large flow incidence approaching inlet lip.

Fig. 4.6. These scale-model inlet test data show that the angle of attack capability of a given installation is a function of engine corrected flow demand (i.e., inlet throat Mach number) and flight speed. The selection of an inlet design must consider both the projected flow incidence, including the actual pitch angle plus local flowfield effects such as wing upwash, as well as the engine flow demand range (e.g., flight idle to takeoff) over which these incidence levels may be encountered. If necessary, inlet design modifications can be made to increase the inlet capability at the expense of a larger nacelle. These design modifications would involve increasing inlet nose bluntness and/or internal lip contraction (highlight-to-throat area ratio A_i/A_{th}). As shown in Fig. 4.6, inlet nose bluntness is generally a function of the design variables D_{HL}/D_{th}, a/b, D_{HL}/D_{max}, and X/D_{max}. Similar considerations apply to the crosswind takeoff condition.

The fourth condition of Fig. 4.4, takeoff climb with an engine out, involves the possibility of external inlet surface flow separation. It is mentioned here because it can be a key constraint on sizing the external inlet cowl for a commercial transport. This situation arises because of the need to certify a minimum climb gradient capability with an engine out in order to obtain sufficient altitude to dump fuel and return to the airfield landing pattern if necessary. The incremental drag that can result from an external inlet unable to maintain attached flow at the extremely low mass flow ratios characteristic of a windmilling engine (say, 0.3) can preclude the attainment of this required minimum climb gradient. The scale-model test data of Fig. 4.7 show that this sudden cowl drag rise can amount to as much as approximately 10% of total aircraft drag. This situation is

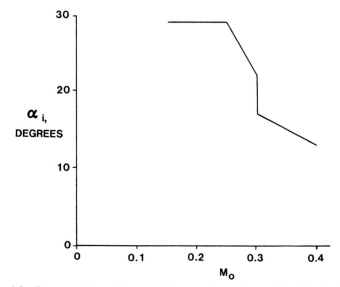

Fig. 4.5   Representative design requirements for subsonic inlet flow incidence.

unacceptable during the critical period of minimal excess engine thrust. As indicated, the data of Fig. 4.7 were acquired at a tunnel total pressure of 4 atm. Significant pressurization is necessary with model scales typical of tests on large turbofan installations in order to achieve Reynolds numbers sufficiently close to those of the full-scale design to avoid unduly pressimistic results in terms of incidence for separation. For example, data acquired at a tunnel total pressure of 1.26 atm and a 70% lower Reynolds number indicate flow separation at 10 deg less incidence than the Fig. 4.7 value. Inlet forebody design, especially the diameter ratio D_{HL}/D_{max}, which directly impacts D_{max}, is the main variable used to achieve required flow incidence capability.

Other Operating Conditions
Affecting Inlet/Engine Integration

Depending upon the particular installation, one or more of the following conditions may impact the inlet/engine interface:

(1) Inlet noise suppression requirements can significantly affect inlet design and installed performance characteristics. Sound-absorbing material is used along the diffuser walls of the current wide-body commercial aircraft. Depending upon the magnitude of suppression required, the need for acoustic absorption material may drive the inlet length, weight, and losses beyond those required solely to fulfill the aerodynamic diffusion function.

(2) Foreign object ingestion is a topic of past and current interest that can relate to the inlet/engine installation via the need to control the

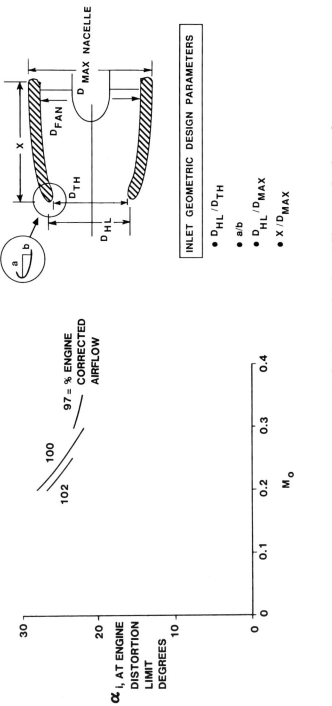

Fig. 4.6 Typical format of subsonic inlet/engine operational capability at angle of attack.

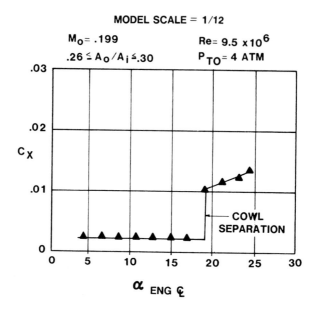

Fig. 4.7 Effect on nacelle drag of upper cowl lip flow separation during takeoff climb with an inoperative engine.

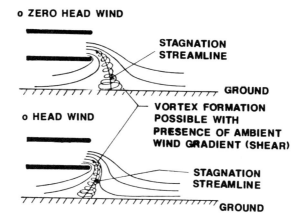

Fig. 4.8 Schematic representation of inlet/ground vortex formation.

inlet/ground vortex formed at low forward velocities. When present the ground vortex can cause ingestion of dust and heavier objects, thereby eroding the compression system, which in turn degrades performance and component life and increases maintenance costs. Ground vortex formation requires wind shear and is schematically represented in Fig. 4.8. Various devices, generally involving the use of compressor bleed air directed outward from the lower inlet lip in some fashion, have been evaluated for their potential in eliminating or reducing the vortex ingestion characteristics. These devices are directed toward eliminating the stagnation point, which typically occurs on the ground, that is inherently required for vortex formation. The Boeing 737 has such a system available and in use in relatively remote areas with unimproved runways. In addition, there have been indications of interest for other existing installations to aid in engine performance retention via controlling compressor blade contour erosion.

Inlet flow distortion occurring during *thrust reverser operation* on ground deceleration can be controlled to some extent by inlet/engine placement, although the reverser type and discharge flow pattern and the aircraft operating procedure (e.g., cutoff aircraft speed and engine power setting) that control the penetration characteristics of the reversed gases are usually the primary means of avoiding excessive distortion. The engine flow disturbances experienced in the reverse thrust mode result from ingestion of the reversed exhaust gases (either self-ingestion or cross-ingestion from an adjacent engine) and, therefore, can include variations in total temperature as well as pressure. Both of these properties can also experience fluctuations with time, owing to a chaotic, turbulent motion of the ingested flow. A simplified reingestion pattern is schematically shown in Fig. 4.9.

Other Considerations
Affecting Inlet/Engine Installations

Additional factors that are not directly associated with the functioning of the inlet/engine, but that must be considered in any overall installation study are cost, weight, reliability, and maintainability.

4.4 DEFINITION OF SUPERSONIC
INLET/ENGINE OPERATIONAL REQUIREMENTS

Although, in principle, considerations of many subsonic applications also apply to transonic and supersonic installations, an additional complexity is

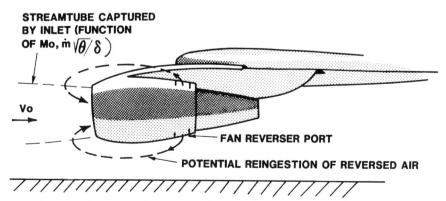

Fig. 4.9 Simplified schematic of potential reingestion during ground thrust reversal of a subsonic transport.

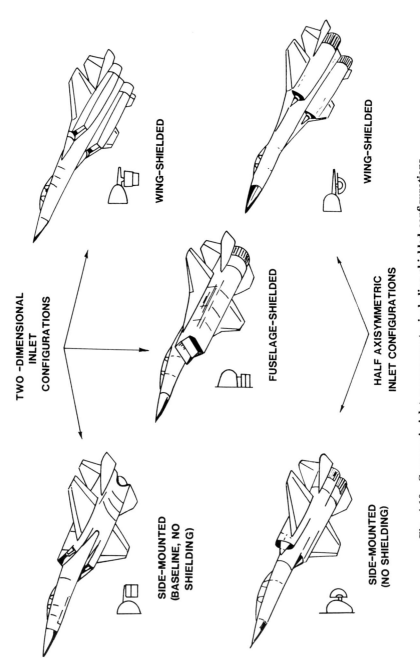

Fig. 4.10 Supersonic inlet arrangements, including shielded configurations.

realized in practice by the latter due to the more extensive flight placard and to the impact of the inlet shock structure on flow stability and inlet/engine flow matching. Additionally, the unshielded local supersonic inlet flowfield includes considerable upwash and outwash as a result of the inlet's proximity to the fuselage and wing and is considerably more sensitive to inlet placement than a typical subsonic installation. Consequently, a number of inlet shielding arrangements have been devised to minimize the effect of aircraft maneuvers on local inflow incidence and thereby improve both the performance (higher pressure recovery) and operability (lower distortion) of a given basic inlet design.

Figure 4.10 illustrates some possible configurations that use either the aircraft fuselage or wing for shielding for both two-dimensional and half-axisymmetric designs. Figure 4.11 shows an example of the amount of local upwash that can exist, even with a shielded inlet, during operation at high Mach number and high angle of attack. This implicitly suggests the greater upwash magnitude of an unshielded inlet. The impact of angle of attack on inlet pressure recovery is demonstrated in Fig. 4.12. Integration of the inlet design, including scheduling of the variable inlet compression surface(s), with the aircraft flowfield is required for superior high-speed performance. There are also some additional factors given increased impor-

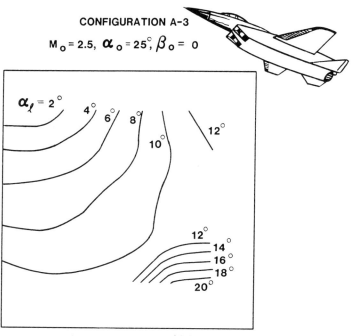

CONFIGURATION A-3

$M_0 = 2.5$, $\alpha_0 = 25°$, $\beta_0 = 0$

$\alpha_l = 2°$ $4°$ $6°$ $8°$ $12°$

$10°$

$12°$ $14°$ $16°$ $18°$

$20°$

FUSELAGE STATION 68.7

Fig. 4.11 Local angle-of-attack contours at high aircraft angle of attack for a fuselage with a shielded inlet (from Ref. 2).

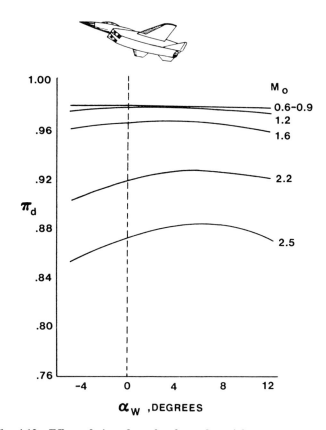

Fig. 4.12   **Effect of aircraft angle of attack on inlet pressure recovery.**

tance by the generally closer integration of the supersonic inlet and engine with the aircraft itself, such as boundary-layer diversion to minimize the amount of low-energy airflow entering the inlet. Finally, the predominantly military nature to date of transonic/supersonic aircraft imposes other requirements on the inlet/engine combination.

Inlet/Engine Operational Placard

The flight placard dictated by the particular aircraft mission has a major effect upon inlet/engine operation. The typical tactical fighter placard shown in Fig. 4.13 provides a framework for discussion of the various demands that can be placed on the system. Flight along line A requires that the aircraft operate at or near peak lift coefficients, due to the low dynamic pressure q available to produce lift. This results in relatively large aircraft angles of attack and inlet lip flow incidence angles that can produce high inlet pressure distortion. Generally, the engine must be able to tolerate this distortion, since the supersonic inlet lip is usually less blunt than on typical subsonic inlets to allow reasonable supersonic drag. The lip shape is a

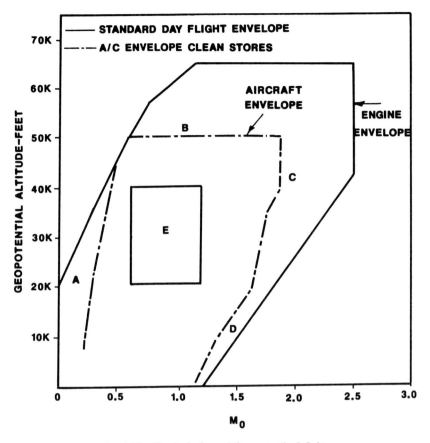

Fig. 4.13 Typical placard for a tactical fighter.

compromise between drag and distortion during maneuvering. This situation can affect compression system blade design and may influence the ultimate selection of the compression system operating line to allocate sufficient margin for surge-free operation. The intersection of lines B and C, or peak Mach number, is generally the inlet design point. Here, inlet recovery and drag, as well as other determinants of engine thrust, are given much emphasis and tailored in conjunction with the aircraft drag polar to ensure that the vehicle can meet or exceed its design speed. Lines C and D define the locus of maximum dynamic pressure conditions and, therefore, are significant to the design of both the inlet and engine structure and the engine cycle, since they represent pressure and/or temperature extremes. The aircraft combat arena is denoted by the box E, which in practice may be fairly extensive. In this region, severe angles of attack and yaw can confront the inlet (e.g., Fig. 4.14). These aircraft attitudes must not be exceeded in order to avoid engine instability or surge.

As previously mentioned, the fuselage and wing can produce local inlet flow angles that differ significantly from the aircraft flight path angles of

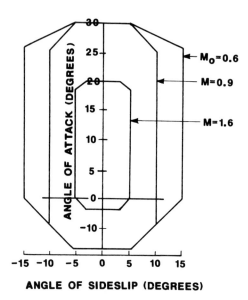
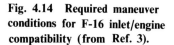

Fig. 4.14 Required maneuver conditions for F-16 inlet/engine compatibility (from Ref. 3).

attack and yaw. For this reason, supersonic inlet verification tests are usually conducted with at least a partial forebody and wing simulation to produce inlet flowfields typical of the actual aircraft operation.

Additional Demands on Inlet Flow Capacity

There are several potential sources of airflow demand, other than the engine requirement, that can affect inlet design and operation. These flow rates are typically small relative to the engine's and are often termed "secondary" flows for that reason and for distinction. Examples include engine periphery and/or exhaust nozzle cooling, electronic equipment cooling, and boundary-layer bleed from the inlet ramp or spike and internal cowl surfaces (for prevention of separation in the adverse pressure gradients and especially at shock wave impingement locations). The extent and manner in which these flows are removed from the inlet and ultimately exhausted can greatly affect the efficiency, uniformity, and stability of the engine inflow. This subject must, therefore, receive commensurate attention during development of the propulsion system and will be discussed later in this chapter.

Considerations Related to Shock Structure Stability

Supersonic inlets may each be classified into one of three types—external, mixed, or internal compression—according to whether the supersonic diffusion occurs external to the inlet duct, partly external and partly internal to the duct, or entirely internal to the duct, respectively. Figure 4.15 compares the three inlet types. Figure 4.16 illustrates the achievable

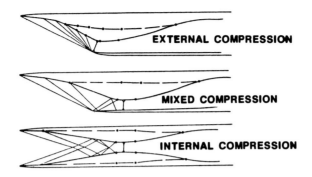

Fig. 4.15   Comparison of supersonic inlet diffuser types.

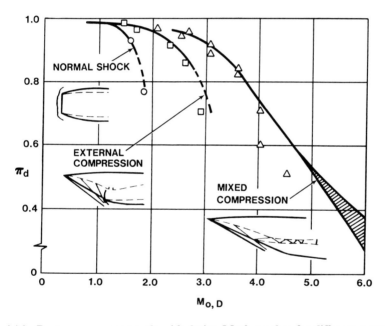

Fig. 4.16   Pressure recovery trends with design Mach number for different types of supersonic inlets.

pressure recovery trends with design flight Mach number for each of these types. Although the mixed compression inlet provides higher recovery at supersonic conditions, its increased weight, increased control complexity, and increased bleed requirements must be balanced against the improved supersonic recovery. For a multimission aircraft, this is seldom a favorable trade.

The shock structure inherently present in a supersonic inlet requires consideration of two phenomena associated with shock stability. One is termed "inlet buzz" and the other is characterized by the axial location of the terminal (normal) shock. Figure 4.17 illustrates these situations in the framework of a pressure recovery/mass flow ratio map; Fig. 4.18 is a corresponding plot of pressure distortion. These figures show that operating at high mass flow ratio, beyond the design point value, reduces pressure recovery (Fig. 4.17) and increases pressure distortion (Fig. 4.18) delivered to the engine. Operating at too low a mass flow ratio, typically well below the design point value, results in inlet buzz, with very low pressure recovery and unstable flow (Fig. 4.17). Either of those two situations is clearly unacceptable.

Inlet buzz is a low-frequency, high-amplitude pressure oscillation. It is most frequently linked to shock/boundary-layer and/or shock/shock interaction during relatively low mass flow ratio operation of an external or mixed compression inlet, but may also occur subsonically due to the separation of a diffusing fuselage forebody/inlet flowfield interaction. In any such scenario, the central physical mechanism is downstream choking caused by the flow separation, which requires flow spillage around the inlet. When this spillage occurs, it changes the flowfield that originally caused the flow separation and attendant choking. Then, the normal shock can move aft toward its initial position and the original flow can become re-established, resulting in a buzz cycle. Analytical techniques have been developed that allow approximate determination of the buzz boundary, but experimental investigation of each design configuration is required for detailed definition of buzz onset and magnitude. The obvious concerns about buzz

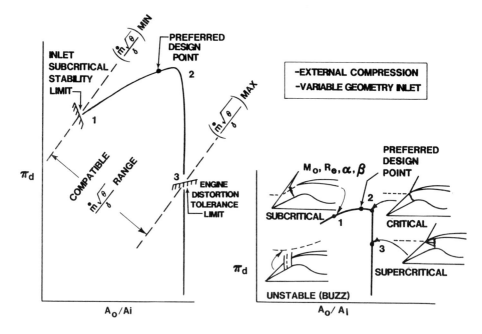

Fig. 4.17 Supersonic inlet operational modes and limits.

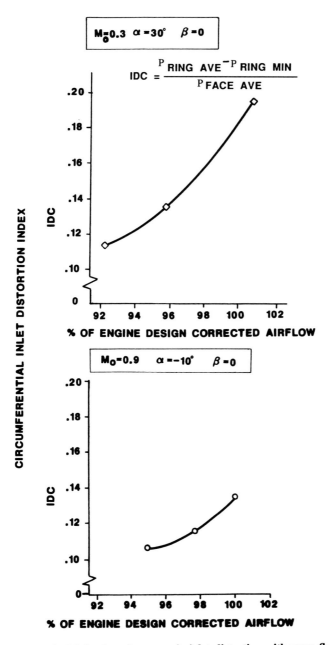

CIRCUMFERENTIAL INLET DISTORTION INDEX

$M_0 = 0.3$ $\alpha = 30°$ $\beta = 0$

$$IDC = \frac{P_{RING\ AVE} - P_{RING\ MIN}}{P_{FACE\ AVE}}$$

IDC

% OF ENGINE DESIGN CORRECTED AIRFLOW

$M_0 = 0.9$ $\alpha = -10°$ $\beta = 0$

IDC

% OF ENGINE DESIGN CORRECTED AIRFLOW

Fig. 4.18   Typical behavior of supersonic inlet distortion with mass flow rate.

are the degraded pressure recovery and (especially) the increased flow distortion, both spatial and temporal. Typically, during inlet development, extensive efforts are made to preclude the buzz regime from intersecting the expected engine/aircraft operating zone. One possible means of achieving that goal is to increase the inlet bypass flow in low mass flow ratio situations where the engine-corrected flow demand is less than the inlet supply.

When operation near the inlet/engine design point is considered, inlet sizing and terminal (normal) shock stability become of critical importance. The desire for peak efficiency, which would occur if a throat Mach number of 1 could be maintained, must be compromised in practice by operating with the normal shock external to the cowl of an external compression inlet. Typically, the terminal shock Mach number will be maintained in the range of 1.2–1.3, whereas the throat Mach number will be in the range of 0.7–0.8. Operation in this mode is termed subcritical operation.

A potential penalty for inlet shock positioning with variable-ramp, external compression inlets is the need for variable geometry and control systems. This need will vary depending on the specific aircraft performance requirements, such as maximum speed, acceleration, and maneuverability. However, such features can include variable ramp angle(s) or spike position (axial or radial movement), variable cowl lip angle, variable throat area and diffuser area distribution, shock position sensor, and variable bypass system, together with the boundary-layer control previously mentioned. The ultimate design will heavily influence the nominal inlet system efficiency and the relative likelihood of encountering situations that tax the engine's available surge margin.

For perspective, we should note that the bulk of supersonic inlets that have been developed have employed all-external compression, with the notable exceptions of the mixed compression designs of the SR-71, XB-70, and United States SST. The mixed compression inlet shows substantial performance advantages over all-external compression inlets when flight at high Mach number ($M > 2.5$) is considered, as shown in Fig. 4.16. Future mixed compression inlet development will probably be restricted to applications whose maximum Mach number and/or time at supersonic cruise can justify the additional complexity, weight, and cost.

All-internal compression inlets have the advantages of excellent pressure recovery at design conditions and little or no cowl drag, since the cowl may be aligned with the local stream flow direction. However, they have several significant practical disadvantages that have made them of largely academic interest. These include (1) complexity, since variable geometry is needed to "start" the inlet (i.e., stably establish the internal shock structure); (2) poor off-design performance; (3) extreme length; (4) high requirement for boundary-layer bleed to enhance performance and stability; (5) need for an active control system; and (6) penalties in maintenance, reliability, and weight accruing from the previous factors.

Military Application Considerations

An application of current military significance is the "stealthy" type of aircraft. This designation generally connotes the suppression of "observables," which include noise, radar cross section (RCS), infrared (IR) radiation, and visual cues. Since inlet noise suppression in connection with subsonic inlets has already been discussed, only RCS reduction will be considered here. The principles involved are somewhat analogous to acoustics, in that the propagation of waves is central to the concept. When radar, rather than acoustics, is considered, however, the frequency range of interest is on the order of gigahertz (10^9 Hz). The RCS reduction is generally accomplished in two principal ways: by exploiting geometrical shapes and by employing absorbent material. The geometrical shaping can be in two forms: placing the inlet opening so as to obscure the inlet cavity in some directions and/or shaping the duct itself to hide the engine face. This special installation and/or duct shaping must be highly integrated with the inlet aerodynamic design and engine airflow characteristic to minimize the installed performance losses and control engine distortion. The use of absorbent material in the inlet cavity opening and duct must also be integrated with the inlet aerodynamics. This can be done potentially by incorporating the radar absorber as a part of the duct structure, as is done with acoustic absorbers. Potential implications of the foregoing RCS suppression concepts for the inlet/engine design integration problem are significant.

The possibility of inflow temperature distortion exists due to several features of some military applications. These include ingestion of exhaust gases during thrust reversal, as well as on V/STOL installations, missile/gun gas ingestion, and steam ingestion for catapult-launched vehicles on naval carriers. Investigation, particularly of the V/STOL situation, has shown hot-gas ingestion to be highly configuration dependent; significant variables include inlet location and shielding potential, wing design (especially vertical position and proximity to the inlet), and number and spacing of exhaust nozzles. In addition to possibly influencing these installation design features, ingestion may impact the compression system via the necessity for interstage bleed and/or a reduced operating line for an additional stability margin.

The subject of temperature distortion has received attention in its own right and is discussed further in Chapter 6. We note here that it can be expressed in terms of an equivalent total pressure distortion and also that it can be a significant factor in designing downstream compression components.

Other Considerations Affecting Operational Requirements

Factors such as cost, weight, reliability, and maintainability assume additional importance as the aircraft design speed increases due to the increased nominal and maneuvering pressure loading, more extensive operational placard, and increased complexity due to need for geometric variability and control systems plus more intimate interaction with the adjacent vehicle flowfield. Optimization of these features with those previously outlined is complex, but it must be addressed on some rational basis, consistent with the needs of the particular application, for the realization of effective mission performance.

4.5 ENGINE IMPACT ON INLET DESIGN

Having just discussed the major factors that may direct the design integration of the inlet and engine, let us now consider the way in which several engine characteristics impact the design of the inlet component. This viewpoint will then be reversed in the following section to address the effect of the inlet on the engine design process.

Inlet Sizing

One of the most fundamental considerations in the design process is selection of the inlet size or intake area. For a subsonic installation, this is a relatively straightforward task because a wide airflow range can be accommodated by a fixed-geometry intake, since the engine compressor can propagate pressure disturbances forward into the freestream as a means of "signaling" the need for a flow rate variation. Thus, the throat, or minimum, area becomes the major determinant of inlet flow capacity. Inlet drag, weight, and cost increase with size, so the objective is to satisfy the maximum engine flow requirement with the smallest possible throat area that is consistent with the desired internal performance (recovery) and stability (pressure distortion). Several ramifications of this statement deserve discussion:

(1) The maximum engine flow demand is influenced not simply by the largest *nominal* corrected flow in the engine operating envelope, but also by engine-to-engine airflow variation, which may run 1.5–2% from nominal. Furthermore, flow increments arising from foreseeable engine thrust growth in a particular installation must be anticipated, so that installation redesign and retooling will be avoided.

(2) Practical considerations preclude sizing the throat to operate at its theoretical maximum flow capacity (Mach 1). Both flow path curvature and boundary-layer growth act to reduce the maximum achievable one-dimensional throat Mach number below unity; a typical value is 0.85–0.90. In addition, some margin from this actual choke point is necessary because, near choking, a small variation in the flow rate produces a disproportionately large change in performance. In practice, a small increase in engine flow demand will cause an abrupt drop in total pressure recovery as well as an abrupt increase in diffuser flow distortion and turbulence. For this reason, existing inlets have generally been sized for maximum throat Mach numbers of 0.75 or less. This situation is represented schematically in Fig. 4.19.

(3) Even this nominal practical value (Mach = 0.75) may need to be reduced to satisfy operating requirements such as the takeoff rotation and

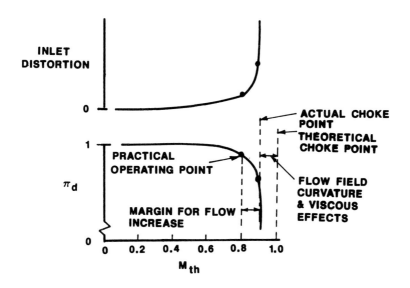

Fig. 4.19   Subsonic inlet sizing considerations.

crosswind takeoff previously discussed, where significant local flow acceleration further reduces the inlet flow capacity or "effective" throat area.

(4) As the throat area is` reduced to achieve a higher throat Mach number, the amount of diffusion required is correspondingly increased. This necessitates a longer diffuser section for a given allowable diffusion rate. Thus, there is a tradeoff between inlet throat diameter and length as it affects nacelle wetted area and inlet pressure recovery. Results of a typical tradeoff study for a subsonic nacelle are shown in Fig. 4.20.

Once the foregoing considerations have been resolved, the required throat area is obtained from the following relation:

$$A_{th} = \frac{(\dot{m}_2\sqrt{\theta_2}/\delta_2)(\delta_2/\delta_{th})}{(\dot{m}\sqrt{\theta}/\delta A)_{M_{th}}} \qquad (4.4)$$

where δ_2/δ_{th} is the reciprocal of the diffuser pressure recovery and may be considered unity for preliminary sizing studies. The throat specific corrected flow $(\dot{m}\sqrt{\theta}/\delta A)$ is a function of the desired throat Mach number, as developed previously and may be obtained from Eqs. (4.2) or (4.3) or from a plot of these, such as Fig. 4.2.

For a supersonic application, inlet sizing is more complex. The inlet shock structure prohibits operation at a mass flow ratio greater than one during supersonic flight. Indeed, variable geometry is required even to approach a unity mass flow ratio across the supersonic range due to the variation in oblique shock angle with Mach number. Therefore, either the shock structure or the throat area can limit the airflow in a supersonic inlet. The supersonic inlet must be able to pass the largest flow demanded by the

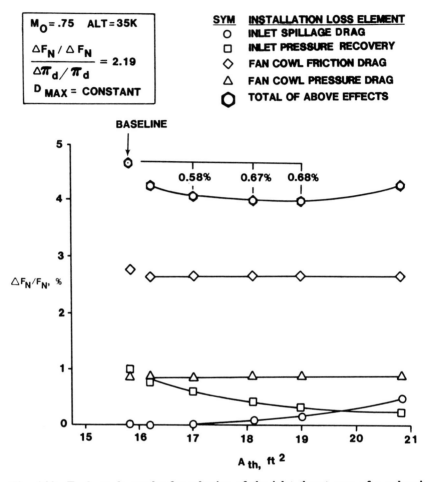

Fig. 4.20 Trade study results for selection of the inlet throat area of a subsonic nacelle.

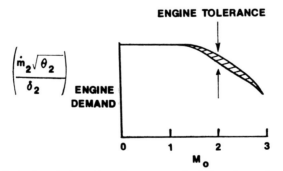

Fig. 4.21 Typical variation of engine corrected airflow demand with flight Mach number.

engine anywhere in the flight envelope. Engine flow vs flight Mach number will typically behave as shown in Fig. 4.21, decreasing at higher M_0 as a result of the reduction in corrected speed $(N/\sqrt{\theta_2})$ that occurs for a given mechanical speed N limit because corrected ram temperature $(\theta_2 = \theta_0)$ rises with increasing Mach number. Again, the total engine flow demand should include engine tolerance and secondary demands such as engine and nozzle cooling flows.

The inlet airflow supply characteristic must then be determined for comparison with engine demand. The inlet supply available to the engine consists of the total flow captured by the inlet minus flow elements such as boundary-layer bleed, subcritical spillage, bypass, and all other aircraft-related flows, such as those required for cooling electronic components, cooling the cockpit, and purging the engine bay, as appropriate. Symbolically, then,

$$A_{0\,\substack{\text{available} \\ \text{to engine}}} = A_i - A_{0\,\substack{\text{supercritical} \\ \text{spillage}}} - A_{0\,\text{bleed}} - A_{0\,\substack{\text{subcritical} \\ \text{spillage}}} - A_{0\,\text{bypass}} - A_{0\,\text{cooling}}$$

$$= A_{0\,\text{inlet}} - A_{0\,\text{bleed}} - A_{0\,\substack{\text{subcritical} \\ \text{spillage}}} - A_{0\,\text{bypass}} - A_{0\,\text{cooling}} \qquad (4.5)$$

These flow elements are illustrated in Fig. 4.22.

During supersonic flight, the flow captured by the inlet will be less than that contained in the projection of its area A_i, unless its shock-generating surface(s) is (are) oriented to focus the oblique shock wave(s) on the cowl lip. Variable-shock generators (planar ramps or conic-section spikes), scheduled with flight Mach number, are used on many aircraft. However, the practical amount of variation is limited by considerations of shock detachment, buzz, and the throat area sizing required for control of the terminal normal shock. Therefore, some supersonic (supercritical) flow spillage will generally occur, at least at off-design flight Mach numbers, as

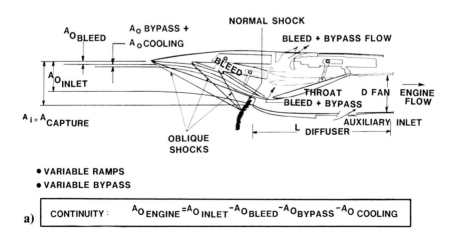

Fig. 4.22 Elements of inlet airflow supply determination.

Continued.

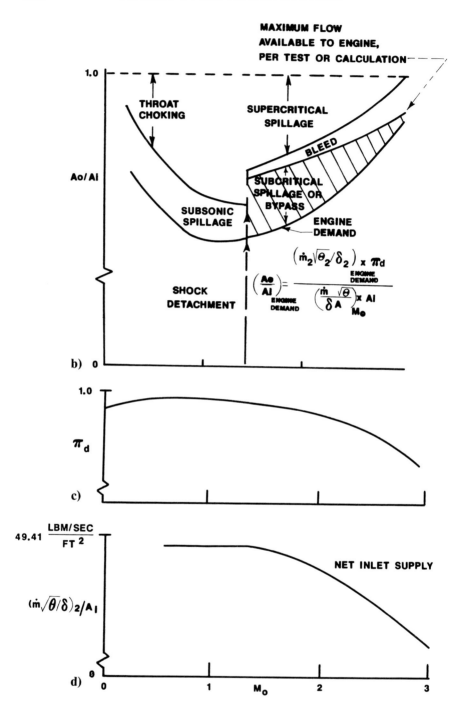

Fig. 4.22 (Continued) Elements of inlet airflow supply determination.

shown in Figs. 4.22a and 4.22b. In addition, a certain amount of flow must be bled supersonically from the inlet compression surface boundary layer to enhance inlet flow stability and pressure recovery. Removing these required supercritical spillage and bleed flows from the capture area results in the maximum inlet flow supply available to the engine, as

$$A_{o\,\substack{\text{maximum} \\ \text{available} \\ \text{to engine}}} = A_i - A_{o\,\substack{\text{supercritical} \\ \text{spillage}}} - A_{o\,\text{bleed}}$$

$$= A_{o\,\text{inlet}} - A_{o\,\text{bleed}} \qquad (4.6)$$

During subsonic and transonic operation at flight Mach numbers below the inlet's oblique shock detachment value, throat choking is the counterpart of supersonic spillage and bleed. This is due to the physical restriction of the inlet's throat area that occurs when the shock generator reaches its mechanical limit and, hence, maximum area.

A representative schedule of engine airflow demand is also shown in Fig. 4.22b to emphasize that any excess flow supply must be either bypassed, as shown in Fig. 4.22a, or spilled subsonically (subcritically) behind the terminal shock, which will move forward of its position in the figure, if necessary, to effect this.

Combining the inlet flow supply term of either Eq. (4.5) or Eq. (4.6), assuming minimum possible spillage, bleed, bypass, and cooling flows, with the typical pressure recovery schedule of Fig. 4.22c, the maximum specific corrected airflow rate available to the engine may be expressed as

$$\left(\frac{\dot{m}\sqrt{\theta_2}/\delta_2}{A_i} \right)_{\substack{\text{maximum} \\ \text{available} \\ \text{to engine}}} = \left(\frac{\dot{m}\sqrt{\theta}}{\delta A} \right)_{M_0} \left(\frac{\delta_0}{\delta_2} \right) \left(\frac{A_o}{A_i} \right)_{\substack{\text{maximum} \\ \text{available} \\ \text{to engine}}}$$

$$= \left(\frac{\dot{m}\sqrt{\theta}}{\delta A} \right)_{M_0} \left(\frac{1}{\pi_d} \right) \left(\frac{A_o}{A_i} \right)_{\substack{\text{maximum} \\ \text{available} \\ \text{to engine}}} \qquad (4.7)$$

where the specific corrected flow $(\dot{m}\sqrt{\theta}/\delta A)$ has been previously discussed. Again, the mass flow ratio term represents only the airflow available to satisfy the engine demand, i.e., it excludes all other applicable flow elements such as spillage, bleed, by pass, and cooling. The resulting airflow supply characteristic [from Eq. (4.7)] will appear similar to that of Fig. 4.22d.

The inlet sizing procedure then consists of determining the required capture area as a function of flight Mach number via the following equation:

$$(A_i)_{M_0} = \frac{\left(\dfrac{\dot{m}_2\sqrt{\theta_2}}{\delta_2} \right)_{M_0}}{\left(\dfrac{\dot{m}_2\sqrt{\theta_2}/\delta_2}{A_i} \right)_{\substack{\text{maximum available} \\ \text{to engine, } M_0}}} = \frac{\text{engine flow demand}}{\substack{\text{maximum available} \\ \text{inlet specific airflow}}} \qquad (4.8)$$

The largest capture area calculated by considering all pertinent Mach number/altitude combinations is the necessary inlet size. A comparison of the resulting inlet supply and engine demand characteristics where the inlet size is set by the maximum Mach number condition is shown in Fig. 4.23. Depending upon the application, it may be feasible to incorporate variable inlet geometry as a way of reducing the extent (Mach number range) and magnitude of the excess flow. The excess flow must be either bypassed, via a duct leading to the engine exhaust nozzle or through overboard exits, or spilled around the inlet. Additional drag forces arise from the failure to employ the captured flow to produce thrust. These remarks also apply to the need for flow dispersal at reduced power settings or in an engine-out situation. Failure to size the inlet large enough to supply the engine with sufficient physical airflow is self-compensating to a certain degree. Since compressors and fans tend to be constant corrected flow devices (changing with speed), the inlet normal shock adjusts itself to provide the desired corrected flow. This is accomplished by drawing the terminal shock to the lip and setting up another normal shock in the diffuser, which lowers the pressure recovery and therefore increases the corrected flow to the required value. Two problems arise when this happens: (1) large pressure distortion levels are usually produced and (2) a loss in net thrust results from an undersized inlet, since engine thrust is a function of engine physical flow and inlet pressure. For these reasons, inlet sizing is undertaken with extreme care to ensure inlet/engine flow matching compatibility and to promote efficient operation (high-pressure recovery) without engine surge.

Although the supersonic inlet is usually faced with a throat area restriction on airflow capacity only while operating subsonically, the lip shapes required for good supersonic external performance can introduce significant internal losses at the high mass flow ratios characteristic of low-speed flight (Ref. 4). If the ramps are already collapsed to their mechanical limit to provide maximum throat area, auxiliary inlets are often utilized to improve performance. These may take the form of discrete slots located around the periphery of the nacelle slightly forward of the engine face plane and incorporate cover doors externally and internally that are actuated, via either pressure differential or mechanically, to expose the required opening.

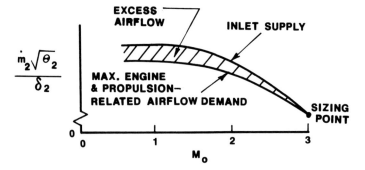

Fig. 4.23   Inlet/engine flow matching for a supersonic cruise sizing point.

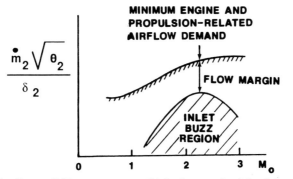

Fig. 4.24   Compatibility assessment of inlet buzz and minimal flow demand.

Figure 4.22a indicates this feature schematically. By utilizing such cover doors, the required corrected airflow can be inducted at a higher overall efficiency, because the auxiliary inlet reduces the flow demand and associated losses for the main inlet, but does not suffer from the losses identified with flow around supersonic lips.

Inlet Buzz Avoidance

Minimum engine airflow demand must also be considered in inlet design and development, in order to preclude inlet buzz occurrence. As discussed previously, this phenomenon occurs at low inlet mass flow ratios and is characterized by low-frequency, high-amplitude pressure fluctuations that are detrimental to engine stability and inlet structural integrity. A goal of the inlet development phase is to arrive at a configuration whose buzz boundary is removed from the minimum propulsion system flow schedule by an appropriate margin, as indicated in Fig. 4.24. To aid this achievement, the inlet bypass system previously discussed and shown in Fig. 4.22a is usually sized to pass sufficient flow to avoid inlet buzz under a failed engine situation.

Inlet Duct Design for Engine Surge Overpressure

The inlet duct's structural design must consider the condition providing the largest internal-to-external surface pressure differential. This situation is likely to arise during engine surge at high dynamic pressure, when a transitory flow reduction occurs, followed by transmission of a strong positive pressure pulse forward from the compression system. This phenomenon is often termed hammershock overpressure, in reference to the suddenness and severity of pressure rise. The initial overpressure pulse is generally followed by successive damped pressure cycles, resulting in application of an oscillatory static pressure loading on the inlet duct surfaces. Figure 4.25 shows a typical test measurement of this phenomenon.

The analytical framework presented in Ref. 5 to describe the overpressure phenomenon indicates that it is, in general, a function of the overall

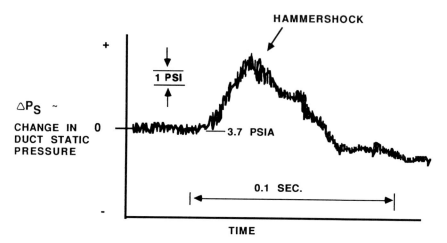

Fig. 4.25 Typical test measurement of a hammershock pressure-time characteristic (first cycle).

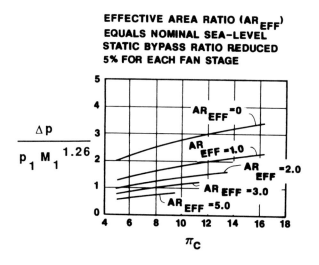

Fig. 4.26 Predicted hammershock overpressure correlation for inlet duct structural design (from Ref. 5).

compressor pressure ratio, inlet duct static pressure and Mach number, and engine bypass ratio. Figure 4.26 shows a typical correlation obtained for the overpressure in terms of a parameter

$$K = \frac{\Delta p}{p_1 M_1^{1.26}} = f(\pi_c, \alpha) \qquad (4.9)$$

where Δp is the pressure rise caused by the passage of the surge-induced wave as it propagates into the inlet duct and p_1, M_1 the duct static pressure and Mach number at engine spinner prior to surge, respectively.

These characteristics are partially based on measurements for bypass ratios α of 0–1.37, with the higher-bypass-ratio curves based on an analytical solution to a transmitted shock wave flow model.

Inlet Control Systems for Flow Stability

Several requirements for a mixed compression inlet control system are briefly summarized here. They arise primarily due to the need to maintain the terminal shock in a stable position inside the inlet diffuser and to prevent boundary-layer separation. Control of shock Mach number value,

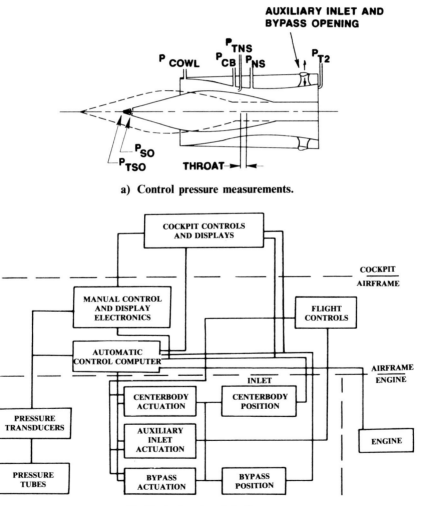

a) Control pressure measurements.

b) Control system interfaces.

Fig. 4.27   Inlet control system pressure measurements and schematic interface diagram.

and hence location, is required for stable and fairly efficient inlet performance and is necessitated by such destabilizing effects as variation in engine power setting/flow demand, intended aircraft flight speed change, atmospheric nonuniformities, and aircraft maneuvers. A flow bypass system, variable inlet geometry, boundary-layer bleed, and throat Mach number sensor are elements of a general control system. Some of these elements are shown in simplified fashion in Fig. 4.27, together with the types of interactions between the cockpit and airframe, for a translating centerbody type of inlet. Use of these particular features depends on the specific vehicle performance requirements previously discussed, such as flight speed and maneuverability, which will strongly influence the inlet type and allowable complexity. Even an external compression inlet generally requires control provision for inlet bleed flow, bypass flow, and ramp scheduling, although the complexity of an active terminal shock positioning system utilized by mixed compression inlets is not required.

4.6 INLET IMPACT ON ENGINE DESIGN

Several aspects of the inlet's behavior are worthy of consideration relative to design of the engine, particularly the compression components. To an extent, they are the complement of the engine factors affecting the inlet design just discussed.

Engine Airflow Demand Tailoring

If the inlet configuration and operating characteristics are defined early enough in the propulsion system design phase, some tailoring of the engine airflow schedule is possible. The incentive for closely matching the inlet supply and engine demand airflows is minimization of inlet spillage and/or bypass system drags, which can contribute to upgraded aircraft performance in terms of maneuverability, range, or acceleration as desired for a particular mission.

Two possible engine/inlet flow matching situations are indicated schematically in Fig. 4.28 for a transonic sizing point. Designated as the

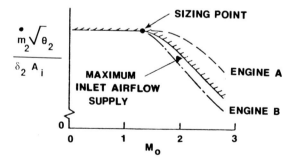

Fig. 4.28 Schematic representation of the typical need for engine airflow tailoring, transonic inlet sizing point.

propulsion-related airflow requirements associated with two engines, A and B, the engine demand will typically differ from the inlet supply. The potential for either over- or undersupply exists (the oversupply case is more common). In Fig. 4.28, flow mismatch occurs in the supersonic regime as a result of the assumed sizing point. Depending on the aircraft mission and associated engine type (turbojet or turbofan), limited flexibility exists to shift the engine demand schedule to fit the inlet supply better through such means as varying the turbine temperature or varying the exhaust area. As developed in Chapter 8 of Ref. 1, if variable engine geometry techniques are employed, more flexibility in airflow matching can result.

In practice, the amount of flow mismatch reduction is constrained by considerations of cycle performance and engine component pressure, temperature, and stress operational limits. For the mixed flow turbofan, an additional barrier is imposed by the physical requirement to match stream static pressure at the juncture of the hot core and cool bypass flows. Accordingly, fan speed and flow must be reduced at higher Mach numbers to achieve the pressure balance, with the possibility of trading subsonic performance for some additional supersonic flow capability via modifications to the basic cycle.

An alternative to accepting the smallest achievable residual supersonic flow mismatch via the means just discussed is as follows. Referring again to Fig. 4.28, engine A's supersonic flow can be reduced via engine speed cutback, as long as full thrust is not required, per the transonic thrust sizing assumption. Engine B's flow schedule can be matched via use of an auxiliary inlet to provide subsonic/transonic flow, together with a down-sized primary inlet. This will reduce the supersonic inlet supply, while maintaining the subsonic/transonic supply.

In addition to the foregoing maximum power consideration, a need exists to reduce the installation losses that arise during part power operation; this is especially important for mixed (supersonic/subsonic) mission aircraft. While the afterburning turbojet and mixed flow-augmented turbofan engines share this problem, the latter type has been selected as the best current powerplant for such applications. The part-power installation losses consist of afterbody drag, due to increased boattail angles associated with reduced area of the variable exhaust nozzle, as well as inlet spillage. These losses are qualitatively indicated in Fig. 4.29 in terms of their associated descriptive parameters. The resulting typical performance decrements are shown in Fig. 4.30 in terms of specific fuel consumption for Mach 0.9 operation of an aircraft with Mach 2.5 capability. Considering that modern fighter engines must be so oversized to meet combat maneuverability requirements that they typically operate at only 40–60% of maximum dry thrust for altitude cruise at Mach 0.8–0.9, significant inlet and nozzle drag penalties can exist. To date, these effects have prevented the fullest realization of multimission capability.

The essential difficulty posed by a mixed-mission aircraft is that it requires propulsion system characteristics that are mutually exclusive in a conventional, fixed-cycle engine. Both the high specific thrust of a turbojet and the low fuel consumption of a turbofan are needed. The particular

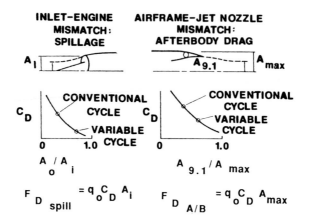

Fig. 4.29 Effect of power setting on installation losses.

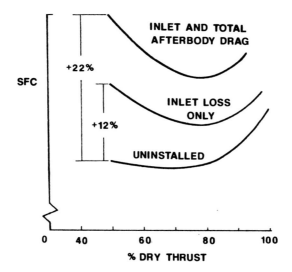

Fig. 4.30 Representative installation losses for Mach 0.9/36,000 ft operation of Mach 2.5 aircraft with a mixed-flow turbofan engine.

features desired may be summarized as follows:

(1) Supersonic cruise, combat maneuver, transonic acceleration: the high specific thrust of turbojet-type engine characteristics including low bypass ratio (~ 0.3), high nozzle pressure ratio, and high turbine inlet temperature.

(2) Subsonic cruise, time-on-station, loiter: low fuel consumption of turbofan-type engine characteristics including moderate bypass ratio (~ 0.8), high cycle pressure ratio, and moderate turbine inlet temperature.

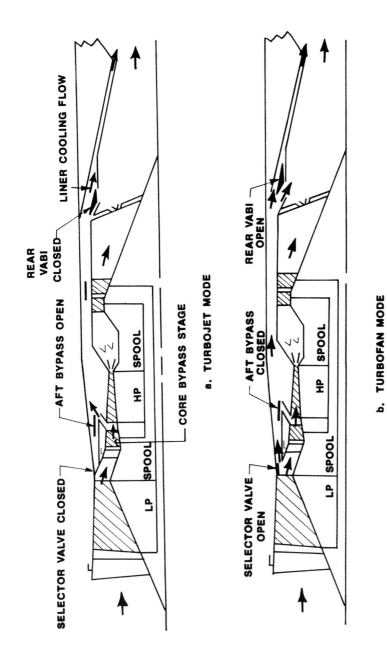

REAR
VABI

LINER COOLING FLOW

CLOSED

AFT BYPASS OPEN

SELECTOR VALVE CLOSED

LP SPOOL

HP SPOOL

CORE BYPASS STAGE

a. TURBOJET MODE

REAR VABI OPEN

AFT BYPASS CLOSED

SELECTOR VALVE OPEN

LP SPOOL

HP SPOOL

b. TURBOFAN MODE

Fig. 4.31   Schematic of a variable-cycle engine concept.

In order to attain these opposing cycle goals to some degree, variable-cycle engine (VCE) concepts have been developed and demonstrated during the past 10–15 years. A key incentive in the inception of these VCE concepts is the reduction of throttle-dependent drag, consisting of both inlet spillage drag (ordinarily associated with reduced engine airflow demand at a low power setting subsonically in a high-thrust-loading aircraft) and afterbody drag created by the large variation in the exhaust nozzle exit area ordinarily required between subsonic and supersonic flight.

One such VCE concept is shown schematically in Fig. 4.31. This dual-mode (turbojet/turbofan) VCE employs a double bypass feature that permits flow to bypass either aft of the compressor first stage for turbojet operation (Fig. 4.31a) or between the fan and compressor for turbofan operation (Fig. 4.31b). A rear variable-area bypass injector (VABI) mixer controls the operating line of the first compressor stage in the turbojet mode and the fan in the turbofan mode, while maintaining the required static pressure balance in the mixed-flow cycle's exhaust system.

In the turbojet mode (Fig. 4.31a), the engine operates as a twin-spool turbojet for high specific thrust. The variable-geometry devices are arranged as follows: (1) the (forward) selector valve is closed, directing all fan air into the core bypass stage; (2) the aft bypass is open to bleed sufficient airflow from the core bypass stage to cool the exhaust system; and (3) the rear VABI is closed to direct the bypass flow outside the augmenter liner and into the exhaust nozzle.

Use of the higher-pressure air for cooling allows very high design turbine inlet temperatures, since a positive static pressure differential is maintained across the augmenter liner, promoting its cooling. In addition, the high turbine inlet temperature increases the tailpipe pressure, producing high specific thrust, as desired. As technology allows turbine inlet temperatures above 3000°F, specific thrusts as high as 100 and 140 $lb_{thrust}/(lb/s)_{airflow}$ can be achieved, respectively for maximum dry and maximum augmented standard sea-level static (SLS) operation.

In the turbofan mode (Fig. 4.31b), the engine operates as a mixed-flow, twin-spool turbofan. The variable features are deployed as follows: (1) the (forward) selector valve is open, allowing flow from the fan [low-pressure (LP) spool] to bypass the core [high-pressure (HP) spool]; (2) the aft bypass is closed, preventing core air from entering the bypass stream; and (3) the rear VABI is open, permitting a larger quantity of lower-pressure air to enter the primary exhaust stream.

With this arrangement, normal matching of the fan and core compressor at part power, coupled with modulation of the core compressor's inlet guide vanes, can provide desired bypass ratios of 0–0.4 for subsonic cruise operation. In addition, the lower pressure ratio in the fan duct reduces the turbine inlet temperature required to balance static pressure at the exhaust system's mixing plane. In this turbofan mode, the combination of lower fan duct pressure ratio, higher bypass ratio, and lower turbine inlet temperature improves the uninstalled specific fuel consumption (SFC). The potential of reduced *installed* SFC is also offered, due to the increased inlet airflow and exhaust nozzle exit area, which reduce the throttle-dependent installation losses.

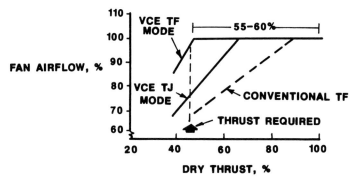

Fig. 4.32   Variable-cycle engine improvement in inlet flow matching at subsonic cruise.

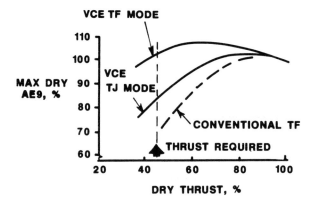

Fig. 4.33   Variable-cycle engine improvement in exhaust nozzle/afterbody closure at subsonic cruise.

The elements of these potential VCE benefits are shown in Figs. 4.32 and 4.33 as a function of thrust at a typical subsonic cruise flight condition, for both turbofan and turbojet modes. For comparison, a conventional (fixed) mixed-flow turbofan, sized for the same mission requirement, is also shown in each figure. In the turbofan (TF) mode, especially, the VCE can hold both the airflow (Fig. 4.32) and nozzle discharge area (Fig. 4.33) over a very wide thrust range down to the low thrust typically required for subsonic cruise with a high-thrust-loading aircraft. Consequently, both inlet spillage drag and afterbody drag will be reduced relative to the conventional turbofan, as schematically indicated in Fig. 4.29.

VCE features such as these allow orienting the basic cycle design toward high supersonic specific thrust, thus minimizing engine size and weight, yet still achieving attractive installed SFC for subsonic cruise, providing more balanced, mixed-mission system capability.

Surge Margin Allocation
for Anticipated Distortion

The general topic of engine stability analysis is discussed in detail in Chapter 6. Accordingly, the objective of this section is to provide an overview of the basic effect of distorted inflow on a compression component and indicate the impact distortion may have on engine design and performance.

A typical compression component aerodynamic performance map may be represented as shown in Fig. 4.34. The operating line is established by back pressure characteristics set by the turbine and exhaust nozzle areas and is selected to achieve the required engine performance levels. It is therefore adjustable at the expense of performance in the case of a reduced (lowered) position. The surge line represents the limit of stable operation and is, by definition, fixed for a given configuration and operating condition.

The nominal operating and surge lines of Fig. 4.34 are subject to various degrading effects that act to decrease available surge margin. Figure 4.35 is a simplified representation of this situation. The stall line degradation results from effects such as inlet distortion, manufacturing tolerances, variable stator accuracy and response time, casing/blade clearance increases, and airfoil contour deterioration with age, as well as Reynolds number effects at appropriate points in the flight placard. The maximum transient operating "line," or locus of maximum operating points, is elevated from the nominal steady-state operating line by inlet distortion, engine component tolerances, fuel control margin required for acceleration and deceleration, compressor bleed effects, afterburner lightoff transients and thrust modulation effects, and control mode effectiveness. The region between the degraded surge line and the maximum transient operating line of Fig. 4.35 represents the excess surge margin or visibility available in the compression system. Visibility, or excess surge margin, is required for assurance of compression system stability, since the destabilizing effects are inherently statistical/probabilistic rather than deterministic.

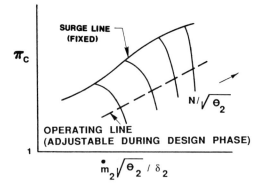

Fig. 4.34 Typical aerodynamic performance map for fan or compressor component.

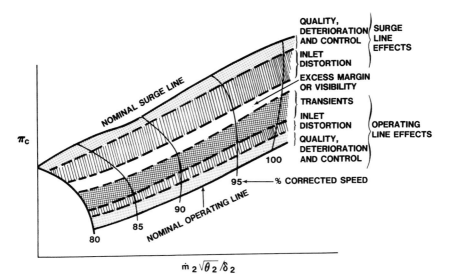

Fig. 4.35 Cumulative representation of degraded surge and operating lines for a compression component, leading to reduced stability margin for inflow distortion.

The fundamental impact of nonuniform flow entering the engine is that a change in the nominal velocity field affects the flow incidence on the blading and may lead to locally stalled regions or, in the limit, to complete surge of a component. Other possible undesirable effects include aeromechanical excitation that can reduce blade life or cause mechanical failure. Depending on the transmission of the inlet distortion through the compression system, creation of a turbine inlet temperature profile that locally exceeds tolerable levels may occur so that either turbine component life is degraded or performance must be derated to maintain life. Actual installed compressor inlet velocity fields can be quite complex, with axial, radial, and circumferential components superimposed on the nominal compressor velocity vector. In practice, flow distortion is expressed in terms of total pressure variation, since this is more readily measurable than components of the entering air velocity.

The detailed effects of pressure distortion on compressor flow stability are complex and not completely understood. They depend heavily upon the flow pattern shape (i.e., magnitude, circumferential and radial extent), location of the distortion (hub vs tip), and compressor geometric and aerodynamic design features such as stage loading, aspect ratio, and solidity. However, the net result of a distorted flow is primarily to effectively lower the undistorted surge line of a compression component; a possible secondary effect is reduction of component efficiency. If this surge line degradation exceeds the available amount, one or more of several options must be exercised in the development cycle. Such options include inlet distortion reduction via modified inlet design or operating point, engine compression system design changes, reduction or reallocation of the other previously mentioned factors degrading the surge margin, redefinition of aircraft operational goals, or acceptance of some type of performance penalty.

The essential feature, then, of surge margin allocation is determination of an amount such that the inlet/engine combination provides both adequate performance and stable compression system operation. An insufficient margin curtails the aircraft's operational envelope and, at worst, can compromise vehicle safety. Alternately, excessive surge margin represents a foregone propulsion potential that prevents efficient realization of the full aircraft performance capability.

Effects of Inlet Buzz, Unstart, and Hammershock Interaction

The transient pressure characteristics of inlet buzz and engine surge-induced hammershock, mentioned previously in connection with inlet duct structural design, are also significant inputs to the engine mechanical design. The inlet buzz pressure frequency spectrum must be carefully defined to ensure that none of the engine's rotating or stationary structural members have resonant frequencies in the range of the predominant buzz frequencies. Similar considerations apply to the case of inlet unstart if a mixed or internal compression inlet is involved. The pressure signature resulting from interaction of the inlet and engine during a hammershock incident must be factored into the compressor design, since it affects the rotor axial force as well as bending and torsional blade loading. These are important engine mechanical design constraints.

4.7 VALIDATION OF INLET/ENGINE SYSTEM

Having set forth the factors that affect the inlet/engine design integration, we can now consider the manner in which integration of the inlet and engine into an effective propulsion system is undertaken and confirmed. This section discusses the basic objective of such an effort, together with a brief historical background of some problems that have arisen from lack of an effective integration program, the current and projected role of analytical methods in aiding inlet and compressor system development, and the progressive complexity and integration of the various types of test programs leading to confirmation of the production vehicle's capabilities.

Background and Basic Considerations

The complexity of inlet/engine interaction is so extensive that, in the past, many unforeseen inlet/engine operability problems have arisen at times in the development schedule that precluded timely, effective resolution. Frequently, resolution of major system operability shortcomings has been forced well into the flight test program. Table 4.1 indicates the difficulties and ultimate solutions for some military vehicles during the 1954–1987 period. In addition, operational problems have occasionally occurred during initial service of commercial transports. This experience has demon-

Table 4.1 Stability Experience of Some Operational U.S. Air Force Systems, 1954–1987 (updated from Ref. 6)

Operational system		Instability problem	Design action to resolve
Fighter	I	Engine stall—level flight and maneuver	Compressor and inlet configuration
	II	Inlet instability, engine stall during maneuver	Compressor and inlet configuration
	III	Engine stall—weapon release, maneuver, inlet cross-feed	Inlet configuration, engine scheduling
	IV	Inlet instability—engine stalls during afterburner lights	Compressor changes
	V	Inlet instability during A/B shutdown	Inlet control
	VI	None	None
	VII	Engine stall—weapons release	Engine control scheduling
	VIII	Engine stagnation stall	Zoned engine augmenter plus fan/core splitter and engine control changes
	IX	Engine stall during severe departed maneuver (i.e., beyond normal A/C flight envelope)	Compliance with A/C operational envelope
Bomber	I	None	None
	II	High power at sea-level static and crosswind takeoff	Inlet configuration
	III	None	None
	IV	Minor inlet instability	Inlet control
	V	Engine stalls due to exhaust nozzle control failures	Control changes
		Nonrecoverable engine stall due to control sensor failure	None
		Nonrecoverable engine stall due to A/C fuel supply problem	Fuel system changes
		Nonrecoverable engine stall during severe departed maneuver	Compliance with A/C operational envelope
Trainer	I	Engine stall with maneuver	Engine control changes
Cargo	I	None	None
	II	None	None
	III	Crosswind takeoff	Inlet configuration
Attack	I	Engine stall due to ingestion of wing flow separation	Wing slats activated by angle-of-attack signal

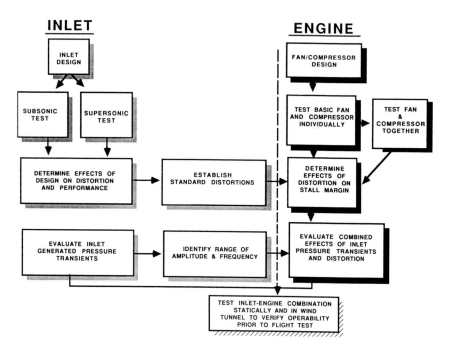

Fig. 4.36   Example of an inlet/engine stability development test program.

strated that it is generally more cost effective to integrate design and development of the engine and its interfacing aircraft components, such as the inlet, through a formally structured series of steps, such that the flight test is merely a final system demonstration built upon subscale and full-scale inlet and compressor component development. The incremental costs associated with more extensive and sophisticated development programs are more than justified by reducing the outlay that would have been created by the aftereffects of a less thorough effort. Such aftereffects may include increased development costs during flight test, delayed system availability, and derated performance capability of the final production aircraft.

An effective integration effort for a supersonic aircraft installation normally requires five or six years, beginning with analytical design studies leading to an increasingly sophisticated series of tests to evaluate inlet and compression component characteristics, singly at first and later together. A general experimental program directed at development of a stable inlet/engine system is shown in Fig. 4.36. These test features are discussed in more detail in subsequent sections of this chapter.

A subsonic aircraft application generally requires a shorter period of less extensive testing following the design phase, depending somewhat upon the particular application and applicability of the various design requirements discussed previously. In the field of subsonic inlet aerodynamic design, computational methods have progressed to the point of accomplishing much of the configuration screening and evaluation tasks, following experiential definition of the preliminary design candidates. Experimental methods are primarily used to determine inlet capability with significant viscous/inviscid flow interactions, such as the separation angle of attack, and where complex, heavily vehicle-integrated inlet geometry is involved. Testing is also conducted to prove concepts and verify performance levels. In the area of engine response to inlet distortion, analytical improvements are needed for detailed understanding of the interrelationship between compressor design variables and distortion sensitivity, component interdependence, and transient performance with distorted inflow. These and other areas are under active development and analytical methods are becoming a progressively more significant factor in the inlet/engine design, integration, and development processes.

The elements and structure of a program to develop and validate an effective inlet/engine subsystem will be discussed in more depth in the following sections.

Analytical Evaluation of Inlet and Engine Operation

To the extent of their capabilities, analytical methods are extremely useful, because they allow evaluation or screening of alternative concepts and facilitate determination of component performance sensitivity to various design parameters of interest. This information can usually be obtained at a lower cost than with an analogous experimental program. The following discussion is intended to outline some current and projected capabilities in three areas: inlet aerodynamic design, modeling compression system response to distortion, and inlet/engine dynamic analysis.

Inlet aerodynamic design analysis. Subsonic inlet design analysis ideally requires methods capable of accurately calculating the details of three-dimensional flows affordably with rapid turnaround times. Two broad classes of solutions are available for this task. The first is a two-dimensional (i.e., axisymmetric or planar geometry) method of the type described in Ref. 7, capable of solving the three-dimensional flowfields generated by operation at nonzero angle of attack. Linear superposition of three independent basic solutions (static, axisymmetric stream flow, and pure crossflow or angle of attack) is employed to achieve the desired solution values of freestream velocity, angle of attack, and mass flow ratio. The second is a fully three-dimensional code (see, e.g., Refs. 8–10). Often, the two-dimensional method is used for design analysis of isolated inlets because of its simpler input, shorter turnaround time, lower cost, greater grid resolution (larger number of smaller panels), and usually adequate accuracy. In that case, the two-dimensional method of Ref. 7 is applied to three-dimensional configurations, where necessary, as described in Ref. 11: i.e., by treating each meridional cut of interest as if it were the profile of an axisymmetric inlet and assuming that the resulting solution is applicable in the immediate vicinity of that meridional cut.

Examples of these analyses are shown in Figs. 4.37 and 4.38. Figure 4.37 compares results of the two-dimensional analysis, an incompressible solution with a compressbility adjustment, with test data for an axisymmetric inlet model simulating takeoff climb with an inoperative engine. Excellent theory/data agreement is shown for both the external surface's peak Mach number and the subsequent recompression. Figure 4.38 compares both two- and three-dimensional analyses with test measurements for the bottom internal and external surfaces of a three-dimensional inlet at takeoff conditions. Reasonable agreement is noted, with little benefit from the more complex three-dimensional code. Since both of these codes provide incompressible solutions with compressibility adjustment, they do not properly model the wavelike nature of the flow on the windward side at nonzero angle-of-attack takeoff conditions, which produce significant transonic flow and, at some angle, a shock/boundary-layer interaction. In that case, the design evaluation must come from a three-dimensional transonic code or from experiment.

Either of the techniques just discussed can be augmented by use of a boundary-layer analysis, such as the one described in Ref. 12, to reflect the viscous effects. The boundary-layer analysis (which is strictly valid only for two-dimensional flows) uses the potential flow solution as the "freestream conditions" it requires to initiate calculations. Results of the boundary-layer solution can be used to adjust the physical body contours to include the effect of the boundary-layer displacement thickness and then recalculate the potential flow and boundary layer. This procedure can be iterated until convergence is obtained. Convergence is determined by a prescribed limit on local velocity changes between successive iterations. Analytical results from this procedure closely duplicate test measurements, provided that flow separation does not occur. Separation onset can be well predicted by this technique, so that the applicable range of the analytical results can be determined prior to testing.

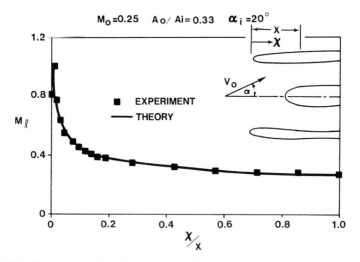

Fig. 4.37 Comparison of axisymmetric analysis with experimental data for the top profile of an axisymmetric inlet at takeoff climb, inoperative engine conditions.

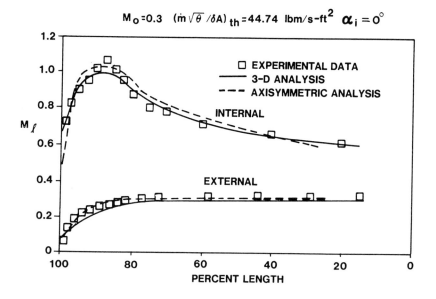

Fig. 4.38 Comparison of three-dimensional and axisymmetric analyses with experimental data for the bottom profile of a three-dimensional inlet at takeoff conditions.

In addition to the foregoing capabilities, continual progress is being made in the development of analytical tools for general three-dimensional potential and viscous flows, with inclusion of interactions between inlet and adjacent aircraft flowfields, shock waves and boundary layers, and inlet turbulence. As an example of current three-dimensional capability, Fig. 4.39 is a comparison of analytical and measured static pressure distributions on the upper inboard side of the nacelle, 30 deg from the top. The complete installation (wing, nacelle, pylon, and fuselage) was analyzed via the three-dimensional panel method of Ref. 10 modified as discussed in Ref. 13. When the pylon is included in the analysis, very good agreement is noted for most of the axial positions. This agreement reflects the absence of significant flow separation in the actual flowfield.

Supersonic inlet design analysis applies to both supersonic and subsonic diffuser components. The former includes the boundary-layer bleed system mentioned previously. Currently, method of characteristics (MOC) solutions continue to be heavily relied upon in the analysis of supersonic diffuser configurations. Interactive viscous calculations are performed and bleed flow effects are considered in subsequent calculations. Use of MOC maintains continuity with the analytical design experience base of past inlet systems and facilitates the design process by screening candidates. Testing currently retains a considerable role in the development of supersonic inlets, especially for considerations related to shock wave/boundary-layer interactions, the boundary-layer bleed system, and inlet stability control, which can impact safety of flight.

The future outlook for capability to analyze supersonic inlets is that the

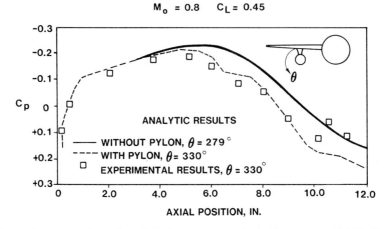

Fig. 4.39 Comparison of analytical and measured nacelle pressure distributions on inboard surface.

recent fully three-dimensional developments, such as refined Euler solvers and parabolized (e.g., PEPSI class) or fully elliptical Navier-Stokes codes able to compute viscous and inviscid interactions plus boundary-layer interactions and bleed effects, will gradually be exercised in parallel with current inlet design procedures, verified with testing, and eventually assume an even more prominent design role. In that connection, computational methods will of necessity direct design of the emerging hypersonic vehicles, because some required flight conditions are incapable of simulation in current or near-term test facilities. However, the flowfield complexity of typical supersonic inlets suggests that their dependence on experimental development methods will continue to be somewhat greater than that of subsonic inlets for the foreseeable future.

Mathematical modeling of compression component/system responses to distortions. In recent years, advances have been made that permit some of the effects of inlet distortion upon compression component stability to be estimated economically using relatively simple models. These models, Refs. 14–16, which are quasi-one-dimensional pitch-line blade-row-by-blade-row or stage-by-stage, parallel-compressor representations of compression components, can predict the effects of circumferential distortions and the effects of planar waves upon the stability of compression components. The parallel compressor concept is discussed more fully in Chapter 6. Distorted flowfield predictions can now be made that include the pumping effect of a compression component upon the distorted inlet flowfield giving rise to tangential velocities and tangential static pressure gradients. For both steady-flow-distorted and spatially undistorted planar wave flows, stability predictions are conducted accounting for the unsteady flow effects associated with the finite time required for the flow surrounding the blades to adjust to the new local flow conditions. Models with these capabilities permit accurately predicting the aerodynamic stability trends and, in many cases, the actual levels associated with the circumferential and unsteady components of inlet distortion. Development of models to handle radial components of distortion in as direct a manner as for the other components of distortion is currently being undertaken.

Inlet/engine control system models. Transient or dynamic simulation models are available for analytically assessing the stability of various propulsion system components or systems. Generally, digital models are preferable to analog, because of their ease of translation/standardization among organizations concerned with the various components and their controls and because of their flexibility in terms of user access and problem setup. To arrive at a useful model during the propulsion system development cycle, exploratory component tests must address dynamic/time-dependent characteristics as well as steady-state performance trends. This facilitates subsequent integration of components into systems.

Initially, the model can be used to roughly approximate response characteristics to other component stimuli; it can then be updated as additional component and subsystem data are acquired. Engine and inlet demonstrator vehicle testing can eventually be used to correlate and fine tune the model, verifying that the model can be used to direct the subsequent demonstrator component scaling and/or minor resizing that are typically required for production hardware.

The dynamic simulation model, then, is a tool that can be used in component and system stability studies as well as general performance evaluation. In terms of inlet/engine integration, for example, it could encompass the inlet installation, inlet geometry control, bypass system, and engine compression system. Successful application of such a tool in its early stages can lead to identification and resolution of aircraft/propulsion system interface problems prior to construction of actual hardware.

Engine Development Test Elements for Inlet Integration

As indicated in Fig. 4.36, an effective military engine development program aimed at achieving compatibility with the aircraft inlet centers on the compression system and proceeds through several phases. Initially, the effort addresses basic fan and compressor operation under clean as well as standardized screen-generated distorted flow conditions and may investigate some blading geometric variables.

As the inlet development proceeds in parallel, the compression system development progresses to evaluating the specific inlet's distortion, both steady state and dynamic, as well as other transient behavior. Finally, the inlet and engine are often tested together in large propulsion wind tunnels, outdoor facilities, or flying test beds for direct interaction effects. This is generally the last step prior to flight testing of the actual experimental or prototype aircraft, which is discussed below. A representative military engine development test sequence oriented toward integration with the inlet system is shown in Fig. 4.40 as a framework for the following discussion of these specific test elements. By comparison, a typical development program for a subsonic commercial engine consists of an abbreviated version of Fig. 4.40. Probable elements include fan and compressor distortion tests and a ground static inlet/engine test.

Compression component testing with distortion.   The test programs shown in Fig. 4.40 typically follow the "exploratory development" or "research and development" phase, which consists of aerodynamic and mechanical design, blade tests in cascade arrangement, and perhaps a limited amount of single stage testing. At this point, as indicated in Fig. 4.40, the fan and compressor will be tested singly, and possibly together in a dual setup, to allow component interactions to be studied. The components are externally powered to desired speed and a variable-area discharge valve is used to throttle them so that the complete operating envelope (i.e., up to surge or mechanical operating limit) can be mapped. The initial testing will be devoted largely to clean (undistorted) inlet flow conditions to obtain basic aerodynamic and mechanical performance data.

Subsequently, fundamental distortion patterns, and/or selected inlet distortion patterns if available from wind-tunnel testing, may be examined. To produce inlet pressure distortion, local area blockage is introduced upstream of the test component via wire screens so that the desired nonuniform total pressure field exists slightly forward of the component's first stage. Some typical types of distortion screens are schematically represented in Fig. 4.41. Fundamental, or standard, pattern types are often used so that a particular design's distortion tolerance can be directly compared with experience on other systems. These include circumferential patterns, termed "one per rev" in the case of the 180 deg contiguous spoiled sector shown in Fig. 4.41 and radial distortion where a complete annulus of low-pressure flow is produced, generally at the tip or hub region of the blade. Refinements of these simple screens are often used, such as variable extent, graded level (multiporosity screening, as opposed to a uniform screen), and "mixed" or complex patterns combining circumferential and radial distor-

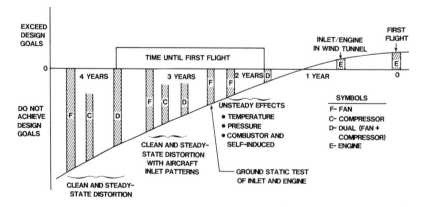

Fig. 4.40   Representative test sequence for inlet integration with a military engine compression system.

tion. The Arnold Engineering Development Center's (AEDC) blowing jet facility[17] is an alternative to screens for distortion testing.

Generally, a second build of the compression system will be required, as shown in Fig. 4.40. This test will be aimed at either mechanical and/or aerodynamic design modifications, as may be required to achieve the required operational capability. Examples of design features, one or two of which may be perturbed at this point, include axial spacing, stage stall margin, stage work, blade solidity, blade aspect ratio, blade tip clearance, radial energy distribution, variable stator scheduling, and interstage bleed. Initial or additional aircraft inlet patterns will be evaluated at this time, generally for key operating conditions like takeoff rotation, crosswind takeoff, subsonic and supersonic cruise, and extreme maneuvering conditions, as applicable.

This type of testing defines the quantitative effect of steady-state distortion on compression component stability and efficiency. Stability or surge margin is determined from knowledge of both the clean and distorted surge line locations, as represented in Fig. 4.42. In this manner, the sensitivity, or reduction in surge margin relative to distortions of various magnitudes, extents, and locations, is directly obtained.

Unsteady flow effects. Until the mid-1960's, the foregoing effort represented the major extent of assessing inlet/engine stability with distorted inflow. However, the introduction of the TF30 augmented turbofan engine into the multimission F-111 aircraft resulted in unexpected engine surges that significantly reduced its operational capabilities or usefulness. After some investigation, the real importance of unsteady pressure distortion was realized. The cause was a complex combination of vehicle flowfield variation with aircraft pitch attitude, boundary-layer ingestion, and shock wave/boundary-layer interaction. This inlet is highly integrated with the aircraft fuselage and wing, as indicated in Fig. 4.43.

Since successful engine operation is commonly achieved despite the known presence of transitory inlet flow effects such as turbulence, vorticity, and swirl, the concept of disturbance frequency is essential to determination of the significance of any unsteady flow phenomenon. Experimental evidence indicates when the period of the disturbance approaches that of one-half to one fan or compressor revolution, the engine will respond to or "see" it as a steady condition, since essentially all its blades will have

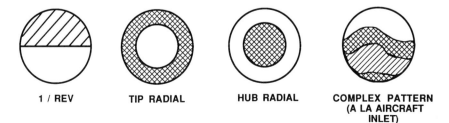

1 / REV   TIP RADIAL   HUB RADIAL   COMPLEX PATTERN
                                                         (A LA AIRCRAFT
                                                             INLET)

Fig. 4.41   Typical types of inlet distortion screens.

experienced the same condition. Conversely, if the disturbance frequency is high enough, very few of the blades will experience it and there will be a negligible effect. Thus, we conclude that there is a critical frequency band in which time-variant inflow can significantly affect the compressor flow stability.

The approach used on the TF30 was to make compressor inlet total pressure surveys with sensors capable of responding to frequencies somewhat beyond the compressor 1/rev rotational value. These data revealed the existence of "instantaneous" distortion well in excess of the corresponding steady-state conditions. When pressure signals (filtered to an appropriate frequency) were used as input to the steady-state distortion index or descriptor, the existence of instantaneous distortion in the TF30/F-111 installation was confirmed.[18]

While this approach facilitated ex post facto identification and resolution of the F-111 problem, its application to the early stages of subsequent integration activities, while certainly warranted, has greatly complicated development testing. Currently, the need exists to measure unsteady pressures during inlet testing and also to determine their effect on compression system stability. While development of analytical compressor modeling techniques is underway, as described previously, use has been made of test devices that simulate inlet dynamic behavior or introduce controlled pressure variations in order to advance understanding of this phenomenon. Such test equipment may use variable ramp or choked plug features or a device to generate variable sinusoidal planar pressure waves. Figure 4.44 depicts a device that produces broad spectrum turbulence. References 19 and 20 address the quantitative assessment of the impact of unsteady pressure distortion on turbomachinery, including the allocation of surge margin, in some detail. This task is quite complex, because of the wide variety of disturbances that may be present, singly or in concert, at the inlet/engine interface.

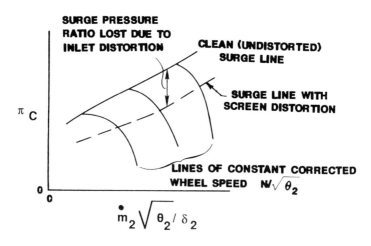

Fig. 4.42 Compressor map showing effect of inlet distortion on surge margin.

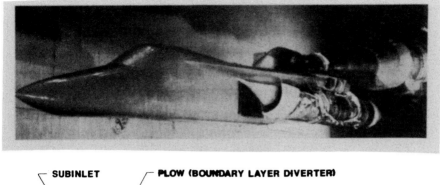

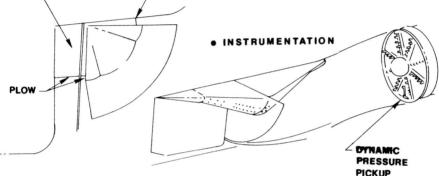

Fig. 4.43   F-111 wind-tunnel inlet model configuration.

Examples of other potential destabilizing transient influences include: (1) ramp-type changes in throttle setting, inlet or engine geometry, or attitude; (2) viscous effects, such as shock wave/boundary-layer interaction, transitory flow separation, and vortex shedding; (3) engine-generated pressure waves caused by rotating compressor stall, combustor instability, aerodynamic mismatch between compression components, and interactions between the engine and its control system; and (4) acoustic wave propagation or resonance within the inlet cavity.

As indicated previously, temperature distortion produced by ingestion of hot gases can also degrade engine operation. Reference 21 describes an investigation of spatial and unsteady temperature distortion conducted by NASA Lewis with a turbofan engine, using a hydrogen-fueled burner as a temperature distortion generator. Figure 4.45 is a sketch of this device. Results showed the high-pressure compressor to be more sensitive than the fan to this type of distortion, with the primary variables affecting stability being the magnitude of spatial temperature distortion and the time rate of change for unsteady temperature distortion.

The foregoing types of compression component or engine testing with unsteady pressure and temperature distortions have been confined mainly to systems for supersonic applications, where the likelihood of occurrence and severity are generally greater than for subsonic systems. Subsonic

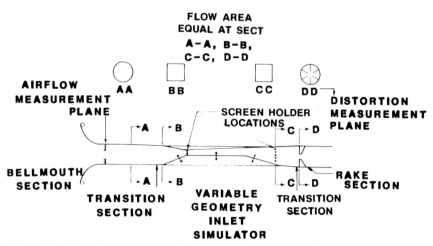

Fig. 4.44 Variable ramp type of unsteady flow generator for engine compression system evaluation.

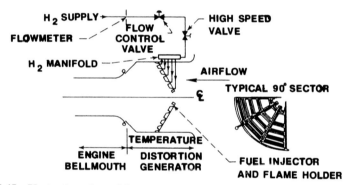

Fig. 4.45 Unsteady and spatial temperature distortion generator for engine compression system evaluation at NASA Lewis Research Center (from Ref. 21).

installation operating conditions producing potentially significant unsteady distortion are generally evaluated initially by defining the inlet dynamic distortion in a model test and later by testing the inlet and engine directly, as discussed in the next subsection.

Inlet/engine testing. Several possibilities exist for evaluating inlet/engine interactions prior to flight testing the actual aircraft. They include propulsion wind tunnels, outdoor static testing, and use of a flying test bed. The role of each of these activities, with some examples, will now be briefly discussed.

Several large propulsion wind tunnels are in place at AEDC and NASA centers to test full-scale inlet and engine installations over a portion of the design flight envelope. Currently, evaluation of pitch and yaw conditions is somewhat restricted due to tunnel blockage considerations. This type of

Fig. 4.46   Crosswind test facility.

Fig. 4.47   DC-10/CF6 installation on B-52 flying test bed.

testing still provides valuable information on installed inlet/engine compatibility and, when conducted roughly one year prior to first flight as shown in the timetable of Fig. 4.40, it allows some lead time for design modifications to resolve any problems that are evidenced. This type of facility is devoted primarily to supersonic applications.

Outdoor static facilities have been used for early evaluation of transport aircraft engines' basic static operation, as well as simulated takeoff rotation, crosswind takeoffs, and compatibility with thrust reverser deployment. The Peebles facility of GE Aircraft Engines is an example of such a test site. It includes a wind generator employing an array of motor-driven fans that produces wind velocities of up to approximately 120 knots. The engine orientation relative to the wind direction is fully variable, allowing simulation of takeoff roll (0 deg orientation), takeoff rotation (15–20 deg), static crosswind (90 deg), and tailwind (180 deg) starts. Figure 4.46 shows a typical crosswind test arrangement for a subsonic installation in this facility.

Flying test beds involve the use of an existing aircraft as a vehicle for mounting and flight testing a developing engine installation, which may either replace one or more of the existing engines or be mounted elsewhere on the aircraft. This test mode has been exercised in the development of many systems, including the DC-10/CF6, C-5A/TF39, 747/JT9D, S-3A/TF34, 727/JT8D, and others. Figure 4.47 shows the DC-10/CF6 installation and its B-52 flying test bed, in which the right inboard pair of existing engines was replaced with a single high-bypass turbofan.

Inlet Development Test Elements for Engine Integration

A representative schedule of supersonic inlet development test programs is shown in Fig. 4.48 as a counterpart of the Fig. 4.40 engine test sequence. In general, two types of testing may be conducted: (1) inlet model tests with engine flow simulated via a vacuum system or propulsion simulator and (2) inlet/engine tests. The latter category was just discussed and will not be further developed here, except to note that static inlet/engine tests are especially meaningful for supersonic installations due to the distortion that generally prevails during takeoff. In addition, the model test discussion in this section is restricted to the internal flow factors that directly impact engine performance and operation, such as recovery, distortion, and buzz. External performance integration is excluded here.

Initial inlet test models are usually small scale to minimize model and wind-tunnel costs. Typical scale range is 5–15%, with a nominal value of 10% in Fig. 4.48. Factors influencing the model scale besides cost and available tunnel size are minimum instrumentation size requirements and requirements for Reynolds number simulation. Both the amount and type of instrumentation desired are significant. Miniaturization of instruments is limited and expensive; therefore, installation access and inlet internal flow area blockage must be considered. Requirements for unsteady pressure sensors and associated electronic acquisition equipment are affected by the

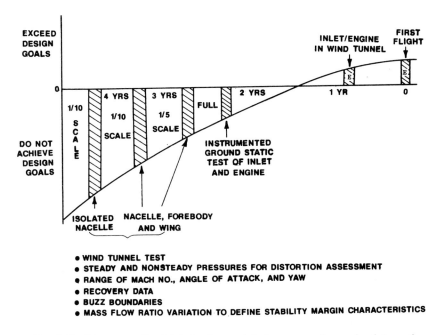

Fig. 4.48 Representative inlet development test sequence for engine integration.

frequency range over which accurate response is needed. As discussed previously, experience has shown that pressure fluctuation frequencies at least as high as the engine rotational speed are significant in terms of engine stability determination. Therefore, for similar geometry and Mach number, similarity requires that a model test must have a frequency response capability that is related to the full-scale value by the reciprocal of the model scale. With typical engine rotational frequencies on the order of 100–200 Hz, a 10% scale model requires response on the order of 1–2 kHz. The trade between the foregoing factors is complex.

The first test is useful in providing an indication of inlet distortion trends that serve to "ballpark" the magnitude of the integration task ahead. Timely information interchange between airframe and engine manufacturers is necessary during this and subsequent test phases to facilitate this evaluation and direct the design modifications required to achieve compatibility between inlet distortion and engine tolerance. In terms of the Fig. 4.48 sequence, which is oriented toward a supersonic application, subsequent tests increase in size and/or complexity as the inlet system becomes increasingly well defined. More of the aircraft components adjacent to the inlet, such as fuselage forebody, wing, and nacelle, if applicable, will be simulated for direct vehicle flowfield interaction effects. This is crucial for evaluation at maneuvering conditions. Other inlet features such as controls, bleed, and bypass systems will be included or modeled more realistically as time progresses. Examples of other supersonic inlet variables that are characteristically evaluated during some portion of the model program include diffuser length and contouring, throat section length, duct axis offset, and auxiliary inlet placement and contouring. Corresponding subsonic inlet test variables include internal lip contraction A_i/A_{th}, leading-edge bluntness, diffuser design, and the auxiliary inlet arrangement. Finally, the amount of unsteady pressure instrumentation generally increases as the model test phase proceeds.

Completion of the inlet model tests should produce a well-resolved inlet aerodynamic design with a good start on definition of distortion characteristics that will be presented to the engine. As the inlet development program progresses, there is a trend of increasing satisfaction of design goals. This trend results from the incorporation of inlet design improvements. Subsequent inlet/engine ground and wind-tunnel tests then serve to extend the inlet model results, within facility limits, by determining the integrated operational characteristics (including buzz) prior to flight testing the actual vehicle. At this point, the bulk of the integration effort has been completed. Achievement of most, if not all, operational and performance goals should be visible, with acceptable trades or minor design modifications already identified to resolve the most marginal area(s).

Flight Testing: The Acid Test

The evaluation of engine/inlet integration is just one of many flight test objectives. If test elements such as those described in the preceding sections have been utilized, the flight test should mainly confirm characteristics already demonstrated. Operational situations either not amenable to prior simulation, or predicted as marginal, represent the most likely problem areas. Examples of the former type of condition could include transient maneuvers and weapon firing. The actual assessment of success in inlet/engine integration is relatively simple. The main criteria are surge-free operation over the complete aircraft flight placard and engine survival without structural damage following an intentional supersonic inlet buzz cycle, providing the aircraft meets its broader goals, such as range/payload and maneuverability. Generally, this entails provision of sufficient diagnostic instrumentation in at least one inlet/engine to allow resolution of any difficulties that may occur during the flight test program. Key instrumentation features include a compression system face total pressure survey, as well as unsteady wall pressure sensors at several stations through the inlet and compression system.

Following the initial flight test, production aircraft experience must be monitored, as manufacturing variations, engine deterioration with age, or altered customer usage may adversely affect some aspect of in-service capability.

Acknowledgment

The authors acknowledge and thank their associates N. O. Stockman, D. J. Lahti, S. J. Bitter, W. G. Steenken, and R. E. Budinger, who provided material used in the preparation of this chapter.

References

[1]Oates, G. C., *Aerothermodynamics of Gas Turbine and Rocket Propulsion: Revised and Enlarged*, Education Series, AIAA, Washington, DC, 1988.

[2]Cawthon, J. A., Crosthwait, E. L., and Truax, P. P., "Forebody Flow-Field Investigation," *Supersonic Inlet Design and Airframe-Inlet Integration Program (Project Tailor Mate)*, Vol. II, Air Force Flight Dynamics Lab., Rept. AFFDL-TR-71-124, May 1973.

[3]Hagseth, P. E., "F-16 Modular Common Inlet Design Concept," AIAA Paper 87-1748, June–July 1987.

[4]Wyatt, D. D. and Fradenburgh, E. A., "Theoretical Performance Characteristics of Sharp-Lip Inlets at Subsonic Speeds," NACA Rept. 1193, 1954.

[5]Marshall, F. L., "Prediction of Inlet Duct Overpressures Resulting from Engine Surge," AIAA Paper 76-758, July 1976.

[6]Tear, R. C., "Approaches to Determine and Evaluate the Stability of Propulsion Systems," Air Force Aero Propulsion Lab., Wright-Patterson AFB, OH, Rept. AFAPL-TR-67-75, Feb. 1968.

[7]Stockman, N. O. and Farrell, C. A, Jr., "Improved Computer Programs for Calculating Potential Flow in Propulsion System Inlets," NASA TM-73728, 1977.

[8]Hess, J. L., Mack, D.-P., and Stockman, N. O., "An Efficient User-Oriented Method for Calculating Compressible Flow In and about Three-Dimensional Inlets," NASA CR-159578, April 1979.

[9]Carmichael, R. L. and Erickson, L. L., "PANAIR—A Higher Order Panel Method for Predicting Subsonic or Supersonic Linear Potential Flows About Arbitrary Configurations," AIAA Paper 81-1255, 1981.

[10]Maskew, B., "Prediction of Subsonic Aerodynamic Characteristics—A Case for Low Order Panel Methods," AIAA Paper 81-0252, Jan. 1981.

[11]Stockman, N. O. and Lieblein, S., "Theoretical Analysis of Flow in VTOL Lift Fan Inlets Without Crossflow," NASA TN D-5065, Feb. 1969.

[12]Harris, J. E. and Blanchard, D. K., "Computer Program for Solving Laminar, Transitional, or Turbulent Compressible Boundary-Layer Equations for Two-Dimensional and Axisymmetric Flow," NASA TM-83207, Feb. 1982.

[13]Dietrich, D. A., Oehler, S. L., and Stockman, N. O., "Compressible Flow Analysis About Three-Dimensional Wing Surfaces Using a Combination Technique," AIAA Paper 83-0183, Jan. 1983.

[14]Tesch, W. A. and Steenken, W. G., "J85 Clean Inlet Flow and Parallel Compressor Models," *Blade Row Dynamic Digital Compressor Program*, Vol. I, NASA CR-134978, 1976.

[15]Reynolds, G. G. and Steenken, W. G., "Dynamic Digital Blade Row Compression Component Stability Model—Model Validation and Analysis of Planar Pressure Pulse Generator and Two-Stage Fan Test Data," Air Force Aero Propulsion Lab., Wright-Patterson AFB, OH, Rept. AFAPL-TR-76-76, Aug. 1976.

[16]Mazzawy, R. S. and Banks, G. A., "Circumferential Distortion Modeling of the TF-30-P-3 Compression System," NASA CR-135124, 1977.

[17]Overall, B. W. and Harper, R. E., "The Air Jet Distortion Generator System— A Tool for Aircraft Turbine Engine Testing," AIAA Paper 77-993, July 1977.

[18]Plourde, G. A. and Brimelow, B., "Pressure Fluctuations Cause Compressor Instability," Airframe/Propulsion Compatibility Symposium, June 1969.

[19]Brimelow, B., Collins, T. P., and Pfefferkorn, G. A., "Engine Testing in a Dynamic Environment," AIAA Paper 74-1198, Oct. 1974.

[20]Collins, T. P., Reynolds, G. G., and Vier, W. F., "An Experimental Evaluation of Unsteady Flow Effects on an Axial Compressor," Air Force Aero Propulsion Lab., Wright-Patterson AFB, OH, Rept. AFAPL-TR-73-43, July 1973.

[21]Rudey, R. A. and Antl, R. J., "The Effect of Inlet Temperature Distribution on the Performance of a Turbo-Fan Engine Compressor System," NASA TM-X32788, June 1970.

CHAPTER 5. VARIABLE CONVERGENT-DIVERGENT EXHAUST NOZZLE AERODYNAMICS

A. P. Kuchar

General Electric Company, Evendale, Ohio

5
VARIABLE CONVERGENT-DIVERGENT
EXHAUST NOZZLE AERODYNAMICS

5.1 INTRODUCTION

The sensitivity of engine net thrust to nozzle performance is higher than any other engine component. For this reason, it is extremely important to obtain the highest possible nozzle performance with due consideration for nozzle cost, weight, complexity, reliability, and maintainability.

There is a wide variety of exhaust system types that fall into two basic categories: (1) fixed nozzles including conic nozzles and high-bypass, separate flow nozzles and (2) variable-area nozzles including converging-diverging nozzles, plug nozzles, and nonaxisymmetric nozzles of all kinds and shapes. The selection of exhaust system type is determined largely by a combination of engine, aircraft, and mission requirements and the appropriate tradeoff studies. A schematic of typical exhaust systems is shown in Fig. 5.1.

Fixed conic nozzles are the simplest. They have no moving parts and are used to exhaust a single engine exhaust flow in a purely converging exhaust duct. They are utilized primarily on subsonic transports, particularly on commercial applications of the older-generation straight turbojet or low-bypass-ratio fan engines. The high-bypass, separate flow converging nozzles are also used in subsonic transport applications. However, they are compatible with the newer generation of high-bypass, separate exhaust engines in which one nozzle is used for fan bypass air discharge and one is used for core engine gas discharge.

Variable-area nozzles are required for engine cycles where large variations in nozzle exit area are necessary. In nearly all cases, this area variation is a result of afterburning engine cycles that require area increases of 50–150% from nonafterburning to full afterburning operation. A few variable nozzle applications employ a converging nozzle only. They are used for applications where aircraft Mach numbers are essentially subsonic. Most variable nozzle applications utilize the axisymmetric, converging-diverging (C-D) type. The C-D nozzle is most common with afterburning engines because, usually, an engine that has afterburning thrust requirements is matched with an airplane that must achieve supersonic Mach numbers. Supersonic flight Mach numbers result in high engine exhaust nozzle pressure ratios and, if a C-D nozzle is not used, the engine exhaust

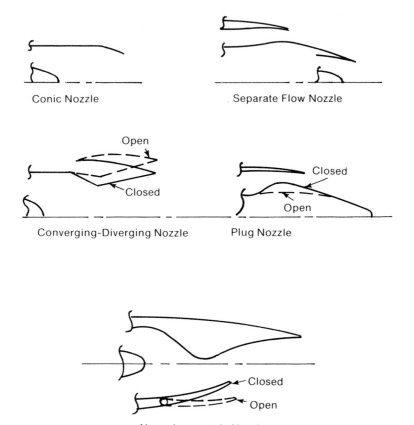

Fig. 5.1   Exhaust system types.

gases cannot be expanded efficiently, resulting in a significant loss in engine net thrust. The axisymmetric nozzles are structurally efficient pressure vessels and, because of their extensive use, have a substantial technology base from which to design.

Even within the class of axisymmetric C-D nozzles, there is a variety of types. The older-generation turbojets still in use are characterized by requirements for bypassing varying amounts of inlet air around the engine to provide engine cooling and good inlet recovery and engine airflow matching. This low-pressure bypass, or secondary, air is injected back into C-D nozzles, which are hence called ejector nozzles. Ejector nozzles can also take air on board from outside the nacelle directly into the nozzle for better overall nozzle installed performance—these are called two-stage ejector nozzles. For very high Mach number aircraft (2.7 and above), the C-D nozzles can be very long with curved diverging sections. For lower supersonic Mach numbers, the diverging section is usually conical. For the newer-generation high-bypass-augmented turbofan engines, secondary air systems are not utilized and the nozzles are simple C-D geometries with conical sections.

In addition to the C-D nozzle, the variable-area plug nozzle has also been considered for afterburning engine applications. These are generally heavier than the C-D nozzles and pose more difficult cooling problems, but because of the kinematic/structural arrangement, the nozzle actuation forces are low. Plug nozzles are characterized by a "flatter" nozzle thrust coefficient curve that gives high performance over a wider range of pressure ratios than a C-D nozzle. They are attractive for very high Mach applications with low Mach requirements.

A final class of nozzles is the nonaxisymmetric variety. These exhaust systems can be fixed or variable geometry and are, by definition, non-axisymmetric. Systems of this type are considered when their need is dictated by unusual requirements imposed on the installation. Examples of such requirements include thrust vectoring, wing lift augmentation, aircraft survivability, or low-drag, blended body exhaust systems. These exhaust systems are becoming more important candidates for future aircraft applications because of additional, more complex aircraft system requirements. These systems are generally heavier and less structurally efficient than axisymmetric nozzles; and, because of their current state of the art, the technology base of nonaxisymmetric nozzles is limited.

In this chapter, the aerodynamic design and analysis procedures for the variable axisymmetric C-D nozzle will be described in detail. This class of nozzle was selected for detailed illustration for several reasons. A prime reason is that this nozzle has associated with it variants of most of the design problems associated with all exhaust systems. Additionally, a substantial data base and high technology level exists, leading to the availability of well-understood design criteria. Because of the additional complexity of the variable C-D nozzle as compared to the fixed convergent nozzle, it was felt that considering the example of the variable C-D nozzle would prove to be the most instructive.

This chapter discusses the aerodynamic design considerations of the variable C-D nozzle that appear most prevalent for current and near-future technology engines. These are turbofan engines with no secondary air and that are compatible with aircraft maximum Mach numbers of approximately 2.5 or less. Included are discussions on how to predict and assess internal performance and aerodynamic gas loads. Considered in the performance aspects are nozzle flow coefficients and thrust coefficients with breakdowns in basic nozzle efficiency, leakage effects, cooling air losses, and flow separation effects. Procedures are provided for estimating aerodynamic loads on the primary nozzle, secondary nozzle, and external flaps. It is noted that the discussions focus on analysis techniques and that specific design curves/criteria are limited.

5.2 NOZZLE CONCEPT

The analysis approaches apply to (but are not necessarily restricted to) variable area C-D nozzles of the type characterized by straight-flap,

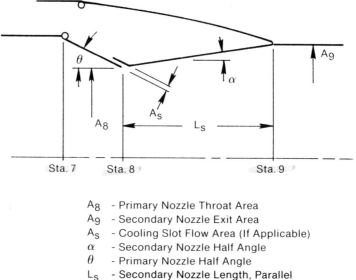

A_8 - Primary Nozzle Throat Area
A_9 - Secondary Nozzle Exit Area
A_s - Cooling Slot Flow Area (If Applicable)
α - Secondary Nozzle Half Angle
θ - Primary Nozzle Half Angle
L_s - Secondary Nozzle Length, Parallel
  to Nozzle Centerline

Fig. 5.2 C-D nozzle geometric patterns.

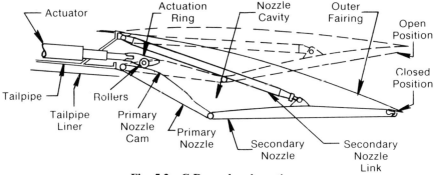

Fig. 5.3 C-D nozzle schematic.

sharp-cornered throats with little or no cooling air introduced at the nozzle throat.

The nozzle flow path geometric parameters referred to are indicated in Fig. 5.2. Station 8 is the nozzle charging station; i.e., all nozzle losses included in the thrust coefficient (Sec. 5.3) are defined to occur between station 8 and the nozzle exit (station 9).

A cross-sectional schematic of a typical single actuation system nozzle is shown in Fig. 5.3. The primary nozzle is most often actuated by a cam and roller system, whereas the secondary nozzle motion is achieved most often by a linkage system. The linkage system shown is a very simple type, but more complex systems can be designed that provide different area ratio schedule characteristics (A_9/A_8 vs A_8). The choice of cam shape and linkage arrangement depends on the requirements of nozzle throat area in combination with the desired area ratio schedule. Some nozzle designs can employ a second actuation system that allows A_9 to be set independently of A_8. This more complicated system is used when good nozzle performance is required over a wide range of nozzle pressure ratios.

5.3 PERFORMANCE PREDICTIONS

Two dimensionless parameters are used to measure exhaust nozzle performance: the thrust coefficient and the flow coefficient. The thrust coefficient is the ratio of actual nozzle gross thrust to the ideal available gross thrust and represents the measure of nozzle efficiency. Referring to Fig. 5.4, the ideal gross thrust available to the nozzle is based on the total airflow supplied to the nozzle, the total pressure and temperature at the nozzle throat, and the fuel-air ratio at the nozzle throat. Thus,

$$C_{fg} = \frac{F_{g\,\text{actual}}}{F_{g\,\text{ideal}}} \tag{5.1}$$

$$F_{g\,\text{ideal}} = f(W_{7\,\text{actual}}, T_{T_8}, P_{T_8}, \text{FAR}_8, P_0)$$

$$= \frac{W_{7\,\text{actual}}}{g} V_{\text{ideal}} = \frac{W_{7\,\text{actual}}}{g} \sqrt{2gJ(H_8 - h_0)} \tag{5.2}$$

where

$W_{7\,\text{actual}}$ = $W_{UL} + W_L$ = total actual flow supplied to the nozzle, lb/s
T_{T_8} = total temperature of gas flow at the nozzle throat (station 8), °R
P_{T_8} = total pressure at the nozzle throat (station 8), psia
P_0 = ambient pressure, psia
FAR_8 = overall fuel-air ratio
g = gravitational constant, = 32.174 ft/s^2
J = Joule's constant, = 778.26 ft-lb/Btu
H_8 = total enthalpy of gas at P_{T_8} and T_{T_8}, Btu/lb
h_0 = enthalpy of gas at fully expanded pressure and temperature, Btu/lb
V_{ideal} = velocity of gas fully expanded to ambient pressure, ft/s

Total pressure losses due to mixing, tailpipe friction, tailpipe liner drag, and afterburner heat addition pressure loss up to the throat are included as part of the mixer/augmentor performance. The nozzle thrust coefficient thus

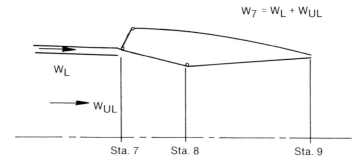

Fig. 5.4   Total flow available to the nozzle.

includes losses due to the following:

(1) Friction (momentum loss due to wall friction in the nozzle).

(2) Angularity (momentum loss due to nonaxial flow at the exit plane of the nozzle).

(3) Expansion (loss due to mismatch of the nozzle exit pressure with ambient).

Depending on the engine cycle/exhaust nozzle performance bookkeeping system selected, the nozzle thrust coefficient can also include losses due to:

(4) Leakage.

(5) Cooling air throttling loss.

The flow coefficient is the ratio of total nozzle flow to ideal nozzle flow,

$$C_{D_8} = \frac{W_{7_{\text{actual}}}}{W_{8_{\text{ideal}}}} \tag{5.3}$$

The ideal flow is calculated from the flow conditions at the nozzle throat P_{T_8}, T_{T_8}, and γ_8, the nozzle physical throat area A_8, and the overall nozzle pressure ratio P_{T_8}/P_0. Using the continuity equation, this can be expressed as

$$W_{8_{\text{ideal}}} = \frac{P_{T_8} A_8 \bar{m}_8}{\sqrt{T_{T_8}}} \tag{5.4}$$

where $\bar{m}_8 = f(\gamma_8, P_{T_8}/P_0)$. The flow coefficient is not a measure of efficiency, but rather a sizing parameter that allows the nozzle physical area to be matched to an engine cycle.

The flow and thrust coefficient parameters are used in engine cycle analysis to complete the engine cycle performance assessment. For a given operating condition, the basic engine turbomachinery determines the flow conditions to the nozzle throat (station 8). These flow conditions include the total flow W_7, total temperature and pressure T_{T_8} and P_{T_8}, and fuel-air ratio FAR_8. From these conditions, the effective flow area at the nozzle throat required to pass the total flow is determined. Using the flow coefficient, the nozzle physical throat area A_8 is calculated, which then "sizes" the nozzle for that operating condition. The nozzle geometry is determined from the selected nozzle area ratio schedule and the thrust coefficient is likewise determined from the geometry and the engine/nozzle operating condition. The thrust coefficient is then used to calculate the actual gross thrust produced by the engine at the operating condition of interest.

It should be noted that, while the flow coefficient definition is essentially universal, the thrust coefficient can be defined in several ways. The variations appear in the definition of ideal thrust. For example, ideal thrust can be based on ideal airflow or it can be based on nozzle exit velocity and pressure rather than fully expanded velocity (or both). The definition selected is usually one that is convenient for the engine cycle or application. Any definition will provide the correct nozzle gross thrust as long as it is consistently used from initial design and analysis through scale-model and full-scale tests. The definition used herein is most commonly used in the aircraft propulsion industry.

Flow Coefficient

As previously stated, the flow coefficient is the ratio of total actual flow through the nozzle to ideal flow. The effective flow area required to pass the actual delivered flow is given by

$$A_{e_{\text{actual}}} = \frac{W_{7_{\text{actual}}} \sqrt{T_{T_8}}}{P_{T_8} \bar{m}_8} \tag{5.5}$$

Combination of Eqs. (5.3–5.5) gives

$$C_{D_8} = \frac{W_{7_{\text{actual}}}}{W_{8_{\text{ideal}}}} = \frac{A_{e_{\text{actual}}} P_{T_8} \bar{m}_8 / \sqrt{T_{T_8}}}{A_8 P_{T_8} \bar{m}_8 / \sqrt{T_{T_8}}} = \frac{A_{e_{\text{actual}}}}{A_8} \tag{5.6}$$

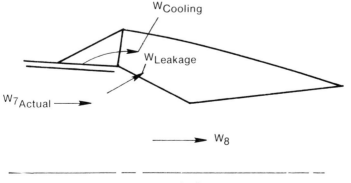

Fig. 5.5   Nozzle flows.

Thus, the flow coefficient is identically equal to the ratio of the effective flow area required to pass the total actual nozzle flow to the nozzle physical throat area. The required effective area is determined from the engine cycle calculations and the flow coefficient is used to calculate the required nozzle physical throat area by substituting the expression for $W_{8_{\text{ideal}}}$ into the flow coefficient equation, as

$$C_{D_8} = \frac{W_{7_{\text{actual}}}}{W_{8_{\text{ideal}}}} = \frac{W_{7_{\text{actual}}}}{P_{T_8} A_8 \bar{m}_8 / \sqrt{T_{T_8}}} = \frac{W_{7_{\text{actual}}} \sqrt{T_{T_8}}}{P_{T_8} A_8 \bar{m}_8}$$

Finally,

$$A_8 = \frac{W_{7_{\text{actual}}} \sqrt{T_{T_8}}}{C_{D_8} P_{T_8} \bar{m}_8} \tag{5.7}$$

Depending on the engine cycle/exhaust nozzle performance bookkeeping system selected, the actual flow in the flow coefficient can include not only the flow passing through the nozzle throat area A_8, but it can also include primary nozzle leakage and, in some cases, bypass cooling air as shown in Fig. 5.5. Thus,

$$W_{7_{\text{actual}}} = W_8 + W_{\text{leakage}} + W_{\text{cooling}}$$

The leakage and cooling flow can be estimated using methods discussed later in this chapter; knowing each of these flows, a total nozzle effective area can be determined based on P_{T_8}, T_{T_8}, and $\bar{m}_8$ as follows:

$$A_{e_{\text{total actual}}} = A_{e_8} + A_{e_{\text{leakage}}} + A_{e_{\text{cooling}}}$$

$$= \frac{W_8 \sqrt{T_{T_8}}}{P_{T_8} \bar{m}_8} + \frac{W_{\text{leakage}} \sqrt{T_{T_8}}}{P_{T_8} \bar{m}_8} + \frac{W_{\text{cooling}} \sqrt{T_{T_8}}}{P_{T_8} \bar{m}_8} \tag{5.8}$$

Thus,

$$C_{D_8} = \frac{A_{e_8} + A_{e_{\text{leakage}}} + A_{e_{\text{cooling}}}}{A_8} \tag{5.9}$$

For this definition, the effective area at the nozzle throat comprises most of the total effective area. The leakage effective area is typically 0.5–1% of the total and the cooling effective area is typically 0–2% of the total.

A typical flow coefficient curve for an isolated converging nozzle with no leakage or cooling bypass flow will appear as shown in Fig. 5.6. When the nozzle is "hard choked," the flow coefficient will reach a maximum value and remain constant with increasing pressure ratio. The level of $C_{D_{8\text{max}}}$ and the pressure ratio at which it occurs is, for all practical purposes, a function of the primary nozzle half-angle as shown in Fig. 5.7.

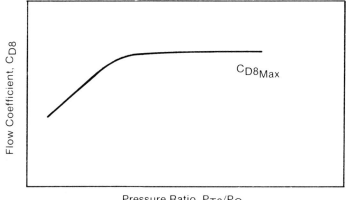

Fig. 5.6   Converging nozzle flow coefficient.

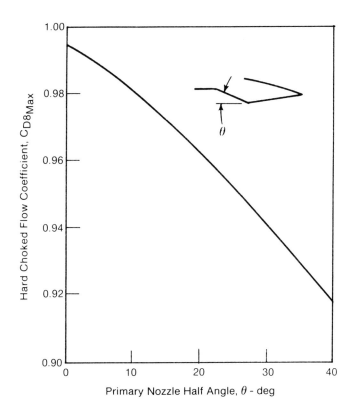

Fig. 5.7   Nozzle maximum flow coefficient.

Since the C-D nozzles in consideration are assumed not to accommodate secondary air, the primary and secondary nozzles are close coupled; there is either a very small cooling slot at the nozzle throat or none at all. With this arrangement, the C-D nozzle will behave as a venturi, and the drop-off in C_{D_8} as pressure ratio is reduced (Fig. 5.6) will not occur. The presence of the secondary nozzle will cause the throat to remain sonic at low pressure ratios (below approximately 1.9) during low engine throttle settings. The flow coefficient is based on overall nozzle pressure ratio. Recall that

$$C_{D_8} = \frac{W_{7_{actual}}\sqrt{T_{T_8}}}{P_{T_8}A_8\bar{m}_8} \quad \text{and} \quad \bar{m}_8 = f\left(\frac{P_{T_8}}{P_0}, \gamma_8\right)$$

The flow coefficient will remain constant down to the sonic pressure ratio and, because of the definition, will rise to higher levels at pressure ratios below sonic as sketched in Fig. 5.8. This occurs because, as in the case with a venturi nozzle, the static pressure in the throat is lower than the ambient pressure. Therefore, while the flow function $\bar{m}_8$ is based on the overall nozzle pressure ratio P_{T_8}/P_0, the actual pressure ratio at the throat P_{T_8}/P_{S_8} is higher than P_{T_8}/P_0. As a result, the throat Mach number is higher than the nozzle exit Mach number and passes more flow than that calculated from P_{T_8}/P_0.

The eventual "flattening" of the C_{D_8} curve at very low pressure ratios is a result of the fact that the amount of static pressure drop in the throat is limited by the nozzle area ratio. This rise characteristic in flow coefficient can be influenced by the size of the cooling slot, the separation characteris-

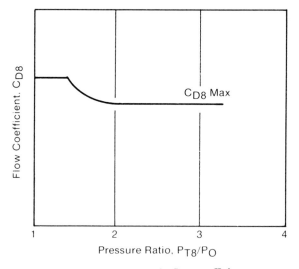

Fig. 5.8   C-D nozzle flow coefficient.

tics of the nozzle, and the secondary nozzle half-angle α, all of which can affect the throat static pressure.

A baseline flow coefficient curve accounting for the throat effective area only is thus defined from Fig. 5.7 and analytical calculations. Since the flow coefficient to be used in the cycle analysis can include the leakage and bypass cooling flows, the final flow coefficient would be that given previously and as shown in Fig. 5.9,

$$C_{D_8} = \frac{A_{e_8}}{A_8} + \frac{A_{e\,\text{leakage}}}{A_8} + \frac{A_{e\,\text{cooling}}}{A_8}$$

It is important to recognize that for a variable-geometry nozzle, the flow coefficient is not a measure of nozzle efficiency; it is a parameter that sizes the nozzle to an engine. The flow coefficient is used simply to determine the nozzle area necessary to pass the required engine flow. For this reason, flow coefficient errors or discrepancies as much as 0.5%, which can be significant for thrust coefficients, have essentially no impact on nozzle performance.

Thrust Coefficient Definition

The thrust coefficient is the measure of nozzle efficiency and accounts for all nozzle losses. As described earlier, these losses include the fundamental friction, angularity, and expansion losses and can include leakage and cooling air throttling losses if desired. The thrust coefficient equation showing these losses can be written as

$$C_{fg} = \frac{\overset{\text{Friction}}{C_V}\,\overset{\text{Angularity}}{C_A}\,\dfrac{W_{7\,\text{actual}}}{g}\,\overset{\text{Expansion}}{\overbrace{V_{9i} + (P_{S_{9i}} - P_0)A_9}}}{\dfrac{W_{7\,\text{actual}}}{g}\,V_s}$$
$$- \Delta C_{fg\,\substack{\text{leakage}\\\text{losses}}} - \Delta C_{fg\,\substack{\text{cooling air}\\\text{throttling losses}}} \qquad (5.10)$$

where

C_V = velocity coefficient accounting for friction losses
C_A = angularity coefficient accounting for angularity losses
V_{9i} = ideal velocity at the nozzle exit based on A_9/A_8, C_{D_8}, T_{T_8}, and FAR$_8$
$P_{S_{9i}}$ = ideal static pressure at the nozzle exit consistent with V_{9i}
V_s = isentropic or fully expanded jet velocity based on P_{T_8}/P_0, T_{T_8}, and FAR$_8$

The nozzle basic aerodynamic efficiency includes the effects of losses due to angularity and friction, while expansion losses are attributable to

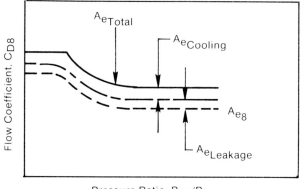

Fig. 5.9   Adjusted C-D nozzle flow coefficient.

"off-design" operation. These three losses establish a baseline C-D nozzle thrust coefficient curve shown in Fig. 5.10. The maximum thrust coefficient is the peak thrust coefficient and occurs at a nozzle pressure ratio where the nozzle exit average static pressure $P_{S_{9i}}$ equals ambient pressure P_0. Equation (5.10), without the leakage or throttling losses, becomes

$$C_{fg} = \frac{C_V C_A \dfrac{W_{7_{actual}}}{g} V_{9i} + (P_{S_{9i}} - P_0)A_9}{\dfrac{W_{7_{actual}}}{g} V_s}$$ (5.11)

At the peak thrust coefficient condition, the ideal exit static pressure is equal to ambient pressure $P_{S_{9i}} = P_0$ and, by definition $V_{9i} = V_s$. Therefore,

$$C_{fg\,peak} = C_V C_A$$ (5.12)

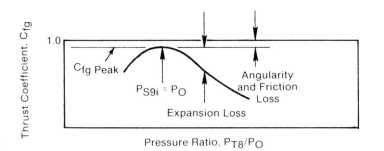

Fig. 5.10   C-D nozzle thrust coefficient.

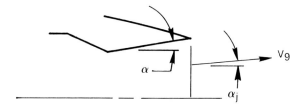

$$V_{9 \, Axial} = V_{9i} \cos \alpha_j$$

$$C_{A_j} = \frac{V_{9 \, Axial}}{V_{9i}} = \cos \alpha_j$$

Fig. 5.11   Nozzle angularity coefficient.

Thus, the peak thrust coefficient is the product of the angularity and velocity coefficients. At any other pressure ratio, the difference between the $C_{fg_{peak}}$ and C_{fg} is the expansion loss.

Angularity Coefficient

The angularity coefficient C_A is the thrust loss due to the nonaxial exit of the exhaust gases from the nozzle. For a small element of flow, this coefficient is essentially the cosine of the local exit flow angle α_j, as diagrammed in Fig. 5.11. Since the exit angle α_j varies from zero at the nozzle centerline to α at the outer radius, the overall nozzle angularity loss coefficient is the integral of $\cos\alpha_j$ across the nozzle exit. For the assumption of constant mass flow per unit area, the flow angularity is

$$C_A = \frac{1}{A_9} \int_{r=0}^{r=R_9} \cos\alpha_j 2\pi r_j \, dr \tag{5.13}$$

For the case of a variable mass flow distribution, the equation would be adjusted to include the mass flow variation. In theory, for a nozzle of sufficient length such that the exit flow angle follows a specific pattern (point source flow), the angularity coefficient is represented by the classical equation,

$$C_A = \frac{1 + \cos\alpha}{2} \tag{5.14}$$

Exhaust nozzles for practical applications, however, are relatively short and do not achieve the level of angularity loss coefficient defined in Eq. (5.14).

Potential flow, analytical studies of the supersonic axisymmetric C-D nozzle inviscid flowfield have allowed the evaluation of the angularity coefficient for a range of practical nozzle geometries. Based on these studies, the angularity loss coefficient can be correlated with nozzle half-angle α and nozzle area ratio A_9/A_8 as shown in Fig. 5.12.

In real nozzle geometries, there is also a slight loss in nozzle aerodynamic efficiency because the exit static pressure is not equal to ambient pressure across the entire nozzle exit area. For practical engineering purposes, this loss is included in the angularity coefficient; thus, the trends shown in Fig. 5.12 represent an angularity coefficient given by the equation

$$C_A = \frac{1}{A_9} \int_{r=0}^{r=R_9} \cos\alpha_j 2\pi r_j \, dr + \frac{1}{\dfrac{W_{7_{actual}}}{g} V_{9i}} \left[\int_{r=0}^{r=R_9} (P_{S_{9j}} - P_{S_{9i}})2\pi r_j \, dr \right.$$

$$\left. + \int_{r=0}^{r=R_9} (V_{9j} - V_{9i})\rho_{9j}V_{9j}2\pi r_j \, dr \right] \qquad (5.15)$$

The second and third terms in this equation are the static pressure variation loss and velocity variation loss (due to pressure variation) and are functions of the nozzle geometry only. The dashed line region of the angularity coefficient map represents nozzle geometries that, although an analytical C_A calculation can be made, are very short and are most probably separated making the analytical prediction invalid.

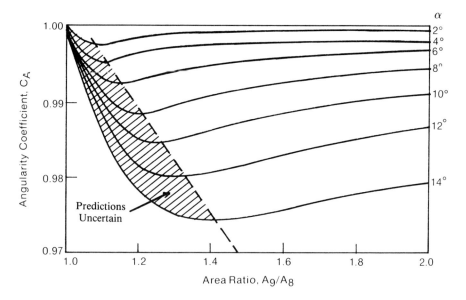

Fig. 5.12 C-D nozzle angularity coefficient.

Velocity Coefficient

The velocity coefficient accounts for the effects of boundary-layer momentum loss caused by friction in the nozzle. Because it is related to friction, this coefficient is essentially a function of the secondary nozzle surface area, Mach number near the wall, and Reynolds number. The Reynolds number effects are generally very small, less than $\pm 0.1\%$ for typical ranges of full-scale nozzle operating conditions. Since the wall Mach number distribution and surface area are a function of nozzle geometry, the velocity coefficient can be expressed uniquely as a function of geometry only. The velocity coefficient, like the angularity coefficient, can be presented as in Fig. 5.13 as a function of nozzle area ratio A_9/A_8 and half-angle α. This characteristic has also been determined analytically by inviscid flowfield analyses and boundary-layer calculations.

It should be recognized that, while angularity and velocity coefficients can be determined separately by analytical studies, it is extremely difficult and virtually impractical to determine the individual coefficients by experimental methods. The velocity coefficients are typically 0.992–0.997 and any reasonably accurate experimental analysis to determine these coefficients will require very precise pressure and flow angle measurements in a supersonic flowfield. However, as stated earlier, since the peak thrust coefficient is the product of C_V and C_A, this product can be determined directly by experiment with a carefully controlled scale-model nozzle test. Scale-model tests using strain gage force balance systems to measure nozzle thrust can provide accurate measurements of thrust coefficients. Comparison of peak scale-model thrust coefficients with the predicted peak coefficient using the analytical $C_V \cdot C_A$ product has shown agreement better than $\pm 0.25\%$.

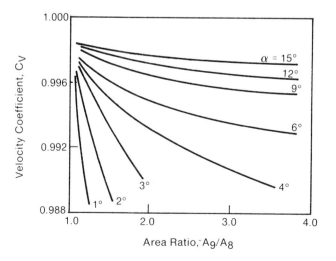

Fig. 5.13 C-D nozzle velocity coefficient.

Thrust Coefficient Curves

The angularity and velocity coefficients define the nozzle peak efficiency or maximum thrust coefficient, as discussed earlier. Once the nozzle peak thrust coefficient has been defined, either from C_V and C_A curves or from a scale-model test, a basic nozzle thrust coefficient curve can be calculated using one-dimensional fluid dynamics. The equation for thrust coefficient (less leakage and cooling air losses) may be manipulated assuming constant specific heat as follows:

$$
C_{fg} = \frac{C_V C_A \dfrac{W_{7_{\text{actual}}}}{g} V_{9i} + (P_{S_{9i}} - P_0)A_9}{\dfrac{W_{7_{\text{actual}}}}{g} V_s} = \frac{C_V C_A C_{D_8} \dfrac{W_{8i}}{g} V_{9i} + (P_{S_{9i}} - P_0)A_9}{C_{D_8} \dfrac{W_{8i}}{g} V_s}
$$

$$
= \frac{C_V C_A C_{D_8} \dfrac{P_{T_8} A_8 \bar{m}_8}{\sqrt{T_{T_8}}} V_{9i} + (P_{S_{9i}} - P_0)A_9}{C_{D_8} \dfrac{P_{T_8} A_8 \bar{m}_8}{\sqrt{T_{T_8}}} V_s}
$$

$$
= \frac{C_V C_A C_{D_8} \dfrac{\bar{m}_8}{\sqrt{T_{T_8}}} V_{9i} + \left(\dfrac{P_{S_{9i}}}{P_{T_8}} - \dfrac{P_0}{P_{T_8}}\right) \dfrac{A_9}{A_8}}{C_{D_8} \dfrac{\bar{m}_8}{\sqrt{T_{T_8}}} V_s} \tag{5.16}
$$

where

$$\bar{m}_8 = f(\gamma_8), \text{ assuming the nozzle is fully choked}$$

$$V_{9i} = f\left(\frac{A_9}{C_{D_8} A_8}, T_{T_8}, \gamma_8\right)$$

$$P_{S_{9i}}/P_{T_8} = f\left(\frac{A_9}{C_{D_8} A_8}, \gamma_8\right)$$

$$V_s = f\left(\frac{P_{T_8}}{P_0}, T_{T_8}, \gamma_8\right)$$

Since $\gamma_8 = f(T_{T_8}, \text{FAR}_8)$, the thrust coefficient can be defined as a function of pressure ratio from the following six parameters:

Geometry: $A_9/A_8, C_V, C_A, C_{D_8}$
Engine cycle: T_{T_8}, FAR_8

The resulting thrust coefficient curve, as shown in Fig. 5.10, includes the losses due to friction and angularity in the velocity momentum term $C_V C_A$

and the losses due to over- or underexpansion in the pressure area term (nozzle pressure ratio and gas properties) with the correct, full-scale engine gas properties. The resulting thrust coefficient curve assumes constant geometry (A_9/A_8) and constant gas property conditions $(T_{T_8}$ and $FAR_8)$.

Nozzle throat area A_8 is set by the engine cycle and operating condition. Exit area A_9 can be established in either of two ways depending on the engine cycle and the application and mission of the aircraft. The exit area can be set kinematically by a linkage system as shown in Fig. 5.3. In this case, there is a single value of A_9 for each A_8, A_9 is dependent on A_8, and the nozzle is said to have a fixed area ratio schedule. In the second method, A_9 is set by its own linkage and actuation system independently of A_8 and is said to have a variable area ratio schedule.

To provide nozzle performance characteristics for an engine cycle deck with a fixed nozzle area ratio schedule, a sufficient number of C_{fg} curves should be defined to cover the nozzle operating range of interest. For example, a typical area ratio schedule for a C-D nozzle is shown in Fig. 5.14, along with the operating ranges of interest and the minimum number of geometries at which C_{fg} curves should be defined. With this type of schedule, nozzle performance is compromised at some operating conditions because the area ratio A_9/A_8 is not properly matched to the nozzle pressure ratios P_{T_8}/P_0 and an expansion thrust loss will occur.

For exhaust systems with independent A_9 control, the area ratio can be set to provide maximum internal performance at any operating pressure ratio and the area ratio is "scheduled" as a function of nozzle pressure ratio as shown in Fig. 5.15. In this case, nozzle maximum performance is characterized as a function of pressure ratio (Fig. 5.16) by cross plotting a series of C_{fg} curves as shown in the Fig. 5.17 flowchart. The thrust coefficient drops off with higher pressure ratios because of increasing angularity and expansion losses as the area ratio increases.

As previously mentioned, the C_{fg} curves assume a full flowing, fully choked nozzle. For fixed area ratio schedule nozzles, the resulting curve must be adjusted to account for separation at low pressure ratio, over-expanded conditions. In addition, any leakage and cooling air throttling

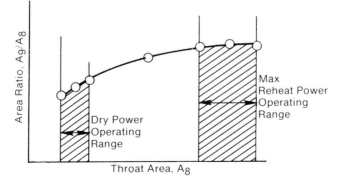

Fig. 5.14   C-D nozzle fixed-area ratio schedule.

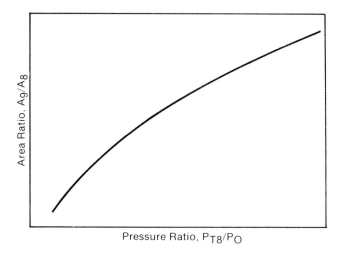

Fig. 5.15   C-D nozzle independent A_9 area ratio schedule.

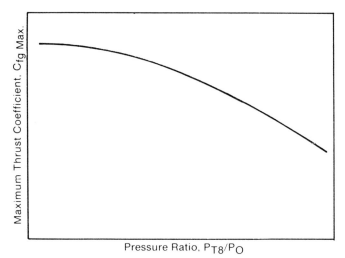

Fig. 5.16   C-D nozzle thrust coefficient characteristic for independent A_9 control.

loss adjustments, if consistent with the nozzle performance bookkeeping system, must be made to these curves. These adjustments are discussed in subsequent sections. This entire procedure is summarized in Fig. 5.18.

Leakage Losses

Nozzle performance losses due to gas leakage out of the nozzle are inherent with variable-geometry nozzles and must be accounted for in performance estimates. Since it is impractical and virtually impossible to

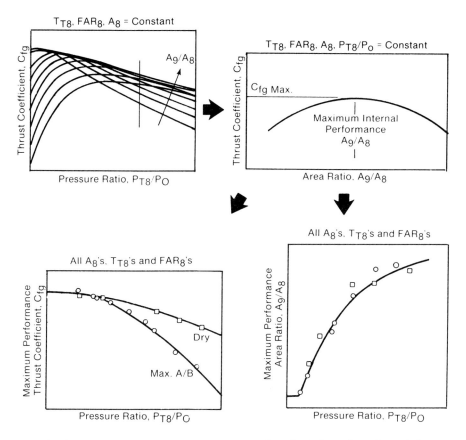

Fig. 5.17   Nozzle performance analysis flowchart for independent A_9 control.

simulate leakage areas in a scale-model test accurately, these thrust losses must be analytically calculated. The calculations require the use of nozzle wall pressure distributions to obtain the gas pressure loading and a knowledge of the effective leakage areas in the nozzle.

The various nozzle leakage paths encountered are defined in Fig. 5.19. Note that, with the exception of the primary nozzle, the leakage flows can be in either direction depending on the nozzle type and operating condition. If a control volume is drawn inside the nozzle as shown in Fig. 5.20, only those leakages that cross the control volume are included in the performance accounting. These include leakages 1–4 (see Fig. 5.19). The reason for this is that, for a nozzle with no cooling slot at the throat the nozzle cavity is most likely unpressurized and is essentially equal to ambient pressure and the external nozzle will not be designed to seal. Thus, leakages 5–7 are of no consequence. With a cooling slot at the throat: (1) the nozzle cavity is probably pressurized with cooling air; (2) there is no throat hinge leakage path 3 as it is substituted by a cooling slot; and (3) although only leakages 1, 2, and 4 need to be specifically calculated, leakages 5–7 contribute to nozzle performance because they affect the cavity pressure level and the fraction of total cooling airflow that comes back into the nozzle through the cooling slot as opposed to going overboard.

Although leakages 3 and 4 are shown to occur in either direction, any flow leaking back into the nozzle is generally neglected. The reason for this is that, for a nozzle with no throat cooling slot, the flow will re-enter the nozzle at a pressure that is at or very near ambient pressures and the thrust gain will be negligible. For a nozzle with a throat cooling slot, there will be conditions where a small amount of flow, which is significantly above ambient pressure, can leak back into the main gas stream through the secondary flaps and seals. However, since this leakage will be coming into the main stream in a circumferential direction, losses will occur in turning the flow axially. Thus, for a pressurized cavity nozzle, a theoretical thrust

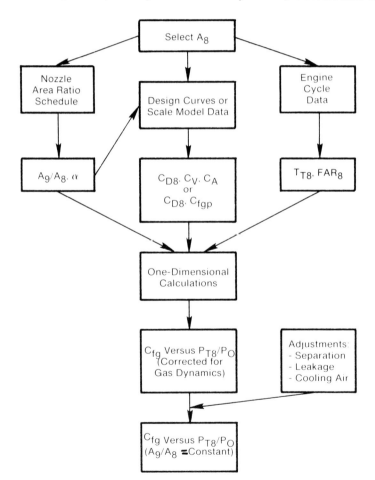

Fig. 5.18 Calculation procedure flowchart for full-scale nozzle performance curves.

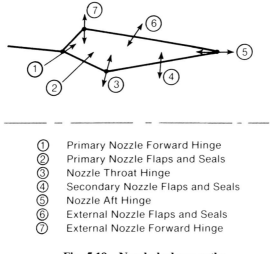

①   Primary Nozzle Forward Hinge
②   Primary Nozzle Flaps and Seals
③   Nozzle Throat Hinge
④   Secondary Nozzle Flaps and Seals
⑤   Nozzle Aft Hinge
⑥   External Nozzle Flaps and Seals
⑦   External Nozzle Forward Hinge

Fig. 5.19   Nozzle leakage paths.

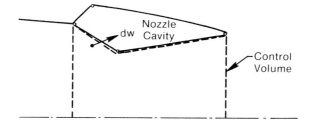

Fig. 5.20   Nozzle leakage control volume.

gain due to leakage flow from the cavity into the nozzle will be very small and can be assumed to be zero.

In order to calculate thrust losses due to leakage, the effective area of the leakage flow paths must be determined. Since this is extremely difficult and impractical to calculate analytically, the leakage characteristics of various sealing arrangements are obtained by tests of full-scale nozzle hardware. A typical setup for a leakage test of a primary nozzle is shown in Fig. 5.21. By measuring the airflow, pressure, and temperature in the nozzle and the ambient pressure, the effective leakage area can be calculated as

$$A_{e_{\text{leak}}} = \frac{W_{\text{leak}}\sqrt{T_N}}{P_N \bar{m}} \tag{5.17}$$

where $\bar{m} = f(P_N/P_0)$. The effective areas can be plotted as a function of the $\Delta P(P_N - P_{\text{amb}})$ across the nozzle.

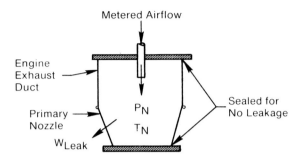

Fig. 5.21   Nozzle leakage test setup.

In general, the performance loss due to leakage is calculated as follows. Leakage flow W_{leak}, which crosses the control volume, is considered to be completely lost to the nozzle with no thrust recovery. The loss in thrust coefficient is the ideal thrust of the leakage flow divided by the nozzle ideal thrust; i.e.,

$$\Delta C_{fg\,\text{leakage}} = \frac{F_{gi\,\text{leakage}}}{F_{gi}} \tag{5.18}$$

Assuming constant specific heat in the expansion process, Eq. (5.2) can be rewritten as

$$F_{gi} = \frac{W}{g}V_i = \left(\frac{F}{W\sqrt{T_T}}\right)W\sqrt{T_T} \tag{5.19}$$

where

$$\left(\frac{F}{W\sqrt{T_T}}\right) = \sqrt{\frac{2\gamma R}{g(\gamma - 1)}\left[1 - \left(\frac{1}{P_T/P_0}\right)^{(\gamma - 1)/\gamma}\right]} = \bar{F} \tag{5.20}$$

Substituting into Eq. (5.18) and rearranging leads to

$$\begin{aligned}
\Delta C_{fg\,\text{leakage}} &= \frac{\bar{F}_{\text{leakage}}W_{\text{leak}}\sqrt{T_{T\,\text{leak}}}}{\bar{F}_8 W_{7\,\text{actual}}\sqrt{T_{T\,8}}} \\[6pt]
&= \frac{\bar{F}_{\text{leakage}}\bar{m}_{\text{leakage}}P_{T\,\text{leak}}A_{e\,\text{leak}}}{\bar{F}_8\bar{m}_8 P_{T\,8}C_{D\,8}A_8} \\[6pt]
&= \left(\frac{\bar{F}_{\text{leakage}}}{\bar{F}_8}\right)\left(\frac{\bar{m}_{\text{leakage}}}{\bar{m}_8}\right)\left(\frac{P_{T\,\text{leak}}}{P_{T\,8}}\right)\left(\frac{A_{e\,\text{leak}}}{C_{D\,8}A_8}\right)
\end{aligned} \tag{5.21}$$

Since leakage losses are generally less than 1% and because the flow function $\bar{m}$ will generally be choked or near choked, assuming $\bar{m}_{\text{leak}} = \bar{m}_8$ will create an insignificant error in the calculation. Therefore,

$$\Delta C_{fg_{\text{leakage}}} \cong \left(\frac{\bar{F}_{\text{leakage}}}{\bar{F}_8}\right)\left(\frac{P_{T_{\text{leak}}}}{P_{T_8}}\right)\left(\frac{A_{e_{\text{leak}}}}{C_{D_8}A_8}\right) \qquad (5.22)$$

Since leakage flow emanates from the boundary layer, the total pressure of the leakage flow is approximately equal to the wall static pressure. The leakage thrust function $\bar{F}_{\text{leak}}$ will be based on this pressure expanded to ambient pressure; i.e.,

$$\bar{F}_{\text{leakage}} = f\left(\frac{P_{T_{\text{leak}}}}{P_0}\right) = f\left(\frac{P_{S_{\text{wall}}}}{P_0}\right)$$

A typical leakage loss calculation is thus performed in the following manner:

(1) For the specific operating condition of interest, nozzle pressure ratio P_{T_8}/P_0, nozzle effective area $C_{D_8}A_8$, and the gas constant γ can be obtained from engine cycle data.

(2) A nozzle geometry is determined consistent with the cycle and wall pressure distributions are estimated as discussed in Sec. 5.4.

(3) From the wall pressure distribution and cavity pressure, a pressure loading is determined, $\Delta P = P_{S_{\text{wall}}} - P_{\text{cav}}$, and a leakage pressure ratio is obtained, $P_{S_{\text{wall}}}/P_0$.

(4) From the pressure loading ΔP and appropriate leakage test data, an effective leakage area is determined, $A_{e_{\text{leak}}}$.

(5) From the nozzle pressure ratio P_{T_8}/P_0 and leakage pressure ratio $P_{S_{\text{wall}}}/P_0$, the thrust functions $\bar{F}_8$ and $\bar{F}_{\text{leak}}$ are determined. The leakage flow γ can be estimated with negligible error.

(6) The leakage thrust loss is then calculated from Eq. (5.22) and subtracted from the thrust coefficient curve defined by Eq. (5.11).

In cycle deck bookkeeping systems where the leakage loss is not included in the thrust coefficient, leakage flows 1 and 2 are calculated and subtracted from the nozzle incoming flow W_7 to obtain the net flow available to the nozzle W_8. This net flow is then substituted for $W_{7_{\text{actual}}}$ in Eq. (5.11). The leakage flow is generally calculated as a ratio using the fundamental flow equation (5.17) and the appropriate subscripts, i.e.,

$$\frac{W_{\text{leak}}}{W_8} = \left(\frac{P_{T_{\text{leak}}}}{P_{T_8}}\right)\left(\frac{\bar{m}_{\text{leak}}}{\bar{m}_8}\right)\left(\frac{A_{e_{\text{leak}}}}{C_{D_8}A_8}\right)\sqrt{\frac{T_{T_8}}{T_{T_{\text{leak}}}}}$$

$$\cong \left(\frac{P_{S_{\text{wall}}}}{P_{T_8}}\right)\left(\frac{A_{e_{\text{leak}}}}{C_{D_8}A_8}\right)\left(\frac{T_{T_8}}{T_{T_{\text{leak}}}}\right) \qquad (5.23)$$

Generally speaking, W_{leak}/W_8 will be a function of A_8 and can be modeled as a curve or family of curves in the cycle deck. For this accounting system, the remaining leakage flows 3 and 4 must be estimated as a thrust loss and subtracted from the nozzle thrust coefficient as previously described.

Cooling Air Losses

Some C-D nozzles incorporate a cooling slot at the nozzle throat to provide additional film cooling air for the secondary nozzle during afterburning operation. The cooling air can be extracted from the tailpipe liner cooling flow upstream of or near the primary nozzle hinge, dumped into the nozzle cavity, and exhausted through the cooling slot as sketched in Fig. 5.22.

Since the pressure level of the liner flow can be significantly higher than the nozzle cavity pressure, the cooling flow that is removed from the liner experiences a pressure drop and is put back into the nozzle at a lower total pressure and, thus, lower available thrust. Note that the liner bleed W_{LB} will dump into the cavity and will mix with primary flap leakage air (and possibly secondary flap leakage). As mentioned previously, some of the flow will leak overboard depending on the sealing qualities of the external flaps and seals. In the design of the nozzle, the cooling slot must be sized to provide the desired amount of cooling air W_C at the estimated cavity air conditions.

The cooling air throttling loss should be estimated as the difference in the ideal thrust of the flow between the condition where it is extracted from the

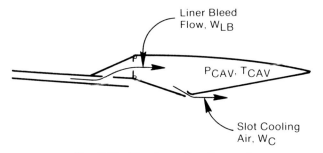

Fig. 5.22   Nozzle cooling flow path.

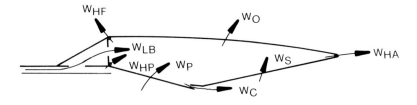

Fig. 5.23   Nozzle cavity flows.

liner and the condition at which it is put back into the nozzle; i.e.,

$$\Delta C_{fg\,\substack{\text{cooling air}\\\text{throttling}}} = \frac{F_{gi\,\substack{\text{liner}\\\text{bleed}}} - F_{gi\,\substack{\text{slot}\\\text{flow}}}}{F_{gi}}$$

$$= \frac{\left(\dfrac{F}{W\sqrt{T_T}}\right)_{7L} W_{LB}\sqrt{T_{TL}} - \left(\dfrac{F}{W\sqrt{T_T}}\right)_{\text{cav}} W_C\sqrt{T_{TC}}}{\left(\dfrac{F}{W\sqrt{T_T}}\right)_8 W_{7\,\text{actual}}\sqrt{T_{T8}}}$$

$$= \left(\frac{\bar{F}_{7L}}{\bar{F}_8}\right)\left(\frac{W_{LB}}{W_{7\,\text{actual}}}\right)\sqrt{\frac{T_{TL}}{T_{T8}}} - \left(\frac{\bar{F}_{\text{cav}}}{\bar{F}_8}\right)\left(\frac{W_C}{W_{7\,\text{actual}}}\right)\sqrt{\frac{T_{T\,\text{cav}}}{T_{T8}}} \quad (5.24)$$

Since the total pressure of the liner bleed flow is very near mainstream total pressure, it can be assumed that $\bar{F}_{7L} = \bar{F}_8$; thus,

$$\Delta C_{fg\,\substack{\text{cooling air}\\\text{throttling}}} = \left(\frac{W_{LB}}{W_{7\,\text{actual}}}\right)\sqrt{\frac{T_{TL}}{T_{T8}}} - \left(\frac{\bar{F}_{\text{cav}}}{\bar{F}_8}\right)\left(\frac{W_C}{W_{7\,\text{actual}}}\right)\sqrt{\frac{T_{T\,\text{cav}}}{T_{T8}}} \quad (5.25)$$

where T_{TL} is the gas temperature of the liner flow near the aft end of the liner, $W_C/W_{7\,\text{actual}}$ is specified by design, $W_{LB}/W_{7\,\text{actual}}$ is as required to obtain W_C/W_7, $T_{T\,\text{cav}}$ is an estimated cavity temperature, and

$$\bar{F}_{\text{cav}} = \left(\frac{F}{W\sqrt{T_T}}\right)_{\text{cav}} = f\left(\frac{P_{\text{cav}}}{P_0}, \gamma_{\text{cav}}\right) \quad \text{and} \quad \gamma_{\text{cav}} = f(T_{\text{cav}})$$

Note that if a scale-model test has been conducted with simulation of the cooling slot geometry A_S/A_8, cavity pressure level P_{cav}/P_{T8}, and corrected flow ratio $W_C/W_{7\,\text{actual}}\sqrt{T_{T\,\text{cav}}/T_{T8}}$, the recovered thrust of the cooling flow is already included in the scale-model coefficient, i.e.,

$$C_{fg\,\substack{\text{scale}\\\text{model}}} = \frac{F_{g\,\substack{\text{primary air}}} + F_{g\,\substack{\text{cooling air}}}}{F_{gi\,\substack{\text{primary}\\\text{air}}}} \quad (5.26)$$

In this case, the correction for cooling air throttling is simply

$$\Delta C_{fg\,\substack{\text{cooling air}\\\text{throttling}}} = \frac{W_{LB}}{W_{7\,\text{actual}}}\sqrt{\frac{T_{TL}}{T_{T8}}} \quad (5.27)$$

In the cooling air loss analysis, the nozzle cavity pressure P_{cav} is the most difficult parameter to estimate, primarily because an analysis to determine the pressure can be very complicated. The cavity pressure (and, in fact, the temperature) will be a function of the flows going into and out of the nozzle cavity.

The cavity pressure and temperature can be estimated by performing a flow and energy balance around the nozzle cavity control volume. For the flow directions shown in Fig. 5.23, the flow balance is

$$W_{LB} + W_{HP} + W_P + W_S = W_C + W_{HA} + W_0 + W_{HF} \tag{5.28}$$

and the energy balance

$$W_{LB}H_{LB} + W_{HP}H_{HP} + W_P H_{HP} + W_S H_S$$
$$= W_C H_{\text{cav}} + W_{HA}H_{\text{cav}} + W_0 H_{\text{cav}} + W_{HF}H_{\text{cav}} \tag{5.29}$$

where H is the total enthalpy of gas.

The leakages will be a function of the individual leakage path effective area characteristics, the nozzle wall pressure and temperature distributions, and the nozzle cavity pressure and temperature.

As can be expected, the cooling slot area has a significant effect on the cavity pressure level and cooling slot flow W_C. The cooling slot area schedule, A_S vs A_8, therefore must necessarily be determined by a proper balance of nozzle performance (thrust recovery), nozzle gas loads, and cooling requirements. This is accomplished by continued feedback and interfacing of analyses by mechanical and aerodynamic designers until an acceptable slot area schedule is established. Because of the small size of the cooling slots (1–3% of A_8), hardware tolerances can affect the cooling slot area by a significant percentage and should therefore be included in the analysis. It should be noted that the cooling flow, W_C, in the final analysis may not be a constant percentage of $W_{7_{\text{actual}}}$ because of the performance and gas load trade considerations. Since the additional cooling flow is usually not required for dry operation, it is advantageous to eliminate or shut off the bleed flow, W_{LB}, at these conditions and, hence, to design the slot accordingly.

Separation/Overexpansion Effects

The thrust coefficient curves previously discussed assumed a "full-flowing" nozzle with no internal separation at low pressure ratio. In reality, this condition occurs only at nozzle pressure ratios that are near or higher than the design pressure ratio corresponding to the nozzle area ratio (i.e., the pressure ratio at which the peak thrust coefficient occurs). At pressure ratios below the full-flowing condition, the flowfield within the nozzle will deviate from the design point flowfield and adjust to provide a static pressure at the nozzle exit that is equal to the ambient pressure. This flowfield adjustment is reflected as a static pressure distribution change on the secondary nozzle that, in turn, affects the nozzle gross thrust coefficient, as shown in Fig. 5.24. Thrust coefficient predictions previously defined must be adjusted at lower pressure ratios to account for these pressure adjustment effects.

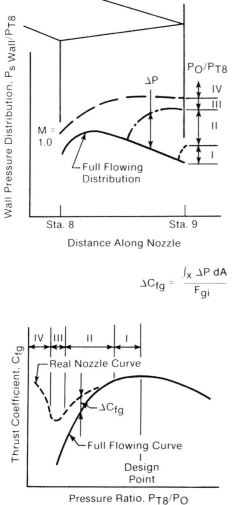

$$\Delta C_{fg} = \frac{\int_x \Delta P \; dA}{F_{gi}}$$

Fig. 5.24   Overexpansion effects on C-D nozzle aerodynamic characteristics.

Referring to Fig. 5.24, as the nozzle pressure ratio is decreased below the design point, a compression wave develops in the nozzle that raises the static pressure at the nozzle exit to the ambient pressure level. As the pressure ratio is further reduced, the location of the compression wave moves forward and its strength increases, affecting more of the nozzle wall pressure distribution. This change in wall pressure distribution can be integrated as a pressure-area term to calculate the effect on C_{fg}. In general, this region of overexpansion can be divided into the following four regions of operation, which are displayed in a simplified fashion in Fig. 5.24:

Region I: In this region, the pressure rise at the nozzle exit is very small. The resulting pressure-area integration is minuscule with virtually no impact on thrust coefficient.

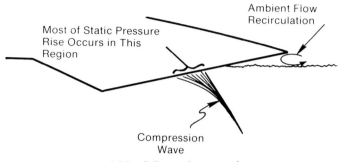

Fig. 5.25   C-D nozzle separation.

Region II: In this region, the compression wave has increased in strength and has moved forward affecting a larger portion of the nozzle. The flow in the forward portion of the nozzle is still supersonic and flow separation within the nozzle may or may not occur as in Fig. 5.25.

Region III: This is a transition region where the location of the compression wave moves rapidly to the throat region during a relatively small pressure ratio change. This region usually occurs around the minimum thrust coefficient pressure ratios. Significant separation can still occur in this band of operation.

Region IV: In this region, the flow is attached and entirely subsonic and the nozzle is behaving as a diffuser.

It is emphasized that these regions are described only in a general sense. They blend and overlap from one to another and should not be mistaken to start and end at discrete pressure ratios. Additionally, each nozzle design will behave differently depending on nozzle geometry. Thrust coefficient predictions must be adjusted at lower pressure ratios to account for these separation effects as shown schematically in Fig. 5.26.

Unfortunately, this region of nozzle operation is the least understood and most difficult to analyze. Experience has shown that real full-scale nozzle separation characteristics can be greatly influenced by nozzle leakage, quantity of slot cooling flow, wall surface quality, and probably local flowfield nonuniformities and Reynolds number. Additionally, little reliable model test data are available for analysis primarily because nozzles normally are not designed to operate at these extremely overexpanded conditions at important design points and little test data have been accumulated in the interest of reducing test costs. Thus, there are no reliable analytical or empirical design methods currently available to accurately predict the real full-scale effects of separation/overexpansion on nozzle performance.

Some scale-model data have been analyzed and correlated to provide the curves shown in Fig. 5.27. This figure will provide a "minimum" thrust recovery estimate. In the real nozzle, the actual thrust recovery will generally be higher since leakage, discontinuities, profiles, etc., encourage more separation. These real nozzle effects will be different for each nozzle design and no design criteria for estimating these effects are available.

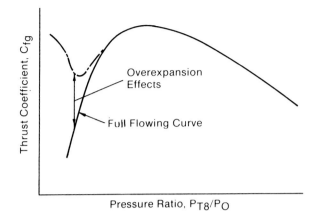

Fig. 5.26 Thrust coefficient curve adjusted for overexpansion effects.

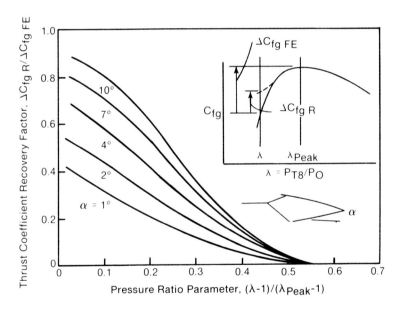

Fig. 5.27 Separation thrust recovery.

5.4 AERODYNAMIC LOAD PREDICTIONS

Nozzle pressure distributions are used by mechanical designers to determine hardware structural loads and actuation system loads. These loads thus effect detailed mechanical design, weight, and cost and are, therefore, very important estimates to be provided by the aerodynamic designer.

Internal pressure distributions can be estimated for the primary and secondary nozzle to cover all nozzle geometries and operating conditions.

Although the external loads on the nozzle are generally much smaller than the internal loads, external pressure distributions need to be estimated at high-load conditions.

Pressure distributions can be obtained either from scale-model tests or from analytical methods. Scale-model tests are more accurate; however, it is not always possible to obtain test data, particularly early in a development program. The following sections define procedures to analytically estimate nozzle pressure distributions.

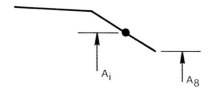

Fig. 5.28   One-dimensional static pressure distribution calculation.

$$(P_S/P_{T8})_i = f\,(A_i/A^*,\ \lambda) = f\,(A_i/C_{D8}\,A_8,\ \lambda)$$

Primary Nozzle

Pressure distributions on a simple conical primary nozzle can be estimated by several methods. There are a number of flowfield analysis computer programs in industry that can provide quite accurate estimates of wall pressure distributions. However, since the flow consists primarily of a subsonic accelerating flow, adequate accuracy for preliminary estimates of the pressure distribution can be obtained by utilizing the assumption of one-dimensional flow. From standard gas flow tables, the local static pressure to total pressure ratio can be obtained as a function of the local flow area to sonic flow area. This technique is applied to the primary nozzle as shown in Fig. 5.28. Pressure distributions can be estimated using this approach over any range of A_8's using the appropriate ratio of specific heats γ.

This approach, however, can produce errors in several ways. A small error will occur at the forward hinge point near the nozzle entrance (station 7) where the local turning of the flow will not be accounted for with the one-dimensional calculation. The actual pressure near the turn will be slightly higher than that calculated assuming one-dimensional conditions, the difference being dependent on the Mach number and turn angle θ' (Fig. 5.29). However, since this is close to the hinge, the impact of this error on actuator loads is usually negligible.

A second and more significant error can occur for nozzle designs with unusually long primary nozzles. The error occurs at afterburning conditions and becomes larger at the high fuel-air ratios. If the gas volume within the primary nozzle is significant relative to the total afterburner burning

Fig. 5.29 Wall-turning effect on static pressure.

P_S Actual $>$ P_S One-Dimensional

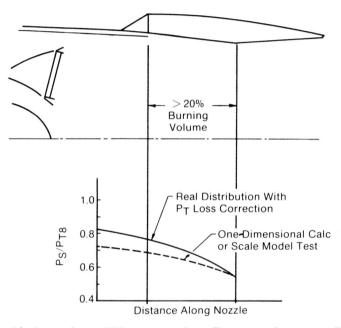

Fig. 5.30 Afterburner heat addition pressure loss effect on nozzle pressure distribution.

volume (approximately 20% or greater), significant combustion can take place within the primary nozzle. Under these conditions, there is a heat addition total pressure loss (Rayleigh line effect) occurring within the primary nozzle. Thus, the reference choke area A^* will be continuously changing along the primary nozzle. As a result, the real static pressure distribution will be significantly higher than the calculated one-dimensional pressure distribution. This can cause a sizeable error as shown in Fig. 5.30.

The correction for reheat burning effects must be applied to any analytical calculation and all scale-model data. For most applications, the primary nozzle is relatively short; thus, this effect is very small, if not insignificant. However, it should always be checked. The correction is a simple Rayleigh line heat addition calculation.

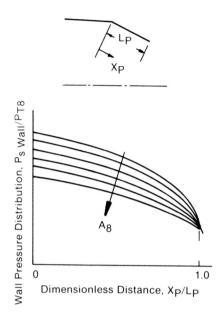

Fig. 5.31   Generalized primary nozzle pressure distribution.

Independent of the method used, pressure load estimates should be provided to the mechanical designer, as shown in Fig. 5.31, at a sufficient number of nozzle A_8 settings to allow easy interpolation for A_8. These curves will be independent of pressure ratio and, because of the sharp corner throat, will apply for very low pressure ratios down to 1.5 or perhaps lower.

Secondary Nozzle

Prediction of the secondary nozzle gas loads is not as straightforward as the primary nozzle for several reasons. Because of the sharp-corner throat, a sudden overexpansion of the flow immediately downstream of the throat causes the pressure distribution to deviate significantly from that predicted by a one-dimensional analysis as shown in Fig. 5.32. The amount of deviation is dependent on both the primary and secondary nozzle half-angles and the amount of cooling slot flow, if present. Thus, reliable analytical predictions can be done by only two methods: semiempirical using test data and theoretical using computerized flowfield equations.

For nozzles with no throat cooling slot, scale-model test data have shown a reasonable correlation with a one-dimensional calculation. This correlation is shown in Fig. 5.33 as the difference between actual pressure minus the one-dimensional calculation vs a nondimensionalized length parameter. Note that the one-dimensional calculation includes a rough estimate of boundary-layer effects in the secondary nozzle (0.995 factor) and the effect of the nozzle flow coefficient C_{D_8}. Inclusion of the flow coefficient accounts for the primary nozzle half-angle effect. Figure 5.33 can be used to estimate gas loads for full-flowing nozzles with no cooling flow at the throat W_C.

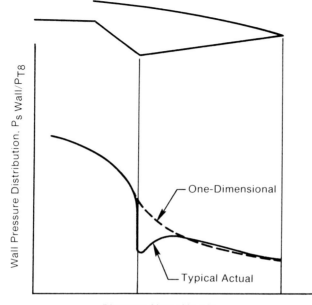

Fig. 5.32   Secondary nozzle pressure distribution.

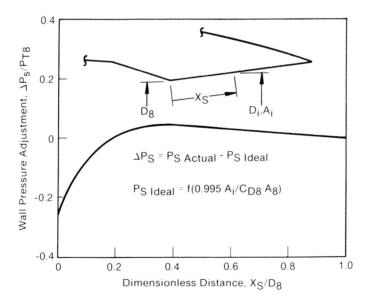

Fig. 5.33   Secondary nozzle gas loads correction factor.

For nozzles with a throat cooling slot, the wall pressure in the forward 20% of the secondary nozzle will be higher than the no-flow case. Additionally, even nozzles with no cooling slot will have some amount of leakage flow at the throat due to hardware stack-up tolerances. An approximate correction due to the cooling flow effect can be made using the curve in Fig. 5.34. This curve presents the change in throat static pressure relative to a no-flow condition based on scale-model test data. The correction is applied as shown in Fig. 5.35. This procedure can be used for an initial estimate of secondary nozzle, full-flowing gas loads.

As with the primary nozzle gas loads, secondary nozzle gas loads should be provided at a sufficient number of geometries to allow easy interpolation by mechanical designers. Reheat total pressure loss corrections made for the primary nozzle are not necessary for the secondary nozzle, since virtually no combustion takes place within this portion of the nozzle. However, if pressure distributions are obtained on a cold flow ($\gamma = 1.4$) basis, as would be the case for scale-model data, an adjustment in secondary nozzle gas load estimates for the effects of changes in specific heat ratio should be made. The error in the trailing-edge pressure for a C-D

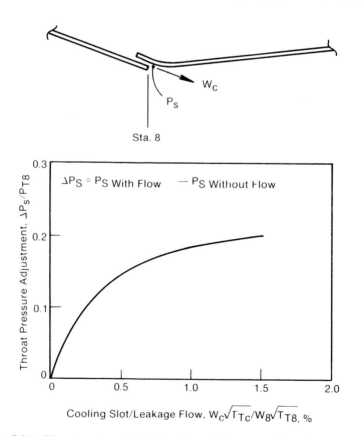

Fig. 5.34   Throat cooling slot/leakage flow effect on nozzle static pressure.

nozzle increases on a percentage basis with increasing area ratio, as noted in Table 5.1. Thus, although the difference in absolute level appears small at first glance, on a percentage basis the correction can amount to nearly 20% at high nozzle area ratios. A mechanical designer must work with pressure differences ΔP and this kind of error could be 20–50% of the ΔP. It is therefore important to provide the most precise load possible. If analytical load predictions are made or, if scale-model tests are conducted, the ratio of specific heat corrections should be made along the entire secondary nozzle. The adjustment is made by (1) converting the "cold-flow" pressure distribution to a Mach number and then an A/A^*, (2) assuming the "hot" A/A^* equals the cold A/A^*, and finally (3) converting the hot A/A^* to a Mach number and then a pressure distribution; i.e.,

$$\frac{P_{S_i}}{P_{T_8}}\bigg)_{\gamma=\text{cold}} \to M_i)_{\gamma=\text{cold}} \to \frac{A_i}{A^*}\bigg)_{\gamma=\text{cold}}$$

$$= \frac{A_i}{A^*}\bigg)_{\gamma=\text{hot}} \to M_i)_{\gamma=\text{hot}} \to \frac{P_{S_i}}{P_{T_8}}\bigg)_{\gamma=\text{hot}} \qquad (5.30)$$

External Loads

Nozzle external pressure distributions will be dependent on many parameters including specific nozzle geometry, propulsion system installation, local aircraft configuration including control surface position, Mach number, and nozzle pressure ratio as well as Reynolds number and aircraft attitude (angle of attack, yaw). Because of the large number of variables, it is virtually impossible to devise semiempirical parametric curves that could be used to provide an initial estimate. External loads are obtained either directly from a specific wind-tunnel test, indirectly by interpolating previous wind-tunnel test data from similar installations, or analytically by computer techniques.

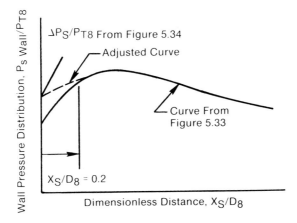

Fig. 5.35   Adjusted secondary nozzle pressure distribution.

Table 5.1 Specific Heat Ratio Effects on Nozzle Pressure
Distribution

A_9/A_8	P_S/P_{T_8}		P_S ($\gamma = 1.3$)
	$\gamma = 1.4$	$\gamma = 1.3$	P_S ($\gamma = 1.4$)
1.2	0.258	0.278	1.08
1.5	0.160	0.176	1.10
2.0	0.093	0.106	1.14
2.5	0.064	0.074	1.16
3.0	0.047	0.056	1.19

By far the most accurate pressure distributions are obtained from wind-tunnel tests of the exact nozzle/airplane configuration of interest at the appropriate operating conditions. Extrapolation or interpolation of these data is often necessary to provide load estimates at all the conditions of interest. In the absence of directly applicable wind-tunnel data, pressure distributions can be estimated by using previous wind-tunnel test data or by analytical methods. When using previous data, configurations and conditions that most closely resemble the nozzle and installation of interest should be used for load estimates. Computerized analytical methods can also be used to predict external loads. This approach gives very good results for two-dimensional or axisymmetric configurations. However, for three-dimensional configurations, which are most usually found in practice, the local geometric variations (which are difficult to simulate in the computer) can have a significant effect on the local nozzle pressure distribution. In either case, engineering judgment will be required to interpret and apply the information to define load estimates properly.

CHAPTER 6. ENGINE OPERABILITY

William G. Steenken
GE Aircraft Engines, Evendale, Ohio

ENGINE OPERABILITY

NOMENCLATURE

AIP	= aerodynamic interface plane
AT	= averaging time
B	= recoverability parameter
BP	= total pressure distortion superposition function
BT	= total temperature distortion superposition function
DGC	= distortion generation coefficient
DTC	= distortion transfer coefficient
EXP	= circumferential total pressure distortion extent function
EXT	= circumferential total temperature distortion extent function
f	= frequency
f_c	= cutoff frequency
F	= total pressure/total temperature superposition function
KP_c	= circumferential total pressure distortion sensitivity
KP_r	= radial total pressure distortion sensitivity
KT_c	= circumferential total temperature distortion sensitivity
KT_r	= radial total temperature distortion sensitivity
K_p	= planar wave sensitivity
MPR_P	= circumferential total pressure distortion multiple-per-revolution distortion descriptor
MPR_T	= circumferential total temperature distortion multiple-per-revolution distortion descriptor
$N\sqrt{\theta}$	= corrected speed
$P3$	= compressor discharge pressure
PR	= total pressure ratio
$P(t)$	= time-varying total pressure
$\overline{P(t)}$	= time-averaged total pressure
$\hat{P}(t)$	= time-varying spatial average of AIP total pressures
$\Delta PC/P$	= circumferential total pressure distortion intensity element
$\Delta PR/P$	= radial total pressure distortion intensity element
ΔPRS	= loss of stability pressure ratio
RSS	= root-sum-square
$\mathrm{SM}_{\mathrm{AVAIL}}$	= available stability margin
ΔSM	= loss of stability margin
$\Delta TC/T$	= circumferential total temperature distortion intensity element
$\Delta TR/T$	= radial total temperature distortion intensity element
WF	= combustor fuel flow
W_c	= corrected flow

ψ_p	= pressure coefficient
ϕ	= flow coefficient
$\phi(f)$	= power spectral density function
$\sum$	= sum
ϵ	= error
θ	= angle
θ^+	= angular extent of total temperature distortion
θ^-	= angular extent of total pressure distortion

Subscripts

i	= ring designator
rms	= root-mean-square
OL	= operating line
SLL	= stability limit line

6.1 INTRODUCTION

Engine operability is the discipline that addresses all the factors having an impact on the installed aerodynamic operation of a gas turbine engine when it is operated either in a steady-state or transient manner over the defined Mach number/altitude and maneuver envelopes. The goal of engine operability is to assure that the engine operates free of instability or with an acceptably small number of recoverable aerodynamic instabilities during the on-aircraft life of the engine. In any case, good design practice would dictate that aerodynamic instabilities do not occur at mission critical points.

While the discipline of engine operability could be the subject of a book in itself, it is the purpose of this chapter to address the major tenets of engine operability and not to be all inclusive. To this end, discussion is focused on the major tool of engine operability — the stability assessment. This method of treatment highlights those factors that are the largest and most variable in the stability assessment and gives a logical framework to the treatment that follows.

Engine operability as a discipline has evolved from considerations that were based on inlet/engine compatibility. Today the spectrum of issues is considerably broader. Although it has long been recognized that engine instabilities result from local compressor blade and vane velocity triangles being distorted to the point that incidence angles become unacceptably large, it was found that acceptable correlations could be developed that related the loss of stability pressure ratio (based on ratios of total pressure) to nonuniformities in the entering total pressure profiles (differences in total pressure), thus obviating the need to accomplish difficult-to-make local velocity measurements. This fortuitous finding has greatly simplified the task of the engine operability engineer and made possible the significant advances that have taken place in this discipline.

An early quantification of the nonuniformity of steady-state inlet total pressure profiles was undertaken by Alford.[1] Subsequently, it was found that engine stability was affected not only by the steady-state profile, but that unsteadiness of the total pressure as characterized by root-mean-square values of the total pressure contributed additionally to the loss of stability pressure ratio.[2] The current state of inlet/engine compatibility owes much to the work of Plourde and Brimelow,[3] who developed the first methodology to quantify the effects of time-varying inlet total pressures on engine stability. It is from these foundations that the discipline of engine operability has evolved with the concomitant need to assess all those factors affecting engine stability.

The current state of the art in engine operability to a large degree is based on the work of SAE S-16, Turbine Engine Inlet Flow Distortion, an industry committee composed of representatives from engine manufacturers, airframe manufacturers, and civil, regulating, and defense governmental agencies. In the material that follows, liberal use has been made of this committee's published works[4,5] and its soon to be published working documents in the areas of total temperature distortion and planar waves as well as the forerunner of this chapter.[6] Extensive bibliographies and references to the literature can be found in Refs. 7–10.

As the reader progresses through this chapter, it will help him to understand that this author explicitly and often implicitly is assuming a two-spool turbofan engine with a fuel control using fuel flow divided by compressor discharge pressure ($WF/P3$) acceleration schedules. This assumption by no means limits the utility of the material of this chapter and was implemented to prevent distracting digressions that amount to nuances. The techniques that are discussed are readily extendable to other gas turbine engine configurations and engine control modes.

6.2 DEFINITIONS

As a result of advances that have taken place, especially as a result of the nonrecoverable stall (stagnation stall) programs conducted by both Pratt and Whitney Aircraft and General Electric Aircraft Engines in the early 1980's, it has become necessary to clarify and sharpen the focus of the terminology used in engine operability studies. The definitions presented in this section and later in this chapter represent a more modern view and are consistent with current usage. It will be helpful if the reader takes the time to familiarize himself with the terms and their definitions as they will occur frequently throughout the chapter.

When presenting definitions and terminology, it is important to determine clearly the point of view that one is adopting. In our case, the point of view taken is that of an engine systems person looking at the overall behavior of a compression component or a compression system. This point of view is adopted because it is tied to those parameters that are most easily measured, that is, annulus average total pressure, corrected speed, corrected flow, and turbine or a related temperature.

Aerodynamic Instability

Aerodynamic instabilities occur as a result of the explicit (e.g., closing of an exit area) or implicit (e.g., distorted inflow conditions) throttling of an axial flow compression component to the point at which the compression component cannot sustain an increase in pressure ratio and/or decrease in corrected flow (loss of pumping capability) without the compression component incurring a sharp drop in discharge pressure. Unsteadiness within the compression component will accompany these aerodynamic flow instabilities. The surge and rotating stall instabilities are the two known compression component instabilities.

Aerodynamic Stability Limit Line

The aerodynamic stability limit line is defined by the locus of points in terms of annulus average total pressure ratio vs corrected flow obtained by slowly throttling a compression component until an aerodynamic instability occurs.

The aerodynamic stability limit line is generally obtained from rig or component tests by throttling the compression component at constant corrected speed until it loses pumping capability due to the occurrence of an aerodynamic instability. (See Fig. 6.1.) It should be noted that for the purposes of conducting stability assessments, the onset of undesirable mechanical stresses may occur prior to the occurrence of an aerodynamic instability in some speed ranges and would, therefore, represent the limiting condition.

In some low-speed regions, a clearly discernible aerodynamic instability may not occur, but rather heavy turbulence indicative of separated flow may develop and a subjective decision has to be made as to the usable pressure ratio/corrected flow range for the purposes of establishing the aerodynamic stability limit line.

In the past, the aerodynamic stability limit line often has been called either the surge line or the stall line, with both terms being used synony-

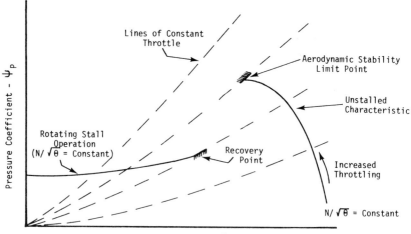

Fig. 6.1   Component characteristic representation.

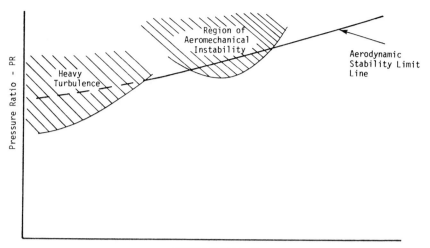

Fig. 6.2 Definition of stability limit line.

mously. Because either rotating stall or surge may occur or because additional constraints may set the limit locus of points over part of the speed range, it is desirable that a more encompassing and descriptive term be used; hence, the term *aerodynamic stability limit line.* (See Fig. 6.2.)

Stall

The initiation of an aerodynamic instability often is characterized by a drop in discharge pressure that occurs when the compression component loses the capability to pump flow at the immediate values of pressure ratio and corrected flow. This characteristic initiation of an aerodynamic instability is called *stall.*

Stall leads to either transient or fully developed aerodynamic instabilities (rotating stalls or surge cycles). (See Fig. 6.3.)

Rotating Stall

Rotating stall is an aerodynamic instability characterized as a local blockage to the axial flow within a compression component that rotates circumferentially in the direction of the rotor rotation at rotational speeds equal to approximately one-half of the rotor speed. Operation with fully developed rotating stall results in a stable annulus average operating point that lies on the stalled characteristic of the component pressure coefficient/flow coefficient representation. (See Fig. 6.3.)

Rotating stalls can occur as full- or part-blade span (hub, midspan, or tip) rotating stalls and may be multicelled. While rotating stalls are generally single celled, the occurrence of two cells is not uncommon. Other investigators have observed rotating stalls with as many as seven cells.[11]

Fully developed rotating stalls generally occur at low speeds for high-pressure-ratio compressors. Low-pressure-ratio (fan and booster) compression components often only exhibit the rotating stall instability throughout their operating speed ranges.

Surge

Surge is an aerodynamic instability characterized by a breakdown in the flow that results in more or less planar waves traveling in the axial direction of the compression component. Fully developed surge cycles will be characterized by alternate cycles of stall, depressurization, and repressurization (recovery). Often, the depressurization portion of the surge cycle exhibits some reverse flow. (See the component pressure coefficient/flow coefficient representation in Fig. 6.3.)

Fully developed surge generally occurs at high speeds for a high-pressure-ratio compressor. For medium and large turbofan engines, surge cycle frequencies will tend to lie in the 5–20 Hz frequency range.

Available Stability Margin

The available stability margin for a compression component is the difference between the pressure ratio at the stability limit line and the pressure ratio at the operating line nondimensionalized by the pressure ratio at the steady-state operating line, all taken at the same corrected flow. With reference to Fig. 6.4, the available stability margin can be written in

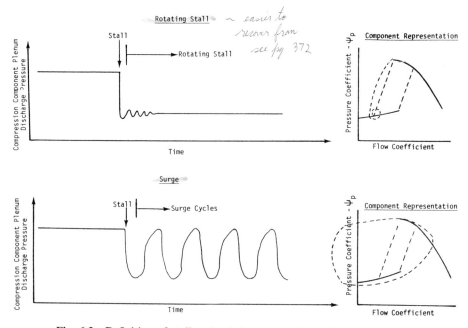

Fig. 6.3 Definition of stall and relation to rotating stall and surge.

equation form as

$$SM_{AVAIL} = \frac{PR_{SLL} - PR_{OL}}{PR_{OL}}\Bigg|_{W_c = const} \tag{6.1}$$

The available stability margin is often quoted as a percentage. In the past, the terms available stall margin or available surge margin have been used interchangeably. The use of available stability margin removes any chance for ambiguity.

Loss of Stability Margin

The loss of stability margin is the difference between the pressure ratio at some assessment point AP1 and another lower assessment point AP2 nondimensionalized by a reference operating line pressure ratio, all taken at the same corrected airflow. Values of the loss of stability margin are generally expressed as a percentage for discussion purposes. However, calculations are carried out in decimal form. In equation form and with reference to Fig. 6.4, the loss of stability margin is defined as

$$\Delta SM = \frac{PR_{AP1} - PR_{AP2}}{PR_{OL}}\Bigg|_{W_c = const} \tag{6.2}$$

Often the assessment point AP2 is located on the operating line and, hence, $PR_{AP2} = PR_{OL}$.

In the past, the term loss of stability margin has been known as loss of stall margin or loss of surge margin. The use of the term loss of stability margin removes any ambiguity caused by the interchangeable use of the words stall and surge.

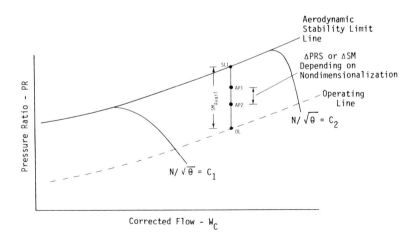

Fig. 6.4   Relationship of pressure ratios at points used in operability definitions.

Loss of Stability Pressure Ratio

The loss of stability pressure ratio is the difference between the pressure ratio at some assessment point AP1 and some lower assessment point AP2 nondimensionalized by the stability limit line pressure ratio, all taken at the same corrected airflow. Values are generally expressed as a percentage for discussion purposes, but for calculational purposes, the values are used in decimal form. In equation format and with reference to Fig. 6.4, the loss of stability pressure ratio is defined as

$$\Delta PRS = \frac{PR_{AP1} - PR_{AP2}}{PR_{SLL}}\bigg|_{W_c = \text{const}} \tag{6.3}$$

Often the assessment point AP1 is located on the stability limit line and hence, $PR_{AP1} = PR_{SLL}$.

In the past, the term loss of stability pressure ratio has been known by the terms loss of stall pressure ratio and/or loss of surge pressure ratio. Use of the term loss of stability pressure ratio removes any ambiguity caused by the interchangeable use of the words stall and surge.

Through use of the definitions for available stability margin, loss of stability margin, and loss of stability pressure ratio [Eqs. (6.1–6.3)], it can be readily shown that the three definitions are related as follows:

$$\Delta SM = \Delta PRS(1 + SM_{AVAIL}) \tag{6.4}$$

This equation has particular utility when transforming stability limit line or operating line effects to a common base for use in conducting stability assessments.

6.3 STABILITY ASSESSMENT

The engine designer is faced with significant dilemmas during his studies to provide adequate available stability margin throughout the aircraft flight envelope, that is, to optimally balance requirements for high thrust levels, low fuel consumption, low engine weight, long life, and low cost with adequate available stability margin. The operability engineer's tool for accounting for the factors for which he is responsible in contributing to this optimal balance is called the *stability assessment*. The stability assessment is a method for accounting in an orderly manner for all the destabilizing factors that consume stability margin during the operational on-aircraft life of the engine at flight conditions of interest.

Through use of the stability assessment, the operability engineer can achieve an appropriate balance between the available stability margin and the stability margin required by the destabilizing factors. If the desired balance is achieved, then the compression components of an engine will operate instability free. If the destabilizing factors should require more stability margin than is available, then aerodynamic instabilities such as surge or rotating stall may occur. Momentary occurrence of these instabilities should be and usually are of little importance. However, if these aerodynamic instabilities should last for extended periods (approximately greater than 0.1 s), then the mechanical life of the engine may be impaired, overtemperature of the turbine may occur, or flameouts may occur. Surge has been known to cause mechanical damage in high-pressure-ratio compressors with flexible blades as a result of the reverse flow causing "tip clang." In this situation, the compressor blades elastically deform during the reverse flow portion of the surge cycle and impact adjacent blades at the tip, causing the inelastic deformation that gives rise to stress concentrations.

The stability assessment is used during the conceptual and preliminary design phases of an engine program to help the cycle designer place the operating line of each compression component in relation to the objective stability limit line that the fan and compressor aerodynamic designers estimate will be achieved. During the development phase of the engine program, the stability assessment is used to ensure that adequate available stability margin is maintained in critical regimes of the flight envelope, while during the operational phase, it is used to diagnose those factors that may be increasing beyond the design allowances—hence, it is another element of maintaining engine quality.

The destabilizing factors that contribute to losses of available stability margin are divided into two groups: internal and external factors.

Internal Destabilizing Factors

Losses of stability margin that result from features internal to the engine are designated as being due to internal destabilizing factors. Examples of these factors are engine-to-engine manufacturing tolerances, deterioration, control tolerances, power (accel/decel/bode) transients, and thermal transients. These factors can impact both the stability limit line and the operating line.

Engine-to-engine manufacturing tolerances. Blade and vane quality, rotor tip clearances, hub leakage paths, and buildup tolerances are considered to be manufacturing tolerances that cause variations in the stability limit line of a compression component. Tolerances in exhaust and turbine nozzle areas and variations in component quality that affect efficiency are examples of manufacturing tolerances having an impact on the operating line of a compression component.

Deterioration. Deterioration items that manifest themselves as degraders of the stability limit line are blade and vane erosion due to sand, deposit buildups on blades and vanes due to dirty atmospheres, opening of hub seal clearances due to wear, and opening of tip clearances due to rubs during rapid acceleration and deceleration. Deterioration of the turbines due to tip clearance wear as well as other factors that would cause the turbine efficiency to decrease will cause increases in the operating lines of compression components.

Control tolerances. Control tolerances exhibit impact on the stability limit line and operating lines through the accuracy with which variable geometry can be positioned and fuel schedules repeated from one time to the next. Examples of such variable geometry include inlet guide vanes (IGV's), ganged IGV's and variable stators, and booster (low-pressure compressor) bypass doors. Fuel schedule repeatability depends on the accuracy with which sensed control parameters can be measured and these signals then being transformed via control schedules to actuation of valves that control fuel flow to the main burner.

Power transients. Turbofan engine power transients that result in the acceleration or deceleration of an engine will have their main impact upon the compressor operating line, while augmenter transients (lightoff and zone transitions) will have their main impact upon the fan operating line.

As an example of a compressor power transient, the trajectory of an acceleration operating line is shown in Fig. 6.5. Some of the available stability margin is consumed, resulting in a loss of stability margin due to the acceleration of the compressor from idle to intermediate power as depicted on the pressure ratio/corrected flow compressor map. Since the steady-state operating line represents balanced torque points (the work produced by the turbine equals the work consumed by the compressor), an acceleration can be accomplished only by having the turbine produce more work than the compressor consumes. This is accomplished by adding excess fuel in the combustor, which results in an increased compressor discharge pressure and, hence, an increase in compressor pressure ratio as depicted by the dashed line labeled "acceleration path" in Fig. 6.5.

The acceleration time of an engine can be increased by increasing the area bounded by the acceleration trajectory and the steady-state operating line. If power transient operating points should occur and cause unacceptable amounts of stability margin usage, "trimming" of the acceleration fuel schedule can result in more optimum usage of stability margin.

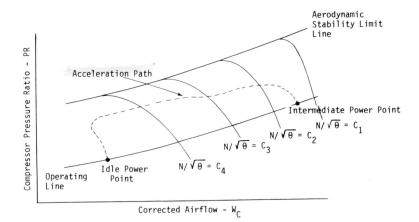

Fig. 6.5 Compressor map acceleration path.

It should be noted that the power transient allowance is developed based on an engine operability philosophy such as the usage associated with a given power transient at the corrected flow of interest or, as is the author's choice, to run many transients of all types and to encompass the maximum transient stability margin usage as a function of corrected airflow by a locus that ensures the engine can handle any transient.

Thermal transients. Accelerations and decelerations give rise to two types of thermal transients: (1) thermal mechanical expansions and contractions that affect clearances and (2) heat transfer to the gas path that results in compressor rear-end stage mismatching. Modern-day compressors are the result of sophisticated thermal design techniques and materials matching to cause the expansion (or contraction) of the compressor case to match in an optimal manner the expansion (or contraction) of the rotor.

External Destabilizing Factors

Losses of stability margin that result from the environment external to the engine are termed external destabilizing factors, examples of which are Reynolds number, inlet total pressure distortion, inlet total temperature distortion, and planar waves. The latter three items may appear in combination and their effect is often accounted for in a combinatorial manner.

In general, Reynolds number effects are accounted for by cycle deck predictions, that is, Reynolds number effects that cause the cycle to rematch and cause changes in the operating lines due to changes in engine inlet conditions are handled automatically by the cycle deck. Similarly, the stability limit line will be properly represented if empirical scalars and adders are included to modify the stability limit line as the engine inlet conditions (Reynolds numbers) change. In this manner, a cycle deck will provide correct estimates of the available stability margin.

Because inlet total pressure distortion, inlet total temperature distortion, and planar waves can consume large amounts of the available stability margin of compression components as functions of engine corrected airflow, aircraft Mach number, and aircraft attitude (angle of attack and angle of sideslip), they play a major and often a dominant role in any stability assessment procedure. The techniques that have been developed for estimating the impact of these external destabilizing influences upon the stability of a compression component are reviewed later in this chapter. It suffices to say here that techniques do exist to permit the operability engineer to estimate values of the loss of stability pressure ratio for a compression component due to these destabilizing influences.

Assessment Technique

The stability assessment technique that has evolved over the last two decades is designed to provide the engine operability engineer with an accounting of all the factors affecting the stability of an engine at a particular power setting and aircraft Mach number/altitude operating condition.

It is important to note that stability assessments are accomplished at given or defined engine corrected airflow settings in keeping with the definitions of Sec. 6.2. Engine corrected airflow is the parameter of choice for correlations because it promotes communication between the airframe manufacturer and the engine manufacturer and because inlet distortion and planar waves are often primarily functions of engine corrected airflow. Although the engine manufacturer generally accomplishes his component testing as a function of corrected speed, the transition to corrected airflow can be made with relative ease. As a result, engine speed changes during the engine development program due to cycle rematching and control schedule changes are transparent to the inlet designer.

With experience, the operability engineer will learn to identify the flow settings and Mach number/altitude conditions at which "pinch" points occur. Pinch points are those combined engine and aircraft operating points at which the remaining stability margin (available stability margin less the stability margin consumed by the sum of the external and internal destabilizing factors) is a relative minimum. It is important to engine and aircraft development programs to identify the pinch points and to conduct stability assessments at them. In conducting a stability assessment at a pinch point, each compression component of the engine should be examined.

Thus, having identified in principle the points at which stability assessments are to be conducted, we will now examine in some detail the manner in which a stability assessment is conducted for a given condition and for one compression component—a compressor. A compressor has been chosen for stability assessment because it illustrates most aspects of the instability assessment procedure. Stability assessments for other compression components would be conducted in a similar manner.

One approach to conducting a stability assessment is illustrated in Fig. 6.6. First, important information about the point selected is gathered. From the aircraft point of view, this information will include Mach number, altitude, engine power setting, engine airflow, angle of attack, angle of sideslip, inlet recovery, and inlet distortion. From the engine point of view, the important information will include engine power setting, Mach number, altitude, customer bleed requirements, and customer horsepower requirements. With this information, the cycle deck can be run to provide the cycle operating points, component corrected speeds, component corrected flows, and the available stability margins. It is for the purposes of obtaining the available stability margins of each compression component that the information just described is tabulated at the beginning of each assessment.

Because loss of stability margin or stability pressure ratio for the destabilizing factors is generally correlated as a function of corrected speed or corrected flow, the cycle output data permits the engine operability engineer to enter his correlations, whether they are based on historic and/or empirical data, test data, or analytical studies to obtain the impact on stability of each destabilizing item that has been identified to be important at this operating condition. Typical values of the loss of stability pressure ratio are shown for each destabilizing influence. Note that influences on the operating line are often correlated in terms of loss of stability margin and

CONDITION:	MO	60 KIAS
	ALT	7340'
	POWER	TAKE OFF
	AIRFLOW	130 LB/SEC
	DISTORTION CONDITION	30 KNOT CROSSWIND
	ESTIMATED DYNAMIC DISTORTION	$\dfrac{\Delta PC}{P} = 0.08$

STABILITY MARGIN CALCULATIONS ⁻ΔPRS	Σ	RSS
OP LINE VARIATION		.008
SURGE LINE VARIATION		.020
STATOR TRACKING ERROR		.004
ACCEL FUEL SCHEDULE VARIATION		.012
DETERIORATION SURGE LINE	.020	
THERMAL ACCEL TRANSIENTS (NO BODE)	.029	
POWER TRANSIENTS	.021	
DISTORTION	.012	.020
TOTAL ΔPRS REQUIRED	.082 ± .032	
CLEAN SM AVAILABLE	13.5	
TOTAL ΔSM REQUIRED = ΔPRS $(1 + SM_{avail})$	12.9	
NET SM REMAINING	0.6	

Fig. 6.6 Example of a compressor stability assessment.

will have to be transformed to loss of stability pressure ratio using Eq. (6.4) in order to maintain the accounting in the same type of units.

Examination of Fig. 6.6 will reveal that there are two columns for each destabilizing factor. The left-hand column is used for those factors that are deterministic in nature and are to be summed directly. The right-hand column is for those items that are statistically distributed and are root-sum-squared. Control and manufacturing tolerances fall into this latter category. In carrying out the RSS process of the right-hand column, it is generally assumed that each individual destabilizing factor is a Gaussianly distributed statistic. Hence, it is important to ensure that the contribution of each destabilizing influence represents the same number of standard deviations. For these purposes, two standard deviations are often used.

If the right-hand column is root-sum-squared and the resulting value is added to the direct sum contribution from the left-hand column, then the loss of stability pressure ratio for a statistically worse case engine is obtained. This value, when transformed to stability margin units, is then subtracted from the available stability margin. If the result is zero or positive, then current interpretations hold that no aerodynamic instability would be expected to occur during the on-aircraft life of the engine. If the remaining stability margin should be negative, then instabilities may occur and, with greater probability, the more negative the value of the remaining stability margin will be.

It is interesting to note that stability assessments can be designed for a specific set of circumstances; that is, a new engine can be represented by setting the deterioration allowances to zero or to some small levels to represent break-in losses or, alternatively, fixed throttle operation can be represented by setting the power transient allowance to zero.

The next subsection provides a discussion of the manner in which the concept of the stability assessment can be extended to broaden its utility.

Probability of Instability Occurrence

The stability assessment illustrated by Fig. 6.6 was accomplished for a compressor with 1.2% customer bleed taken from the compressor discharge and resulted in positive remaining margin. If a stability assessment was accomplished for zero customer bleed and hence a higher operating line, then the available stability margin is reduced and the remaining stability margin results in a negative amount equal to -0.6%. It is appropriate to ask what proportion of the number of times this combined set of flight and engine conditions will result in engine aerodynamic instabilities.

The proportion of occurrences can be estimated using the assumption of Gaussian statistics. The procedure is illustrated by Fig. 6.7. The probability of the occurrence of an aerodynamic instability at these conditons is noted to be 4.3%.

It should be noted that the SAE S-16 Turbine Engine Inlet Flow Distortion Committee currently has a subcommittee investigating the possibility of extending the above limited statistical analysis to one that is able to estimate the number of occurrences per flight or per aircraft lifetime using a Monte Carlo simulation technique and the occurrence distributions for the many environmental factors, flight envelope encounters, angle of attack and sideslip encounters, power lever transients, augmentation selection, etc. This effort is in the latter stages of development and will soon be entering a validation phase. The successful development of an automated statistical assessment technique will give the operability engineer a very powerful tool—a tool that will allow a quantitative assessment to be made of the importance of a negative remaining margin at a flight envelope condition and to balance the estimated number of occurrences at this condition vs the program constraints of budget and time required to change this negative margin to zero or positive remaining stability margin.

6.4 AERODYNAMIC INTERFACE PLANE

The common instrumentation plane that is chosen for making measurements necessary for inlet/engine compatibility is crucial to achieving a successfully integrated propulsion system. This instrumentation plane, called the aerodynamic interface plane, is used to make all performance and distortion measurements and should remain invariant throughout the propulsion system life cycle for all subscale, full-scale, and flight tests. The aerodynamic interface plane should be located as close as practical to the engine face in a circular or annular section of the inlet duct downstream of all auxiliary air systems so that only engine airflow passes through it.

The rake probe instrumentation array located at the aerodynamic interface plane should also remain invariant throughout the propulsion system life cycle and should be designed such that the engine performance and stability characteristics are unchanged by its presence. A typical rake probe array for measuring inlet recovery, total pressure distortion, total temperature distortion, and planar waves uses eight equiangularly spaced rakes with five probes per rake located on the centroids of equal areas, as shown in Fig. 6.8.

The importance of selecting an aerodynamic interface plane with its rake probe array such that both remain invariant throughout the life of the program cannot be overemphasized if successful propulsion system integration is to be achieved.

The next few sections of this chapter address obtaining estimates for the loss of stability pressure ratio due to some of the most variable and complex destabilizing factors, that is, the external destabilizing factors known as total pressure distortion, total temperature distortion, and planar waves. This chapter concludes with a discussion of those factors that promote recovery from stall should an aerodynamic instability be encountered in spite of following good design practices and a discussion of some of the more important analytical techniques that can be used to support engine operability analysis efforts and to provide, validate, or give insight into the quantitative values used in stability assessments.

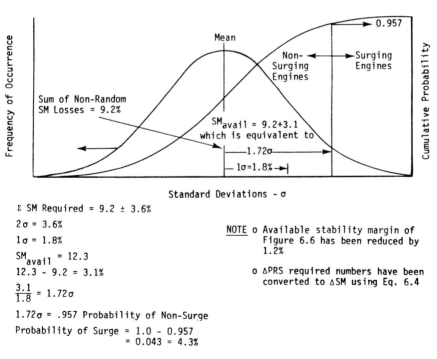

Fig. 6.7   Estimate of probability of stall.

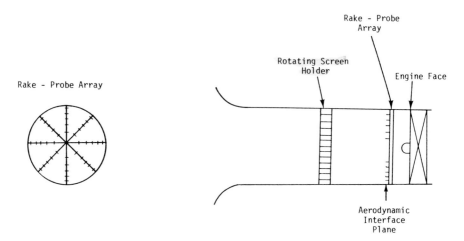

Fig. 6.8   Aerodynamic interface plane and relationship to engine face and distortion screen.

6.5   TOTAL PRESSURE DISTORTION

Engine-face total pressure distortion can be produced in many ways. Sources of total pressure distortion internal to the inlet are high local duct curvatures and rates of diffusion and incorrect design of boundary-layer bleeds and diverters. Sources of total pressure distortion external to the inlet are flowfield angularities due to angle of attack and angle of sideslip, shock/boundary-layer interactions, and protuberances such as landing gears, weapons bay doors, pods, armaments, and antennas. See Chapter 4 for a more complete discussion.

It is these inlet total pressure distortions that are to be measured by the instrumentation located at the aerodynamic interface plane and are to be transformed into estimates of loss of stability pressure ratio for use in the stability assessment procedure. The process for developing an estimate of the loss of stability pressure ratio for total pressure distortion is shown pictorially in Fig. 6.9. This figure illustrates the inlet data flow path from the rake probe instrumentation array through the data filtering process that approximates the engine dynamic response characteristic through the calculation of the distortion descriptors and through the loss of surge pressure ratio calculation, where inputs on the sensitivity of the engines compression components to the various elements of distortion are required. The major steps of the just described data flow process are described in the following paragraphs.

Engine Dynamic Response Characteristics

The ability to predict the loss of stability pressure ratio due to inlet total pressure distortion within acceptable tolerances was developed based on the work of Plourde and Brimelow.[3] In this work, the concept of "averaging time" was introduced, the purpose of which was to cause the peak

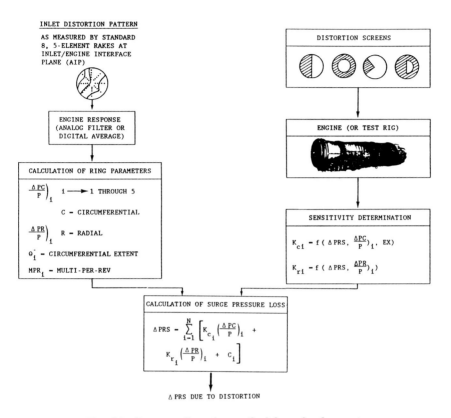

Fig. 6.9 Pressure distortion methodology development.

distortion that caused stall initiation to be larger than previous or subsequent peaks prior to stall. The averaging time has to cause the "right" peak to stand out above all others consistent with three items: (1) the time it takes for the distorted flow to travel from the aerodynamic interface plane to the stalling stage, (2) the time it takes for a stall to develop in that stage, and (3) the time it takes for the stall pulse to travel upstream to the aerodynamic interface plane. Other investigators also have found and corroborated the finding that the distortion must persist for a significant portion of a compression component rotor revolution if the peak distortion is to cause stall.

It is not possible to define one "averaging time" for all compression components even in nondimensional terms because, as many investigators have found, the averaging time is inextricably entwined with the loss-of-stability pressure ratio computation algorithm being used. From another point of view, it is one more parameter to be used in obtaining loss of surge pressure ratio correlations due to the effects of distortion within the desired accuracy. See the subsection below on stability pressure ratio correlation.

Determination of the optimum averaging time for a compression component, given a loss of the stability pressure ratio computation algorithm, can be accomplished best by comparing the measured loss of stability pressure ratio vs the calculated loss of stability pressure ratio and varying the averaging time to determine the value that gives the least error between the measured and calculated values of loss of stability pressure ratio. The inlet distortion for this study should have significant unsteady (dynamic) content and the pattern contours should be similar to those that will be encountered in aircraft applications. One study[12] reported in the literature used a two-dimensional shock/boundary-layer interaction produced by a device known as a random frequency generator to create the dynamic distortion patterns. For the "critical" compression component (the two-stage fan), the optimum averaging time was determined to be equal to one rotor revolution. Figure 6.10 illustrates a typical presentation of results for determining the optimum averaging time. While, ideally, the minimum error should equal zero if the averaging time and the distortion algorithm are perfect correlators of the data, this rarely happens in practice as illustrated in the figure.

Implementation of the averaging time has been accomplished with equal success using either digital or analog filters. A relationship between the averaging time AT in seconds of a digital running average (average M samples, advance one) and the cutoff frequency f_c (-3 dB point) of an equivalent five-pole linear phase analog filter was derived[13] with the following result:

$$f_c = 0.45/\text{AT} \tag{6.5}$$

where the averaging time AT is equal to the number of samples M being averaged divided by the sampling rate N in samples per second.

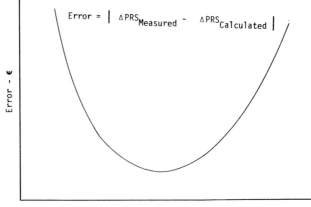

Error = | $\Delta\text{PRS}_{\text{Measured}}$ − $\Delta\text{PRS}_{\text{Calculated}}$ |

Error – ϵ

Averaging Time - AT

Fig. 6.10 Definition of optimum averaging time.

Total Pressure Distortion Descriptors

Investigators have determined that the loss of stability pressure ratio of a compression component depends on the manner in which the total pressure distribution deviates from uniformity at the engine face, that is, how the profile varies in the circumferential direction (circumferential distortion) and in the radial direction (radial distortion).

A closer examination of the effects of circumferential distortion reveals that the loss of stability margin will be a function of the magnitude of the distortion (magnitude of the total pressure deficit relative to some average), the circumferential extent of the low total pressure region, and the number of low total pressure regions per revolution (multiples per rev). Radial distortion produces losses of stability pressure ratio that are functions of the magnitude of the distortion and its location (hub or tip). In an excellent paper, Cotter[14] presents data that illustrate the effects of inlet patterns with the above types of total pressure defects (also see Ref. 15). Each engine manufacturer has developed its own procedure for correlating all the abovementioned effects of total pressure distortion. Unfortunately, the many different methods have led to confusion, impeded communication, and increased the amount of numerical computations an inlet designer might have to make to quantify results from inlet scale-model wind-tunnel tests. (Note that issues associated with the scale-model testing are addressed in Refs. 4 and 5.) In order to standardize the manner in which communication takes place between the engine manufacturer and the airframe manufacturer relative to quantifying the effects of inlet total pressure distortion, the SAE S-16 Inlet Turbine Engine Flow Distortion Committee was formed. While the reader will find details in Refs. 4 and 5, inlet distortion descriptors and the manner in which they may be used to correlate the loss of stability pressure ratio for a compression component are presented in the subsequent paragraphs to illustrate some of the issues that must be addressed in constructing a methodology for correlating the loss of stability pressure ratio for a compression component.

The distortion descriptor elements that are presented in this paragraph have been taken from Ref. 4. The rake probe array described in Sec. 6.4 facilitates the formulation of circumferential and radial distortion descriptor elements. The descriptor elements are defined on a ring-by-ring basis, that is, for each of the five rings of probes shown in Fig. 6.8. The ring circumferential profile is defined by the function $P(\theta)_i$, which is formed by a linear fit between successive probe values for each ring. Circumferential distortion is described by *intensity*, *extent*, and *multiple-per-revolution* elements. Radial distortion is described by an *intensity* element alone.

The descriptor element which describes the magnitude of the circumferential total pressure defect for each ring is called the intensity and is defined by the following equation (also see Fig. 6.11):

$$\text{Intensity} = \left(\frac{\Delta PC}{P}\right)_i = \frac{(P_{\text{AV}})_i - (P_{\text{AV,LOW}})_i}{(P_{\text{AV}})_i} \tag{6.6}$$

where the ring-average total pressure is defined as

$$(P_{AV})_i = \frac{1}{2\pi} \int_0^{2\pi} P(\theta)_i \, d\theta \tag{6.7}$$

and the average total pressure of the ring low total pressure region is

$$(P_{AV,LOW})_i = \frac{1}{\theta_i^-} \int_{\theta_i^-} P(\theta)_i \, d\theta \tag{6.8}$$

The ring distortion-descriptor element that defines the extent of the total pressure defect region is expressed in analytical form as

$$\theta_i^- = \theta 2_i - \theta 1_i \tag{6.9}$$

For a distortion pattern with only one total pressure defect per revolution, the ring distortion-descriptor element that describes the number of total pressure defects per revolution—the multiple-per-revolution element—is equal to one, that is

$$(MPR)_i = 1 \tag{6.10}$$

If a distortion pattern should contain more than one low-pressure defect per revolution, the pattern is termed a "multiple-per-rev" pattern. Although one-per-rev patterns occur most often, two-per-rev patterns are not uncommon. In the case of multiple-per-rev patterns, the intensity and extent are defined to describe the worst (in terms of loss of stability pressure ratio) total pressure defect and the multiple-per-revolution element is defined in terms of the number of equivalent low-pressure regions to give a continuously varying function rather than giving integer jumps (1,2,..., etc.). For details in treating patterns with multiple-per-revolution total pressure defects, the reader is referred to Refs. 4 and 5.

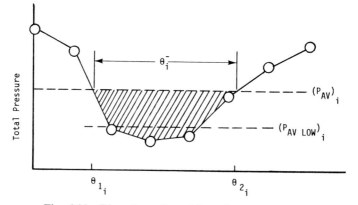

Fig. 6.11   Ring circumferential total pressure profile.

The radial distortion intensity-descriptor element describes the manner in which the ring-average radial profile changes in going from the hub of the engine face to the tip. This description is defined by a value that is a measure of the ring-average total pressure relative to the face-average total pressure. In algebraic form,

$$\left(\frac{\Delta PR}{P}\right)_i = \frac{(P_{F,AV}) - (P_{AV})_i}{(P_{F,AV})} \qquad (6.11)$$

where the face-average total pressure is defined as

$$(P_{F,AV}) = \frac{1}{N} \sum_{i=1}^{N} (P_{AV})_i \qquad (6.12)$$

That is, since it is the average of all 40 probes in the rake probe instrumentation array, it can be shown to be the average of the ring average values. Positive values of $(\Delta PR/P)_i$ reflect a ring average that is below the face-average total pressure.

It should be noted that if any probe value is missing due to misreading or malfunctioning pressure sensors, then values must be substituted for the

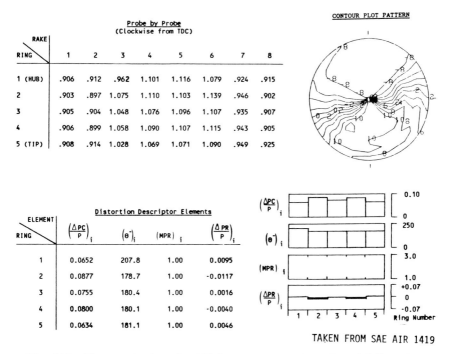

TAKEN FROM SAE AIR 1419

Fig. 6.12   Characterization of a 180 deg one-per-rev circumferential distortion pattern.

missing probe value. Techniques for accomplishing probe value substitution can be found in Ref. 5.

The results of carrying out the calculation of circumferential and radial distortion-descriptor elements are shown for a one-per-rev pattern in Fig. 6.12. The probe values from the rake probe array, a contour map (pattern), the values of the descriptor elements for each ring, and a graphic presentation of these elements are presented.

Typically, the AIP dynamic data required for input to distortion-descriptor calculations at a given condition is recorded over a 15–30 s period. Under most circumstances, a data sample of this length will provide a good estimate of the maximum distortion that will be seen over the operational life of the aircraft at this condition. However, it is good practice to make estimates of the maximum expected distortion[16] based on the maximum measured peak using extreme value statistics. Following this procedure will ensure that data samples of appropriate length will be acquired and the peak observed distortion thus representative of the expected maximum distortion.

Development of Total Pressure Distortion Sensitivities

The next step in defining the loss of stability pressure ratio of an aircraft inlet distortion pattern, having now decomposed the inlet total pressure distortion pattern into elements to which a compression component is sensitive, is to define the magnitude of these sensitivities.

Definition of sensitivities is often accomplished using distortion screens, that is, mounting wire mesh of varying shapes and porosities to a supporting structure (backing screen) that will create total pressure loss patterns similar to idealized design intents or to measured inlet patterns. See Fig. 6.13 for photographs of typical backing and distortion screen that will produce a distortion pattern similar to an aircraft pattern. Typically, screens are designed to produce distortion-descriptor values similar to those produced by a full-scale application, that is, those dynamic patterns that persist sufficiently long to cause loss of stability pressure ratio.

For typical test setups, the distortion screen is mounted in a rotating screen holder to give better pattern definition, since a fixed rake probe array is being used to define pattern effects relative to fixed sensor positions in the engine installations. It is located approximately one engine face diameter upstream of the engine face (see Fig. 6.8).

Other methods such as the airjet distortion generator[17] turbulators (turbulence generator[3,18]) and distortion valves[18] have been developed for producing distortion patterns. Screens suffer several drawbacks: (1) a new combination of mesh shapes and sizes is required for each different pattern and (2) a screen produces significant distortion levels over only a limited flow range (the magnitude of the circumferential distortion intensity element varies roughly as the square of the flow). In spite of these disadvantages, the screen remains the method of choice by most investigators for defining distortion element sensitivities.

The definition of circumferential distortion element sensitivities begins by choosing a reference pattern, in most cases the 180 deg one-per-rev pattern.

Fig. 6.13   Backup and distortion pattern screens.

In the early days of engine operability, 180 deg square wave patterns were often used. However, the steep gradients produced by square-edged patterns are not often found in "real" inlet patterns. Hence, the 180 deg reference pattern has evolved to a 180 deg graded pattern so that the edge effect (gradients) between the distorted and undistorted regions are more representative of naturally occurring patterns.

The sensitivity to 180 deg circumferential distortion is determined by throttling (raising the discharge pressure to increase the operating line) a compression component to instability at each corrected speed of interest. Typically, the average operating point at the initial occurrence of instability is below the aerodynamic stability limit line of clean inlet flow. By dividing the loss of stability pressure ratio due to inlet distortion by the circumferential distortion intensity at the point of instability, the 180 deg one-per-rev graded pattern sensitivity is obtained as

$$KP_{c,180} = \frac{\Delta \text{PRS}}{(\Delta PC/P)_{\text{max}}} \qquad (6.13)$$

It has been found by some investigators that the circumferential distortion sensitivity is best correlated when the maximum circumferential distortion sensitivity is based on the average of the largest two-adjacent ring values of the circumferential distortion intensity elements as follows:

$$\left(\frac{\Delta PC}{P}\right)_{\text{max}} = \text{max of } \frac{1}{2} \sum_{i=1}^{N-1} \left[\left(\frac{\Delta PC}{P}\right)_i + \left(\frac{\Delta PC}{P}\right)_{i+1} \right] \quad \text{for } i = 1,2,3,4$$
$$(6.14)$$

The sensitivity to the angular extent of circumferential distortion is found by testing one-per-rev screens of varying angular extents and forming the ratio of the loss of stability pressure ratio for an extent less than 180 deg to the 180 deg loss of stability pressure ratio. Hence, for a screen of angular extent θ^-, the extent function is given by

$$EXP(\theta^-) = \frac{\Delta PRS(\theta^-)}{\Delta PRS(180)} \quad (6.15)$$

The extent function typically varies from zero in the range of 0 to 45–60 deg to one at 180 deg. As with the 180 deg one-per-rev sensitivity, the extent function also varies with corrected speed.

For a one-per-rev pattern, the multiple-per-rev function is equal to one. The sensitivity to the multiple-per-rev element is obtained by testing with screens with multiple extent equal to a one-per-rev extent—that is, if two-per-rev sensitivity was required for two 90 deg extents separated by 90 deg, a screen with this characteristic would be tested and the resulting loss of stability pressure ratio would be referenced to the one-per-rev 90 deg pattern loss of stability pressure ratio. Hence,

$$f(MPR_P) = \frac{\Delta PRS(\theta^-, MPR)}{\Delta PRS(\theta^-)} \quad (6.16)$$

The sensitivity to radial distortion is determined using graded hub or tip radial screens. Thus,

$$KP_r = \frac{\Delta PRS}{(\Delta PR/P)_{max}} \quad (6.17)$$

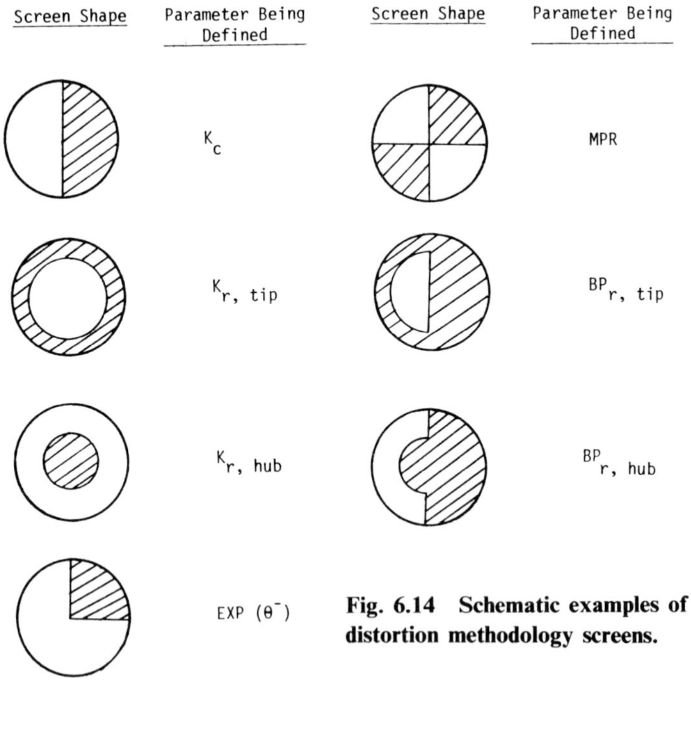

Fig. 6.14 Schematic examples of distortion methodology screens.

where $(\Delta PR/P)_{max}$ is determined from the ring that produces the maximum value of radial distortion, usually ring 1 (hub) or ring 5 (tip). It should be noted that, in some instances, the radial distortion sensitivity will be negative due to the point of aerodynamic instability being above the clean-inlet-flow stability limit line. This result occurs when a compression component has been designed for a radial profile differing from uniform flow. Thus, relative to the compression-component design point with a radial profile, the flat clean-inlet-flow profile is, in essence, a distortion.

Although the distortion descriptors treat the loss of stability pressure ratio due to circumferential distortion as being independent of the loss of stability pressure ratio due to radial distortion, they are, in fact, not independent. To account for the fact that coupling does occur, a parameter called the superposition factor has been developed to account for this coupling effect. In general, the superposition factor is determined by testing screens that have graded 180 deg one-per-rev content superposed with graded tip or hub radial pattern content. This superposition factor BP for total pressure distortion has been found to be a function of the ratio $(\Delta PR/P)_{max}/(\Delta PC/P)_{max}$ and corrected speed.

Figure 6.14 illustrates the conceptual nature of the screens that can be used to determine the necessary sensitivities and the superposition factor.

Total Pressure Distortion Loss
of Stability Pressure Ratio Correlation

The loss of stability pressure ratio due to the total pressure distortion associated with an arbitrary inlet pattern can be written in the following form:

$$\Delta PRS = \underbrace{BP}_{\substack{\text{superposition} \\ \text{factor}}} * \underbrace{f(\text{MPR}_{\text{P}}) * EXP * KP_c * \left(\frac{\Delta PC}{P}\right)_{max}}_{\substack{\text{circumferential} \\ \text{distortion} \\ \text{component}}} + \underbrace{KP_r * \left(\frac{\Delta PR}{P}\right)_{max}}_{\substack{\text{radial distortion} \\ \text{component}}} \quad (6.18)$$

Examination of Eq. (6.18) reveals that the loss of stability pressure ratio of an arbitrary pattern is composed of the loss of stability pressure ratio due to the radial distortion intensity element and to the circumferential distortion-decriptor elements modified only for the coupling between the circumferential and radial distortions by the superposition factor.

This type of distortion correlation equation has been used with significant effectiveness for the compression components of many engine programs. By industry standards, an acceptable degree of correlation has been achieved when the predicted loss of stability pressure ratio can be correlated with the measured loss of stability pressure ratio to within ± 0.02 units of loss of stability pressure ratio. An example of this degree of correlation is shown in Fig. 6.15. It should be noted that it is often necessary to test several aircraft patterns and to use the results obtained

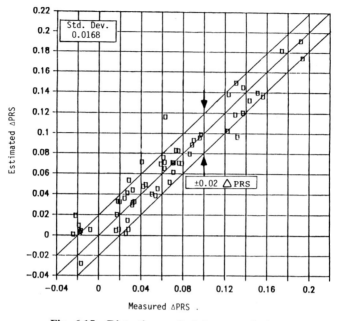

Fig. 6.15   Distortion methodology correlation.

from them to "tune" the results obtained by testing the pure and stylized patterns of Fig. 6.14 used to develop the distortion methodology.

Total Pressure Distortion Transfer and Total Temperature Distortion Generation Coefficients

As a distortion pattern passes through a compression component, the total pressure distortion is generally attenuated due to work of the fan adding total pressure to the deficit regions. However, in the process of accomplishing this attenuation, total temperature distortion is created. Thus, in order to predict the loss of stability pressure ratio of one compression component downstream of another—for example, a compressor downstream of fan in a turbofan engine—and using engine face distortion descriptor elements, it is necessary to estimate the distortion-descriptor elements at the entrance to the downstream compression component. This is accomplished through the use of the total pressure distortion transfer coefficient (DTC) and the total temperature generation coefficient (DGC), defined as

$$\text{DTC} = \frac{(\Delta PC/P) \text{ compressor inlet}}{(\Delta PC/P) \text{ engine face}} \qquad (6.19)$$

and

$$\text{DGC} = \frac{(\Delta TC/T) \text{ compressor inlet}}{(\Delta PC/P) \text{ engine face}} \qquad (6.20)$$

The total temperature distortion descriptor intensity element will be defined in Sec. 6.6. As has been found by many investigators, it is not necessary to address the transfer of radial total pressure distortion or generation of radial total temperature distortion because a fan tends to produce the same radial exit profile independently of either the inlet radial or circumferential distortion. Hence, as long as the fan exit radial profile is reproduced during testing of the downstream component, for example, during rig testing of a compressor, the impact of the fan exit radial profile will be inherent to the stability limit line of the downstream compression component and no further accounting will need to be made.

The loss of stability pressure ratio for a compressor downstream of a fan would then take the following form based on the measured engine face total pressure distortion:

$$\Delta \text{PRS} = F * f(\text{MPR}_\text{P}) * EXP * KP_c * \text{DTC} * \left(\frac{\Delta PC}{P}\right)_{\text{max}}$$

$$+ f(\text{MPR}_\text{T}) * EXT * KT_c * \text{DGC} * \left(\frac{\Delta PC}{P}\right)_{\text{max}} \quad (6.21)$$

where $f(\text{MPR}_\text{P})$, EXP, and KP_c are the total pressure circumferential distortion element sensitivities for the compressor; $f(\text{MPR}_\text{T})$, EXT, and KT_c the total temperature circumferential temperature distortion element sensitivities for the compressor; and F the superposition function for combining total pressure and total temperature distortions.

Synthesis of Dynamic Total Pressure Distortion Descriptors and Patterns

Sometimes the operability engineer is faced with only steady-state inlet data or steady-state data and the root-mean-square (rms) values from a few dynamic total pressure probes and the need to make stability assessments that require knowledge of dynamic inlet data. There are several techniques that can be used to provide estimates of the dynamic distortion.

One method is based upon correlations of dynamic distortion levels as a function of steady-state distortion levels. For a circumferential distortion parameter similar to $(\Delta PC/P)_{\text{max}}$, the ratio varies 1.5–2.2. Further, it has been found that dynamic radial distortion values of $\Delta PR/P$ tend to equal the steady-state value as a result of the averaging nature of the parameter. Note, however, that this type of correlation provides no information about the engine face inlet contour map pattern.

Many attempts have been made to synthesize dynamic distortion levels and patterns from steady-state probe readings and the rms values from a few dynamic probes located at the AIP. Since these methods inherently lack information about the manner in which each probe is correlated, in the statistical sense, with any other probe in the rake probe array at the AIP, limitations exist. These limitations reveal themselves at higher distortion levels where separated flow strongly influences the distortion levels and the probes become less correlated with each other. Two of the more recent methods that have enjoyed some success are due to Borg[19] and Sedlock.[20] Borg's method, which uses the AIP steady-state probe values, a few rms levels, and input from a random number generator produces only dynamic distortion levels (no dynamic patterns) and provides good correlation with measured dynamic distortion levels up to the point that separated flow begins to become significant. Sedlock's method provides estimates of both intensity distortion descriptors and contour maps using the steady-state probe values, rms values from a few dynamic probes, and a random number generator. Results using this method show that good agreement between measured and predicted values of both the distortion descriptors and between maps is achieved for turbulence (rms) levels up to 4%.

Techniques, such as those mentioned above and employed with caution, can be used to provide estimates of the total pressure distortion levels necessary for conducting stability assessments and pattern contours.

6.6 TOTAL TEMPERATURE DISTORTION

In many respects, the effects of total temperature distortion are addressed in a manner similar to that employed for total pressure distortion except for accounting for transient operating line effects that occur due to rapid changes in the engine face average total temperature as in the case of ingesting hot gases. While total pressure distortion mainly affects the stability limit line, total temperature distortion can affect both the stability limit line and the operating line. The operating line effect results from corrected speed changes (Sec. 6.9) and front-to-back thermal mismatch. The operating line effects are best estimated using predictions made from transient cycle decks that have provisions for sensor and actuator responses as well as thermal storage effects. It should be noted that the SAE S-16 Turbine Engine Inlet Flow Distortion Committee expects to publish a working document in late 1988 or early 1989 as a prelude to an SAE Aerospace Informational Report that will attempt to do for total temperature distortion what ARP 1420[4] and AIR 1419[5] have done for total pressure distortion. Some of the more accepted tenets of the work are included in the following paragraphs as an aid to students of engine operability practices.

Engine Response Characteristics

To this author's knowledge, there has been no definitive work that defines the response characteristic of a compression component to time-varying (dynamic) total temperature distortion. This stems from two factors: total temperature distortion is less common than total pressure distortion and the point measurement of rapidly varying total temperatures under laboratory conditions is difficult, requiring special instrumentation with compensating networks—to say nothing of making 40 such measurements in an inlet environment.

In the prediction of the loss of stability pressure ratio of a compressor located downstream of a fan, circumferential total temperature distortion

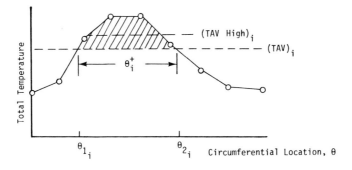

Fig. 6.16   Ring circumferential total temperature profile.

can be estimated at the compressor inlet using Eq. (6.20). If time-varying inlet circumferential total pressure distortion is present at the fan inlet, then time-varying circumferential total temperature distortion will be present at the compressor inlet. Thus, if the inlet total pressures have been filtered to the one-per-rev frequency equivalent, then implicitly so have the total temperatures at the compressor inlet.

Hence and because no cogent counterargument has been offered, the practice has been to assume that the analog filter characteristic or digital running average used to identify the engine face time-varying total pressure distorton that causes the maximum loss of stability pressure ratio for a compression component can be used to identify the time-varying engine face total temperature distortion that causes the maximum loss of stability pressure ratio.

However, the question as to the proper frequency to filter inlet time-varying total temperature data is probably moot. The reason for this statement is that in order to have thermocouples sufficiently durable to withstand flight or engine test cell environments, the thermocouple must be large with its concomitant low-frequency response. Hence, even with corrections, the frequency response of the thermocouples is significantly less than the one-per-rev frequency of the compression component and the operability engineer is faced with constructing a correlation with the available data.

Total Temperature Distortion Descriptors

In the case of total temperature, it is the portions of the distortion patterns where the total temperatures are greater than the average total temperature that cause compression component loss of stability pressure ratio. Thus, while the development of total temperature distortion descriptors parallels that for total pressure distortion descriptors, it will differ in this important respect.

The circumferential distortion descriptor intensity element is defined with respect to Fig. 6.16 as

$$\text{Intensity} = \left(\frac{\Delta TC}{T}\right)_i = \frac{(T_{\text{AV,HIGH}})_i - (T_{\text{AV}})_i}{(T_{\text{AV}})_i} \qquad (6.22)$$

and where the ring average total temperature is defined as

$$(T_{AV})_i = \frac{1}{2\pi} \int_0^{2\pi} T(\theta)_i \, d\theta \tag{6.23}$$

The average of the high-temperature region is defined as

$$(T_{AV,HIGH})_i = \frac{1}{\theta_i^+} \int_{\theta_i^+} T(\theta)_i \, d\theta \tag{6.24}$$

and the total temperature distortion descriptor extent element is defined as

$$\theta_i^+ = \theta 1_i - \theta 2_i \tag{6.25}$$

For a one-per-rev total temperature distortion pattern, the multiple-per-rev distortion descriptor element is equal to one, that is,

$$(MPR_T)_i = 1 \tag{6.26}$$

Patterns with more than one high-temperature region per revolution are termed multiple-per-rev patterns. Definition of the circumferential total temperature distortion descriptor elements for multiple-per-rev patterns will be found in the forthcoming SAE S-16 publication previously mentioned.

The radial total temperature distortion intensity descriptor element defines the magnitude by which the ring-average total temperature deviates from the face-average total temperature. In algebraic form

$$\left(\frac{\Delta TR}{T}\right)_i = \frac{(T_{AV})_i - (T_{F,AV})}{(T_{F,AV})} \tag{6.27}$$

where the face-average total temperature is based on the 40 probe average and is defined as

$$T_{F,AV} = \frac{1}{N} \sum_{i=1}^{N} (T_{AV})_i \tag{6.28}$$

Positive values of radial temperature distortion represent annuli where the ring-average total temperature exceeds the face-average total temperature.

Total Temperature Distortion Loss of Stability Pressure Ratio Correlation

The development of sensitivites to the total temperature distortion descriptor elements is accomplished conceptually in the same manner as for total pressure distortion-descriptor elements. However, the total temperature distortion is often created by injecting hot air into the engine inlet airstream or through the use of hydrogen burners. Methods for creating total temperature distortion and some of the results obtained during tests of compression components and engines has recently been reviewed by Biesiadny et al.[21]

For the purpose of this chapter, it suffices to say that the loss of stability pressure ratio due to inlet total temperature distortion can be correlated in a manner similar to that for total pressure distortion [Eq. (6.18)], that is,

$$\Delta PRS = BT * f(MPR_T) * EXT * KT_c * \left(\frac{\Delta TC}{T}\right)_{max} + KT_R \left(\frac{\Delta TR}{T}\right)_{max} \quad (6.29)$$

where the terms of the equation have significance similar to Eq. (6.18). The distortion sensitivities and the superposition factor are determined in a manner similar to that described in Sec. 6.5 for total pressure distortion through use of total temperature distortion patterns similar to the total pressure distortion patterns illustrated in Fig. 6.14. However, creation of these patterns using injected hot air or hydrogen burners is considerably more difficult than creating total pressure distortion patterns.

6.7 PLANAR WAVES

Another external destabilization item that can result in the loss of stability pressure ratio is planar waves. Planar waves can be generated by supersonic buzz, subsonic low-flow inlet instabilities, and protuberances located forward of the inlet such as landing gears, weapons bay doors, antennas, pods, and armaments.[22]

Until recently, a methodology for estimating the loss of stability pressure ratio due to planar waves did not exist. In a recent work by Steenken,[23] it was proposed that statistical techniques be employed to estimate the rms loss of stability pressure ratio. Then, the maximum loss of stability pressure ratio could be estimated based on knowing whether the planar wave is sinusoidal, harmonic, or random, or some combination thereof.

Planar waves are defined as those inlet total pressure fluctuations that act in unison over the engine face. Thus, the spatial average of all the total pressure probes located in the rake probe array in the aerodynamic interface plane will provide the time-varying planar waveform. Analytically, this definition of a planar wave can be expressed as

$$\hat{P}(t) = \frac{1}{N} \sum_{i=1}^{N} P_i(t) \quad (6.30)$$

The correlated fluctuations acting over the engine face will contribute to the average, while the uncorrelated fluctuations will tend to cancel each other.

Fundamental to this planar wave methodology is the postulate that a loss of stability pressure ratio can be estimated using the power spectral density of the planar wave and a compression component sensitivity to planar waves in a manner analogous to the output of an electrical amplifier being a function of the power spectral density of the input waveform and a gain function. Hence, the rms loss of stability pressure ratio for a compression component due to planar waves can be written as

$$\Delta PRS_{rms} = \left[\int_0^\infty [\phi(f) * K_p^2(f) \, df] \right]^{1/2} \qquad (6.31)$$

where $\phi(f)$ is the power spectral density of the nondimensionalized planar wave $\hat{P}(t)/\overline{P(t)}$ and $K_p(f)$ the compression component sensitivity to planar waves. An example of the sensitivity of a two-stage fan to planar waves is given in Fig. 6.17.[3] This sensitivity is based upon test data[24] and extrapolations based on unpublished computer simulations.

It should be noted that this formulation is independent of the type of waveform that might be present (sinusoidal, harmonic, random, or any combination thereof) and, hence, is applicable to any arbitrary waveform.

In order to estimate the peak loss of stability pressure ratio, its relationship to the rms loss of stability pressure ratio must be established. This can be accomplished by examining some specific waveforms. For example, if the planar wave was a pure sinsuoid, it can be shown that

$$\Delta PRS_{peak} = \frac{K_p(f)}{0.707} * \hat{P}(t)_{rms} \qquad (6.32)$$

If the planar wave was a harmonic wave, then the relationship between the rms loss of stability pressure ratio and the peak loss of stability pressure ratio can be established using Fourier analysis. Depending on the situation, the frequency components of a wave could combine constructively or destructively, or some combination thereof. However, if one were interested in obtaining a bound to the peak loss of stability pressure ratio, then one could assume that the harmonics simultaneously acted in the most constructive manner giving rise to an upper bound that could be readily derived from Fourier analysis.

Due to the nonlinear nature of the compressor component sensitivity $K_p(f)$, it is not possible to use Gaussian statistics in a straightforward

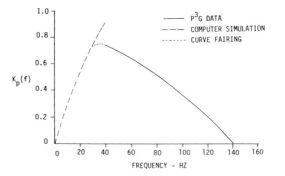

Fig. 6.17   Fan component planar wave sensitivity (from Ref. 23).

manner to estimate a level of peak loss of stability pressure ratio that will encompass some percentage of the peaks as the Gaussian statistics one, two, and three standard deviation levels do. However, the author believes it will be possible to bound the peak loss of stability pressure ratio for random planar waves using the following expression:

$$\Delta PR_{S_{peak}} = C * \Delta PR_{S_{rms}} \tag{6.33}$$

where C is to be determined empirically. The value of C will be influenced significantly by the shape of the planar wave-sensitivity function. It is anticipated that the value required to encompass the peak loss of stability pressure ratio 99% of the time will be considerably less than the value of 3.0 used in Gaussian statistics. A value of 2.0 is recommended until studies can be conducted to provide a more refined value.

6.8 RECOVERABILITY

Recoverability, the ability of gas turbine engines to recover from aerodynamic instabilities following their occurrence, has become of paramount importance following the high incidence of stagnation (nonrecoverable) stalls incurred by early versions of the Pratt & Whitney Aircraft F100 engine. Technology study investigations carried out in the early 1980's by GE Aircraft Engines and Pratt & Whitney Aircraft showed that the end state of nonrecoverable stalls is the aerodynamic instability rotating stall. As a result of these studies, the following definitions of recoverable and nonrecoverable stalls have been adopted:

Recoverable stall—A recoverable stall is a stall that, after initiation by some event, self-recovers from a transient surge or rotating stall aerodynamic instability to an unstalled operating condition.

Nonrecoverable stall—A nonrecoverable stall is a stall resulting in aerodynamic instability that may begin as a surge, but then degenerates into a fully developed rotating stall and does not recover, rather either achieving a steady-state operating condition or resulting in engine operating limits being exceeded or a speed loss to below the operating range. It is often characterized by decreasing speed and increasing turbine temperature. In general, the engine will not respond to control inputs and must be shut down in order to prevent thermal distress to the engine.

Greitzer in his landmark papers[25,26] formulated a nondimensional parameter B that defines a compression component speed known as the aerodynamic critical speed. The aerodynamic critical speed is the corrected speed (or more appropriately the speed range) of a compression component where the type of aerodynamic instability resulting from stall transitions from one type to another, that is, from surge to rotating stall or vice versa. This transition is governed by the recoverability parameter B. To first order, this parameter is defined as

$$B = \frac{U}{2a} \sqrt{\frac{V_p}{L_c A_c}} \tag{6.34}$$

where U is the average wheel speed, a the average acoustic speed within the compression component, V_p the volume to which the compression component discharges, L_c the characteristic length of the compression component, and A_c the characteristic area of the compression component.

When the value of B falls below a certain critical value, the aerodynamic instability *rotating stall* is expected following the occurrence of stall. If the value of B falls above this critical value, then the aerodynamic instability *surge* is anticipated.

B can be plotted as a function of corrected speed for a given compression component as in Fig. 6.18. Because nonrecoverable stall is mainly associated with the compressor in a turbofan engine, it is desirable that the aerodynamic critical speed range be as low as possible—below idle would be ideal. Then any external destabilizing event that was of sufficient magnitude to cause an aerodynamic instability would cause a surge—an instability that is significantly easier to recover from since less unthrottling is required than is required to recover from rotating stall.[25]

It is the author's contention that having the critical speed B occur at low corrected speeds is a necessary condition for an engine to have good recoverability characteristics, but it, in and of itself, is not sufficient. Other compression component factors such as the stability limit line pressure ratio, inlet Mach number, size of the compressor hysteresis region, magni-

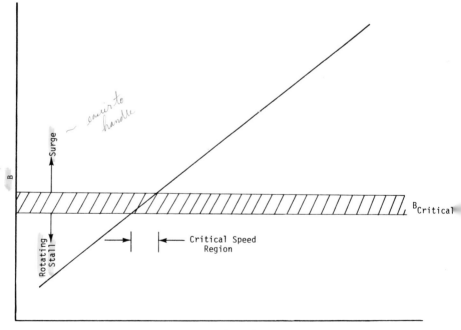

Fig. 6.18   **Relationship of recoverability parameter to compression component corrected speed.**

tude of the separation between the unstalled and the in-stall characteristics, and rate of throttling have been identified as playing a part in establishing good recoverability characteristics.

Engine fuel controls with their effect on compressor throttling and unthrottling by the combustor as well as variable geometry schedules play an important role in ensuring that an engine recovers once an aerodynamic instability occurs. Studies[27,28] have been conducted showing that a transient cycle computer simulation model with the internal volumes and lengths properly represented along with the in-stall and reverse flow characteristics of the compression components can replicate the transient poststall behavior of an engine to an adequate degree[27] and that these computer models can be used to develop stall recovery control strategies.[29]

Computer simulations such as those just described are having great utility during the preliminary design phase when an engine is being configured to ensure that its compresion system fundamental volume and speed relationships are chosen to ensure good recoverability characteristics and during the development phase for developing stall recovery control strategies should an aerodynamic instability be incurred.

6.9 ANALYTICAL TECHNIQUES

Some of the analytical techniques which have proved useful to the engine operability engineer in addition to the steady-state cycle deck and the transient cycle model are reviewed in this section. In particular, the basic foundations of parallel compressor theory are reviewed, an overview of some of the types of compressor component models that have been developed and their uses are presented, and a discussion of the required characteristics of computer simulations employed for stall recovery studies is given.

Parallel Compressor Models

The parallel compressor concept was first discussed by Pearson and McKenzie[30] in 1959. The concept still remains a powerful tool for understanding the manner in which distortion and especially circumferential distortion impacts the aerodynamic stability of a compression component. However, similar arguments could be developed for radial distortion with no difficulty. In this subsection, the application of parallel compressor theory to both circumferential total pressure distortion and circumferential total temperature distortion will be discussed.

Total pressure distortion. In its most elemental form and with reference to Fig. 6.19, parallel compressor theory would view a compressor with 180 deg one-per-rev inlet circumferential total pressure distortion as two parallel compressors, one with a low inlet total pressure of $P_{T,\ LOW}$ and the other with a high inlet total pressure of $P_{T,\ HIGH}$, both operating at the same corrected speed and exiting to a common, uniform static pressure. This latter assumption becomes increasingly better the further downstream

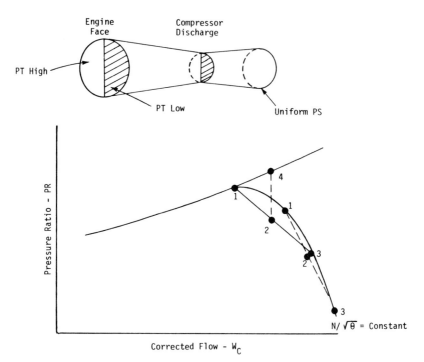

Fig. 6.19   Total pressure distortion parallel compressor model.

of a compression component it is imposed and the lower the Mach number. As indicated by examination of the compressor map of Fig. 6.19, the parallel-compressor sector operating with the low inlet total pressure will operate with the high total pressure ratio denoted as point 1 on the corrected speed line. The parallel compressor with the high inlet total pressure will operate with the low total pressure ratio denoted by point 3. The average operating point of the compressor as a whole is denoted by point 2. If the magnitude of the distortion was increased sufficiently (that is, the low total pressure decreased further) as in the case of an aircraft angle-of-attack maneuver or the average operating point was increased as during an accel power transient, point 1 would reach the clean inlet flow stability limit line first. Under this set of operating conditions, the compressor would incur an aerodynamic instability even though the average operating point of the compressor were well below the clean inlet flow stability limit line. Then, relative to the average operating point of the compressor, the loss of stability pressure ratio for this 180 deg one-per-rev circumferential distortion in keeping with Eq. (6.3) would be

$$\Delta PRS = \frac{PR_4 - PR_2}{PR_4} \qquad (6.35)$$

Thus, with the use of parallel compressor theory and the pressure ratio corrected-flow representation of a compressor, it has been illustrated that low total pressure regions in total pressure distortion patterns lead to compressor instability.

Total temperature distortion.   Now let us consider a compressor operating with 180 deg one-per-rev inlet total temperature distortion. As illustrated by Fig. 6.20, we can view the compressor as divided into two parallel compressors—one operating with a high inlet total temperature of $T_{T, \text{HIGH}}$ and one operating with a low total temperature of $T_{T, \text{LOW}}$ and each with the same inlet total pressure. The two parallel compressors discharge to a common, uniform static pressure where the Mach number is low.

Since each parallel compressor operates at the same physical speed, then and with reference to Fig. 6.20, the sector with the high inlet total temperature operates at a lower corrected speed denoted by point 1 and the sector with the low total temperature operates at a higher corrected speed denoted by point 3, each operating at the same pressure ratio. The average operating point of the compressor is illustrated by point 2. If the total temperature distortion were increased as during a landing of a STOVL (short takeoff, vertical landing) vehicle due to exhaust gas ingestion or the average operating point were throttled toward the stability limit line such

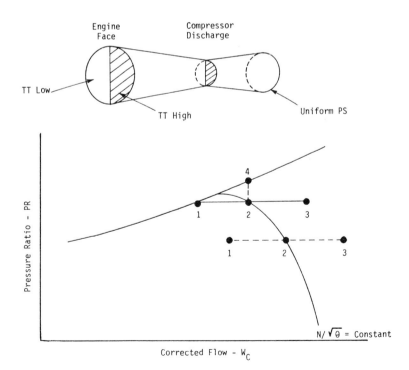

Fig. 6.20   Total temperature distortion parallel compressor model.

as during an engine acceleration, the high total temperature sector denoted by point 1 would reach the clean inlet flow stability limit line first. Then, in keeping with the definition of Eq. (6.3), the loss of stability pressure ratio using the average operating point would be

$$\Delta PRS = \frac{PR_4 - PR_2}{PR_4} \qquad (6.36)$$

Thus, parallel compressor theory illustrates the fact that it is regions of high total temperature in a total temperature distortion pattern that contribute to compression component instability.

Significantly more sophisticated parallel compressor models have been developed that permit multiple sectors and allow for circumferential redistribution of flow upstream and downstream of compression components. Models such as these give improved accuracy over simple parallel compressor theory and provide significantly greater information about the upstream flow, including the magnitude of the swirl velocity component. For further details, the reader is referred to Refs. 31–34.

Compression Component Models

A number of compression component digital computer models have been constructed, the complexity being related to the accuracy or fidelity of the desired results. Where only low frequencies were of interest, the models have successfully achieved solutions using only the continuity and momentum conservation relationships. Where higher frequencies were of interest or the time rate of change of flow parameters was large, then solutions using all three conservation equations (continuity, momentum, and energy or their equivalents) have been employed.

As the desired frequency response increases, then greater care must be taken in the representation of the compression component. In order of providing increasing frequency response with the requisite fidelity, various types of compression component representations are listed in Table 6.1 with appropriate references.

As the frequencies become higher, it has been noted[40] that the use of a single block representation may provide too "hard" a reflection plane since

Table 6.1 Types of Compression Components Models

Type	Reference
Single compressor block stage characteristics (actuator disk)	25
Multiple blocks	35
Stage by stage	36
Blade row by blade row	37, 38, 39

the acoustic impedance change all takes place at one plane rather than at planes (stages) distributed through the compressor. In such cases, it appears that the magnitude of the fluctuations may be larger than measured, although the predicted frequencies are correct. In addition, use of representations that do not represent the first and last stages of a compressor may give rise to inaccurate frequency predictions where differences between simulaton and test vehicle length and/or volume relationships within, upstream, and downstream of the compression component are significant.

While the use of linearized compression component models is limited in engine operability studies due to the magnitude of both time-dependent rates of change and spatial gradients, one study[41] that used a linearized set of equations is important because of the results obtained. This work, which made use of the first method of Liapunov, showed that the prediction of instability in the flow of a compression system occurred at the point of the measured aerodynamic stability limit line when clean-inlet-flow operating points were examined by "throttling" to the stability limit line. The quotation marks surrounding the word throttling are intended to draw attention to the fact that the linearized set of equations allowed examination at discrete points rather than a continuum of states. Thus, this work conclusively showed that the occurrence of mathematical instability at the aerodynamic instability limit line was a physical instability inherent to the aerothermodynamics of the compression component, that is, the system of equations that represents compression components and the specific compression component stage characteristics (pressure and work coefficients) as a function of flow coefficients or their equivalent. It should be noted that for the purposes of this study, the stage characteristics were continuous with no significant changes of curvature in the region of the aerodynamic stability limit line.

The results achieved using the types of models cited here provided the background and paved the way for engine system models that could predict the poststall behavior of turbofan engines as well as their unstalled behavior. This type of model is discussed in the next subsection.

Stall Recovery Models

A number of investigators have constructed digital computer simulations that predict poststall behavior sufficiently accurate to permit definition of the aerodynamic critical speed region and development of stall recovery control strategies should an engine incur either surge or rotating stall instabilities.

These powerful tools are not complex in the computational fluid dynamic (CFD) sense for they make no significant demands on fineness of the spatial grid or on the shortness of the time step. The choice in either case is dictated by the desired frequency response and the numerical stability of the equations and of the solution technique employed. That these simulations have been successful is a result of the flows being essentially one-dimensional or at most two-dimensional (axial-radial upstream of the fan and downstream of the fan in the splitter region) for surge and that the

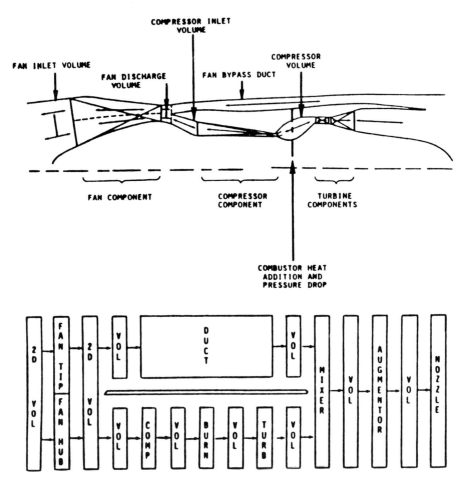

Fig. 6.21   Volume representation of turbofan engine for stall recovery models.

rotating stall *state* can be predicted since thermodynamics is independent of spatial dimension. Thus, there is no need to predict the circumferentially propagating nature of the rotating stall.

Some of the issues that should be addressed in constructing digital computer simulations with poststall capability are presented here. In one scheme of constructing this type of simulation, the basic aerothermodynamic representation of the engine components used in the steady-state cycle forms the starting place. The steady-state cycle is then combined with a transient cycle deck that provides the representations for rigid-rotor dynamics, the control, and the sensor and actuator dynamics. To this point, a standard path for constructing a transient cycle deck has been described. It is at this point that stall recovery technology is implemented. First, it must be decided the manner in which the volumes of the engine will be represented. For the case chosen for discussion and with reference to Fig. 6.21, two-dimensional volumes upstream and downstream of the fan are used that allow for axial and radial interchange of mass, momentum, and energy. The purpose of this type of representation is to allow the dynamics of a compressor surge to be simulated, that is, that compressor reverse flow can enter the fan bypass duct flow around the splitter and to allow the fan hub to surge and the hub-expelled gas to redistribute with some being pumped downstream by the fan tip. All volumes are sized to represent actual lengths and volumes of the engine. All volumes except those downstream of the turbines must permit reverse flow. The only other significant geometric feature to be considered is the modeling of the burner by dividing it into cold and hot volumes commensurate with the volumes outside and inside the burner liner. Burning and the burner pressure drop are assumed to occur at the interface between the cold and hot burner volumes.

The equations chosen to represent the flow in these volumes include the appropriate one- or two-dimensional conservation equations (mass, momentum, and energy), together with a thermodynamic relationship such as a Tds relation[42] to ensure consistency among the thermodynamic variables. It is the author's opinion that inclusion of the thermodynamic relation helps to meet the sufficiency condition for a set of equations since the conservation equations are necessary, but not sufficient in and of themselves.[43] Inclusion of the thermodynamic relation has been shown to avoid inconsistencies in time-dependent flows such as the local static pressure exceeding the local total pressure.

Additional care must be given to the representation of the compression components and the burner. For fans where the bypass ratio is approximately one or greater, hub and tip representations of the compression characteristics should be used. The compression characteristics of the fan hub, the fan tip, and the compressor should include rotating stall and reverse flow characteristics as well as the unstalled characteristic representations. While most investigators have included the hysteresis associated with the pressure coefficient representations or its equivalent, it has not been recognized until recently[44] that the work coefficient or its equivalent should also include a hysteresis effect.

Because under certain conditions the burner will operate fuel rich, the burner temperature rise characteristic must extend well beyond the stoichiometric fuel-air ratio to include the cooling effects seen in fuel-rich regimes. Additionally, the flammability limits of the combustor, especially the rich-side blowout limits, must be represented. While these statements are made in a straightforward manner, obtaining representations based on data is anything but straightforward and requires considerable engineering judgement.

If considerations such as these are addressed, then computer simulations with the requisite capabilities will be available to conduct the necessary recoverability studies. For further detailed discussions of these types of computer models and results, the reader is referred to Refs. 25, 27, 28, and 40.

6.10   SUMMARY

In this chapter, the major issues have been addressed that the engine operability engineer will have to face. References have been provided that should provide the more serious reader with entry to the literature of the discipline.

As the reader reflects on this chapter, it is hoped that he will have gained an appreciation of the many disciplines that contribute to successful inlet/engine integration and ones with which the operability engineer will have to become intimately familiar. These include inlet design, cycle design, performance analysis, transient analysis, controls design and analysis, fan and compressor aerodynamics, instrumentation design, test facilities and techniques, and data acquisition and reduction techniques. It is this myriad of disciplines along with the distinct features of engine operability that make and will continue to make engine operability a challenging field with still many unknowns—some of which have been identified in this chapter to serve as inspiration for a new generation of operability engineers.

References

[1]Alford, J. S., "Inlet Flow Distortion Index," paper presented at International Days of Aeronautical Sciences, ONERA, May 1957.

[2]Kimzey, W. F. and Lewis, R., "An Experimental Investigation of the Effects of Shock-Induced Turbulent In-Flow on a Turbojet Engine," AIAA Paper, June 1966.

[3]Plourde, G. A. and Brimelow, B., "Pressure Fluctuations Cause Compressor Instability," *Proceedings of the Air Force Airframe-Propulsion Compatibility Symposium*, Air Force Aero Propuslion Laboratory, Wright-Patterson AFB, OH, Rept. AFAPL-TR-69-103, June 1970.

[4]SAE Aerospace Recommended Practice, "Gas Turbine Engine Inlet Flow Distortion Guidelines," Society of Automotive Engineers, Warrendale, PA, ARP 1420, March 1978.

[5]SAE Aerospace Information Report, "Inlet Total-Pressure-Distortion Considerations for Gas Turbine Engines," Society of Automotive Engineers, Warrendale, PA, AIR 1419, May 1983.

[6]Oates, G. C. (ed.), *The Aerothermodynamics of Aircraft Gas Turbine Engines*, Air Force Aero Propulsion Laboratory, Wright-Patterson AFB, OH, Rept. AFAPL-TR-78-52, July 1978, Chap. 23.

[7]Hercock, R. G. and Williams, D. D., "Aerodynamic Response," *Distortion-Induced Engine Instability*, AGARD Lecture Series LS72, Paper No. 3, Nov. 1974.

[8]Williams, D. D. and Yost, J. O., "Some Aspects of Inlet/Engine Flow Compatibility," *Journal of the Royal Aeronautical Society*, 1973.

[9]Hercock, R. G., "Effects of Intake Flow Distortion on Engine Stability," *Engine Handling*, AGARD CP-324, Feb. 1983.

[10]Williams, D. D., "Review of Current Knowledge on Engine Response to Distorted Inflow Conditions," *Engine Response to Distorted Inflow Conditions*, AGARD CP-400, March 1987.

[11]Johnson, I. A. and Bullock, R. O., "Aerodynamic Design of Axial-Flow Compressors," NASA SP-36, 1965, pp. 315–316.

[12]Brimelow, B., Collins, T. P., and Pfefferkorn, G. A., "Engine Testing in a Dynamic Environment," AIAA Paper 74-1198, Oct. 1974.

[13]Moore, M. T., "Distortion Data Analysis," Air Force Aero Propulsion Laboratory, Wright-Patterson AFB, OH, Rept. AFAPL-TR-72-111, Feb. 1973.

[14]Cotter, H. N., "Integration of Inlet and Engine—an Engine Man's Point of View," SAE Paper 680286, April-May 1968.

[15]Reid, C., "The Response of Axial Flow Compressors to Intake Flow Distortion," ASME Paper 69-GT-29, March 1969.

[16]Jacocks, J. L., "Statistical Analysis of Distortion Factors," AIAA Paper 72-1100, Dec. 1972.

[17]Overall, B. W and Harper, R. E., "The Air Jet Distortion Generator System—A New Tool for Aircraft Turbine Engine Testing," AIAA Paper 77-993, July 1977.

[18]Van Deusen, E. A. and Mardoc, V. R., "Distortion and Turbulence Interaction, a Method for Evaluating Engine/Inlet Compatibility," AIAA Paper 70-632, June 1970.

[19]Borg, R., " A Synthesis Method for Estimating Maximum Instantaneous Inlet Distortion Based on Measured Inlet Steady State and RMS Pressures," *Aerodynamics of Power Plant Installations*, AGARD CP-301, Paper No. 19, Sept. 1981.

[20]Sedlock, D., "Improved Statistical Analysis Method for Prediction of Maximum Inlet Distortion," AIAA Paper 84-1274, June 1984.

[21]Biesiadny, T. J., Braithwaite, W. M., Soder, R. H., and Aboelwahab, M., "Summary of Investigations of Engine Response to Distorted Inlet Conditions," *Engine Response to Distorted Inflow Conditions*, AGARD CP-400, Paper No. 15, March 1987.

[22]MacMiller, C. J. and Haagenson, W. R., "Unsteady Inlet Distortion Characteristics with the B-1B," *Engine Response to Distorted Inflow Conditions*, AGARD CP-400, Paper No. 16, March 1987.

[23]Steenken, W. G., "Planar Wave Stability Margin Loss Methodology," AIAA Paper 88-3264, July 1988.

[24]Reynolds, G. G., Vier, W. F., and Collins, T. P., "An Experimental Evaluation of Unsteady Flow Effects on an Axial Compressor-P3 Generator Program," Air Force Aero Propulsion Laboratory, Wright-Patterson AFB, OH, Rept. AFAPL-TR-73-43, July 1973.

[25]Greitzer, E. M., "Surge and Rotating Stall in Axial Flow Compressors, Part 1: Theoretical Compression System Model," ASME Paper 75-GT-9, 1975.

[26]Greitzer, E. M., "Surge and Rotating Stall in Axial Flow Compressors, Part 2: Experimental Results and Comparison with Theory," ASME Paper 75-GE-10, 1975.

[27]Hosny, W. M., Bitter, S. J., and Steenken, W. G., "Turbofan Engine Nonrecoverable Stall Computer-Simulation Development and Validation," AIAA Paper 85-1432, July 1985.

[28]French, J. V., "Modelling Post-Stall Operation of Aircraft Gas Turbine Engines," AIAA Paper 85-1431, July 1985.

[29]Hopf, W. R. and Steenken, W. G., "Stall Recovery Control Strategy Methodology and Results," AIAA Paper 85-1433, July 1985.

[30]Pearson, H. and McKenzie, A. B., "Wakes in Axial Compressors," *Journal of the Royal Aeronautical Society*, Vol. 63, No. 58, July 1959, pp. 415–416.

[31]Tesch, W. A. and Steenken, W. G., "Blade Row Dynamic Digital Compressor Program, Vol. I, J85 Clean Inlet Flow and Parallel Compressor Models," NASA CR-134978, March 1976.

[32]Tesch, W. A. and Steenken, W. G., "Blade Row Dynamic Digital Compressor Program, Vol. II, J85 Circumferential Distortion Redistribution Model, Effect of Stator Characteristics, and Stage Characteristics Sensitivity Study," NASA CR-134953, July 1978.

[33]Mazzawy, R. S. and Banks, G. A., "Circumferential Distortion Modeling of the TF30-P3 Compression System," NASA CR-135124, Jan. 1977.

[34]Walter, W. A. and Shaw, M., "Distortion Analysis for F100 (3) Engine," NASA CR-159754, Jan. 1980.

[35]Chung, K., Leamy, K. R., and Collins, T. P., "A Turbine Engine Aerodynamic Model for In-Stall Transient Simulation," AIAA Paper 85-1429, July 1985.

[36]Davis M. W., Jr., "A Stage-by-Stage Dual-Spool Compression System Modeling Technique," ASME Paper 82-GT-184, March 1982.

[37]Tesch, W. A. and Steenken, W. G., "Dynamic Blade Row Compression Component Model for Stability Studies," AIAA Paper 76-203, Jan. 1976.

[38]Reynolds, G. G. and Steenken, W. G., "Dynamic Digital Blade Row Compression Component Stability Model, Model Validation and Analysis of Planar Pressure Pulse Generator and Two-Stage Fan Test Data," Air Force Aero Propulsion Laboratory, Wright-Patterson AFB, OH, Rept. AFAPL-TR-76-76.

[39]Baghdadi, S. and Lueke, J. E., "Compressor Stability Analysis," ASME, Paper 81-WA/FE-18, Nov. 1981.

[40]Steenken, W. G., "Turbofan Engine Post-Instability Behavior—Computer Simulations, Test Validation, and Application of Simulations," *Engine Response to Distorted Inflow Conditions*, AGARD CP-400, Paper No. 13, March 1987.

[41]Tesch, W. A., Moszee, R. H., and Steenken, W. G., "Linearized Blade Row Compression Component Model—Stability and Frequency Response Analyses of a J85-13 Compressor," NASA CR-135162, Sept. 1976.

[42]Obert, E. F., *Concepts of Thermodynamics*, McGraw-Hill, 1960, pp. 130–131.

[43]Steenken, W. G., "Modeling Compression Component Stability Characteristics—Effects of Inlet Distortion and Fan Bypass Duct Disturbances," *Engine Handling*, AGARD CP-324, Paper No. 24, Feb. 1983.

[44]Hosny, W. M. and Steenken, W. G., "Aerodynamic Instability Performance of an Advanced High-Pressure-Ratio Compression Component," AIAA Paper 86-1619, June 1986.

CHAPTER 7. AEROELASTICITY AND UNSTEADY AERODYNAMICS

Franklin O. Carta

United Technologies Research Center, East Hartford, Connecticut

7
AEROELASTICITY
AND UNSTEADY AERODYNAMICS

NOMENCLATURE

Note: Equation numbers refer to the defining relationship or to the first use of the given symbol. Multiple definitions are given for multiple use of symbols.

A	= disturbance amplitude function, Eq. (7.85) and (7.166), or speed of sound, Eq. (7.147)
A_h, A_α	= influence coefficients, aerodynamic lift per unit deflection, Eq. (7.21)
$\bar{A}, \bar{B}, \bar{D}, \bar{E},$	= coefficients in the equation of motion, Eq. (7.25)
A, B	= matrices in Eq. (7.140)
a	= dimensionless distance of elastic axis aft of midchord, semichords, Eq. (7.6), or speed of sound, ft/s, Eq. (7.80)
a, b	= cascade parameters, Eq. (7.127)
B_h, B_α	= influence coefficients, aerodynamic moment per unit deflection, Eq. (7.21)
b	= airfoil semichord, ft, Eq. (7.1)
C	= damping coefficient, lb-s, Eq. (7.194), or force coefficient matrix, Eq. (7.143)
C_c	= critical damping, lb-s, Eq. (7.44)
$C_{F_q}, C_{F_\alpha}, C_{F_w}$	= force coefficient due to given motion, Eq. (7.143)
C_M	= moment coefficient, Eq. (7.169)
$C_{M_q}, C_{M_\alpha}, C_{M_w}$	= moment coefficient due to given motion, Eq. (7.143)
C_m	= complex amplitude of the perturbation unsteady moment coefficient, Eq. (7.154)
$\tilde{C}_m$	= perturbation unsteady moment coefficient, Eq. (7.154)
C_N	= normal force coefficient, Fig. 7.34
C_p	= complex amplitude of the perturbation unsteady surface pressure coefficient, Eq. (7.152), or pressure coefficient, Eq. (7.220)
$\tilde{C}_p$	= perturbation unsteady surface pressure coefficient, Eq. (7.152)
C_w	= work coefficient, Eqs. (7.155) and (7.193)
C_x	= axial velocity, Fig. 7.51
C_α	= aerodynamic damping, lb-s, Eq. (7.42)
$C(k)$	= Theodorsen circulation function, Eq. (7.8)
c	= chord length, ft, Eq. (7.3)
D	= cascade parameter, Eq. (7.163)

D_S/Dt = mean flow convective derivative operator, Eq. (7.147)

$\mathrm{d}s$ = differential vector tangent to blade surface and directed counterclockwise, Eq. (7.154)

e_η = unit vector in the cascade "circumferential" or η direction, Eq. (7.148)

F = aerodynamic force per unit span, lb/ft, Eq. (7.118)

$F(k)$ = real part of Theodorsen function, Eq. (7.8)

F,G = compressible cascade functionals, Eq. (7.167)

f = frequency, cycle/s, Fig. 7.14

$G(k)$ = imaginary part of Theodorsen function, Eq. (7.8)

g = upper limit of integration, Eq. (7.89)

$H_0^{(2)}, H_1^{(2)}$ = Hankel functions, Eq. (7.9)

h = dimensionless bending deflection of elastic axis (or rotation point) in semichords, positive downward, Eq. (7.4)

$\mathrm{Im}\{\ \}$ = imaginary part of $\{\ \}$

$I_\alpha = \int r^2\,\mathrm{d}m$ = mass moment of inertia about elastic axis, slug-ft^2, Eq. (7.5)

$I_{0,n}^\pm$ = influence functions, Eq. (7.164)

K = kernel function, Eqs. (7.100) and (7.135), or semi-infinite series, Eq. (7.166)

K_0 = Possio kernel function, Eq. (7.171)

K_1 = kernel function, Eq. (7.172)

$\bar{K}_E$ = average system kinetic energy, ft-lb, Eq. (7.233)

K_h = bending spring stiffness, lb/ft, Eq. (7.4)

K_α = torsion spring stiffness, ft-lb, Eq. (7.5)

$k = b\omega/V$ = reduced frequency, Eq. (7.1)

k = compressible reduced frequency, Eq. (7.157)

$\bar{k}$ = function defined by Eq. (7.91)

L = lift, positive up, lb, Eq. (7.6), or linear difference operator, Eq. (7.150)

$\mathscr{L}$ = linear differential operator, Eq. (7.150)

L_h, L_α = lift influence functions, Eq. (7.18)

L_1, L_2 = lift functions, Eq. (7.49)

M = moment about elastic axis, positive nose up, ft-lb, Eq. (7.7), or Mach number, Eq. (7.80)

M_h, M_α = moment influence function, Eq. (7.18)

M_1, M_2, M_3 = moment functions, Eq. (7.30)

m = wing mass per unit span, slug, Eq. (7.4), or blade number, Eqs. (7.120) and (7.148)

N = rotor speed, rpm, Fig. 7.14; number of neighbor mesh points, Eq. (7.150); number of oscillations per revolution, Eq. (7.222); or number of nodal diameters, Eq. (7.243)

n = mesh point index, Eq. (7.150); blade number, Eq. (7.172); or number of blades in rotor, Eq. (7.243)

P = zeroth-order or steady pressure, Eq. (7.148), or modified relative pressure, Eq. (7.159)

$P_\mathscr{B}$	= linearized unsteady pressure at the instantaneous position of the reference blade surface, Eq. (7.152)
p	= pressure, lb/ft^2, Eq. (7.61), or complex amplitude of first-order or linearized unsteady pressure, Eq. (7.149)
Q_h	= generalized force in bending mode per unit span, lb, Eq. (7.4)
Q_n	= mesh point, $n = 0$ refers to calculation point, $n = 1,...,N$ refers to neighboring mesh points
Q_α	= generalized force in torsion mode per unit span, ft-lb, Eq. (7.5)
q	= dynamic pressure, lb/ft^2, Fig. 7.38 and Eq. (7.220), or blade vertical translation, ft/s, Eq. (7.137)
q^0	= multiplicative constant, Eq. (7.150)
R	= radius function, ft, Eq. (7.86), or dimensionless function, Eq. (7.175)
R_p	= position vector extending from reference blade axis of rotation to points on the reference blade surface, Eq. (7.154)
Re{ }	= real part of { }
r	= chordwise coordinate relative to elastic axis, positive aft, ft (used in definition of S_α below), or radius, ft, Eq. (7.234)
r, r_0	= radius functions, ft, Eqs. (7.173) and (7.174)
r_α	= radius of gyration about elastic axis, ft, Eq. (7.28)
$S_\alpha = \int r\, dm$	= static moment about elastic axis, slug-ft, Eq. (7.4)
s	= cascade gap, ft, Fig. 7.17, or Laplace transform variable, Eq. (7.105)
T	= disturbance period, s, Eq. (7.2), or time, s, Eq. (7.85)
t	= time, s, Eq. (7.10)
U	= freestream velocity, ft/s, Eq. (7.59), or wheel speed, Fig. 7.51
u	= perturbation velocity, ft/s, Eq. (7.112), or function defined by Eq. (7.101)
V	= freestream velocity, ft/s, Eq. (7.1); auxiliary function, Eq. (7.125); or normal velocity distribution, Eq. (7.166)
V	= zeroth-order or steady velocity, Eq. (7.145)
V_P	= propagation velocity, ft/s, Eq. (7.178)
v	= induced vertical velocity, ft/s, Eq. (7.122)
W	= work per cycle, ft-lb, Eq. (7.168)
W_1	= inlet relative velocity, ft/s, Eq. (7.223)
w	= downwash velocity, ft/s, Eq. (7.59), or transverse gust velocity, ft/s, Eq. (7.137)
w'	= function defined by Eq. (7.91)
$X = (\omega_\alpha/\omega)^2$	= frequency ratio used in flutter solution, Fig. 7.9
X	= position vector, Eq. (7.146)
X_α	= location of center of gravity relative to elastic axis, positive aft, ft, Eq. (7.27)

x	= dimensionless chordwise position in semichords, positive aft, Eq. (7.59)
x_A, y_A	= cascade parameters, Eq. (7.160)
x_n	= distance from blade leading edge to pitching axis, Fig. 7.24
z	= vertical coordinate, positive up, ft, Eq. (7.59)
α	= angular deflection about elastic axis (or rotation point), positive nose up, rad or deg, Eq. (7.4), or complex amplitude of the blade angular displacement, Eq. (7.151)
$\tilde{\alpha}$	= blade angular displacement, Eq. (7.151)
β	= Southwell stiffness parameter, Eq. (7.58); interblade phase angle, positive forward, deg or rad, Eq. (7.120); or compressible function $= \sqrt{1 - M^2}$, Eq. (7.170)
β_1^*	= cascade stagger angle, deg or rad, Fig. 7.27
Γ	= airfoil bound circulation, ft^2/s, Eq. (7.65), or circulation matrix, Eq. (7.140)
γ	= damping ratio, Eq. (7.45); vorticity distribution, ft/s, Eq. (7.60); or specific heat ratio, Eq. (7.145)
ΔC_p	= complex amplitude of the perturbation unsteady pressure difference coefficient, Eq. (7.153)
$\Delta \tilde{C}_p$	= perturbation unsteady pressure difference coefficient, Eq. (7.153)
δ	= logarithmic decrement, Eq. (7.233)
δ_n	= finite-difference coefficient, Eq. (7.150)
ϵ	= free vorticity, ft/s, Eq. (7.109), or small parameter following Eq. (7.146)
ζ	= vertical coordinate, ft, Eq. (7.85)
η	= chordwise coordinate, Eq. (7.122); dimensionless spanwise station, Eq. (7.234); or cascade "circumferential" coordinate, Fig. 7.20
θ	= phase angle between torsion and bending, rad, Eq. (7.11), or cascade stagger angle, deg or rad, Eq. (7.175) and Fig. 7.20
Λ_1	= auxiliary function, Eq. (7.75)
λ	= disturbance wavelength, ft, Eq. (7.2); reduced frequency based on full chord, Eq. (7.130); or function defined by Eq. (7.101), or normalized work, Eqs. (7.240–7.242)
λ, λ_0	= characteristic value of equation of motion and no flow value, s^{-1}, Eqs. (7.39) and (7.40)
μ	= mass parameter, Eq. (7.26); compressible function, $= \sqrt{M^2 - 1}$, Eq. (7.157); or real part of eigenvalue, Fig. 7.71
ν	= integer, Eq. (7.183), or imaginary part of eigenvalue, Fig. 7.71
Ξ	= dimensionless aerodynamic damping, Eqs. (7.155) and (7.207)

ζ	= aerodynamic damping, ft-lb, Eq. (7.203); cascade stagger angle, deg or rad, Fig. 7.17; or dimensionless chordwise coordinate in semichords, positive aft, Eq. (7.60)
ρ	= air density, slug/ft^3, Eq. (7.6)
$\bar{\rho}$	= zeroth-order or steady density, Eq. (7.145)
$\tilde{\rho}$	= nonlinear time-dependent fluid density, Eq. (7.144)
σ	= interblade phase angle, positive forward, deg or rad, Eq. (7.175)
τ	= cascade gap, ft, Eq. (7.170), Fig. 7.20
Φ	= zeroth-order or steady velocity potential, Eq. (7.146)
$\tilde{\Phi}$	= nonlinear time-dependent velocity potential, Eq. (7.144)
ϕ	= disturbance potential, ft^2/s, Eq. (7.102); velocity potential, Eq. (7.150); or complex amplitude of the linearized unsteady velocity potential, Eq. (7.147)
$\tilde{\phi}$	= first-order or linearized unsteady velocity potential, Eq. (7.146)
$\phi_{m,p,\Delta p}$	= phase angle by which a complex response vector (moment, pressure, pressure difference) leads the complex angular displacement vector, Eqs. (7.152), (7.153), and (7.154)
χ	= dimensionless chordwise position, Eq. (7.220)
ψ	= acceleration potential, (ft/s)^2, Eq. (7.80), or modified velocity potential, Eq. (7.158)
Ω	= rotor angular velocity, rad/s, Eq. (7.58), or interblade function, Eq. (7.161)
$\bar{\Omega}$	= auxiliary function, Eq. (7.70)
ω	= oscillatory frequency, rad/s, Eq. (7.1), or reduced frequency based on full chord, Eq. (7.157)
ω_h,ω_α	= natural modal frequencies, rad/s, Eqs. (7.14) and (7.15)
ω_0	= no flow frequency, rad/s, Eq. (7.40)
ω_1,ω_2	= frequencies used in Southwell theorem, rad/s, Eq. (7.57)
$\bar{\omega}$	= compressible reduced frequency, Eq. (7.107)

Subscripts

a	= on airfoil, Eq. (7.60)
B	= due to bending, Eq. (7.235)
C	= due to coupling, Eq. (7.235)
D	= doublet, Eq. (7.92)
I	= imaginary part, Eq. (7.188)
m	= pertaining to mth blade, Eq. (7.120), or at mean rotor passage height, Fig. 7.52
MCL	= referred to mean camber line, Fig. 7.41
n	= mesh point index, Eq. (7.150)
0	= root value, Eq. (7.234)

P	= due to pitch, Eq. (7.235)
q	= due to blade vertical translation, Eq. (7.138)
R	= real part, Eq. (7.186)
S	= source, Eq. (7.88)
SP	= source pulse, Eq. (7.85)
T, TIP	= tip value, Eqs. (7.232) and (7.239)
TOT	= total, Eq. (7.235)
t	= partial derivative with respect to time, Eq. (7.144)
w	= in wake, Eq. (7.60), or due to transverse gust, Eq. (7.138)
α	= due to pitch, Eq. (7.138)
1	= uniform inlet flow condition, Eq. (7.145)
$+$	= just above plane or airfoil, Eq. (7.113)
$-$	= just below plane or airfoil, Eq. (7.113)
∞	= freestream value, Eq. (7.81)

Superscripts

*	= transformed quantity, Eq. (7.105), or dimensionless coordinate, Eq. (7.106)
$(\dot{\ })$, $(\ddot{\ })$	= first and second derivative with respect to time, Eq. (7.4)
$(\bar{\ })$	= amplitude, Eq. (7.10)

Operators

D( )	= substantial derivative, Eq. (7.82)
Δ( )	= difference, upper minus lower, Eq. (7.61)
∇	= gradient operator, Eq. (7.144)
∇^2	= Laplacian operator, Eq. (7.105)
L	= linear difference operator, Eq. (7.150)
$\mathscr{L}$	= linear differential operator, Eq. (7.150)
$\mathscr{L}$	= Laplace transform operator, Eq. (7.105)
$\oint$	= cyclic integral, Eq. (7.186)
$\int$	= Cauchy principal value, Eq. (7.60)

7.1 INTRODUCTION

The object of this chapter is to acquaint the reader with several turbomachinery flutter problems confronting the design engineer and to suggest some of the ways in which the designer can cope with these problems. It is obvious that an in-depth examination of the problem is beyond the scope of this single chapter, and the reader will be directed to the available literature for more details. Nevertheless, an attempt will be made to approach this exposition from the point of view of the designer, so the application of theoretical or empirical methods to overcome design barriers will be stressed wherever possible. To achieve this end, the problem area will first be defined, and following this the reader will be exposed to the basic differences between wing flutter and turbomachinery blade flutter. This will be helpful in establishing the areas in which elementary theories can be used to explain basic principles and will also emphasize the need for advanced theories. Next, a brief discussion of the unsteady aerodynamics of isolated airfoils will lead into the more complicated problems involving cascaded airfoils, and some time will be spent dealing with the dynamic stall problem. Finally, in the section devoted to stability theory, the design methodology for avoiding flutter will be discussed. Here, the introduction of advanced unsteady aerodynamic theories into a basic stability theory involving energy balance will be shown to provide the designer with some of the necessary constraints for flutter-free operation.

It should be emphasized that flutter has never been a "trivial" problem. Several instances of aircraft flutter have been documented in the first chapter of Ref. 1, with the earliest known instance occurring during World War I, indicating that flutter is not a newcomer to the hierarchy of aerodynamic problems. Nor should its severity be underestimated, in view of two known instances of catastrophic wing failures due to flutter on passenger aircraft (c. 1960). Turbomachinery flutter is not as well documented as aircraft lifting surface flutter, but as noted in the following historical overview, has been known to exist since the initial development of the turbomachine. Furthermore, the potential for damage to the aircraft and/or engine is very real and ranges from individual blade failures to the destruction of several stages (if the blades can be contained within the case) and to the ultimate destruction of the engine support structure or the aircraft itself (if the blades cannot be contained or if the unbalanced rotating mass cannot be restrained by the surrounding structure). Clearly, then, this is a problem to be avoided at all cost, and it is to this end that the following sections are directed.

7.2 OVERVIEW OF TURBOMACHINERY FLUTTER

Definitions and Concepts

In general, flutter is the name given to any self-excited oscillation of an aerodynamically lifting surface. The phenomenon is characterized by two necessary requirements: an available energy supply (i.e., the moving airstream) and a zero or negatively damped system. Here, the word "system" applies to the combination of the lifting surface *and* the airstream. By itself, the lifting surface is a system with positive damping. The inclusion of the airstream will either increase or decrease the damping. The mechanism whereby the airstream changes the damping will be discussed in detail below.

First, however, it should be noted that there are many different types of flutter which affect the many different types of lifting surfaces. To name but a few, there are: (1) classical coupled flutter of an aircraft wing or of a wing/aileron combination; (2) single-degree-of-freedom stall flutter of compressor blades, propeller blades, or helicopter rotor blades; (3) coupled flutter of shrouded compressor blades. These few examples, plus all of the other possible types of flutter, share one thing in common—the joint requirement of an available energy supply and a zero or negatively damped system. However, there are many more differences than similarities and a different technique is often required to deal with each of the various types of flutter.

Also shared by all flutter phenomena is a dimensionless parameter called the reduced frequency,

$$k = b\omega/V \tag{7.1}$$

or its inverse, the reduced velocity, $V/b\omega$. Here, b is the blade semichord, ω the frequency of the oscillation, and V the freestream velocity. As will be seen in the work that follows, this parameter appears both explicitly and implicitly in the governing equations for unsteady aerodynamics of oscillating airfoils and also plays an extremely important role in the empirical relationships associated with dynamic stall and stall flutter. Therefore, it is appropriate to begin this chapter with a brief discussion of the physical meaning of k. Consider an airfoil of chord length $2b$ oscillating at a frequency of $\omega = 2\pi f = 2\pi/T$ in a stream moving past it at a velocity V, as shown in Fig. 7.1. A sinusoidal wake will be formed that is imbedded in the freestream and hence also moves relative to the airfoil at a velocity V, with wavelength

$$\lambda = VT = \frac{2\pi V}{\omega} \tag{7.2}$$

If we divide the airfoil chord by this wavelength, we obtain

$$\frac{2b}{\lambda} = \frac{2b\omega}{2\pi V} = \frac{k}{\pi}$$

or

$$k = \pi \left(\frac{2b}{\lambda}\right) = \frac{\pi c}{\lambda} \tag{7.3}$$

Thus, at low reduced frequency ($k \cong 0.05$) the wavelength is very large relative to the chord, while at high reduced frequency ($k \cong 1$) the wavelength is not so large relative to the chord. (Here we cannot talk about the wavelength small relative to the chord because, even at the largest values of reduced frequency encountered in practice, the wavelength is still three times the blade chord.)

Fig. 7.1   Disturbance wavelength of an oscillating airfoil.

Historical Sketch

In the early days of compressor technology, bending flutter near the surge line was a primary design problem. Because it occurred near stall, it was universally referred to as stalling flutter and it was understood that the phenomenon occurred in the first bending or flexural mode of vibration. Pearson[2] specifically states that stalling flutter in the torsion mode was not a problem. A great deal of time and effort was devoted to this problem and a large number of reports were written by British authors and others (cf. Bibliography). Most of these studies dealt with the problem phenomenologically, although a few attempted to explain the physical nature of stalling flutter.

As early as 1945, Shannon[3] found that torsional stall flutter of compressor blades could occur for values of the reduced frequency $k = b\omega_\alpha / V < 0.75$, based on blade semichord. (This is equivalent to a reduced velocity $V/b\omega_\alpha > 1.33$.) However, as stated above, the primary flutter problem in these early days of compressor technology occurred in the bending mode. More recently, however, the problem of flutter near stall has shifted from the bending mode to the torsion mode. Most of this newer work has been done privately and hence has gone largely unreported. In 1960, Armstrong and Stevenson[4] corroborated Shannon's results and cited a critical value of $k < 0.8$ for torsional stall flutter; i.e., if $k < 0.8$ or if $V/b\omega_\alpha > 1.25$, then torsional stall flutter is possible. Sisto and Ni[5] attribute this shift from bending flutter to torsion flutter to the progressive changes that have occurred in the mechanical and aerodynamic design of compressor and fan blades. They state that "the trend toward transonic stages and supersonic blade tips has been at the expense of blade thickness and has been accompanied by a rearward movement of the center of twist along a typical blade chordline. For example, double circular arc airfoil profiles have the elastic axis and center of twist both at midchord as opposed to more forward locations for the earlier 65-series profiles. Supersonic sections of the so-called parabolic or 'J section' variety have even more rearward twist centers. Higher stagger angles and increased solidities have been other concomitant parametric trends that may have aggravated the current susceptibility to severe torsional stalling flutter."

The several flutter phenomena that are most likely to occur in turbomachines are discussed below.

7.3   BRIEF SURVEY OF TURBOMACHINERY FLUTTER REGIMES

A typical plot showing the flutter regions on a compressor performance map is shown in Fig. 7.2. The subsonic stall flutter region, schematically shown near the surge line at part speed, can be regarded as representing either the bending mode or the torsion mode. An equivalent plot of the flutter region on an incidence, Mach number diagram is shown in Fig. 7.3. Superimposed on this plot are two possible operating lines for the compressor. At idle the incidence angle is high and the relative Mach number is low, while at full speed the incidence angle is low and the Mach number is high.

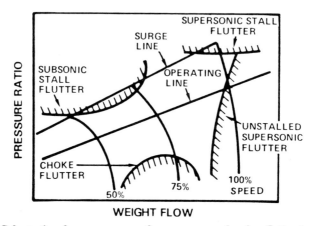

Fig. 7.2   Schematic of compressor performance map showing flutter boundaries for four types of flutter.

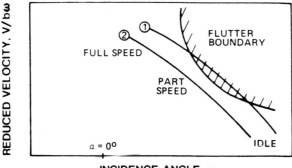

Fig. 7.3   Schematic of stall flutter map with hypothetical operating lines.

Thus, the operating line traverses the plot from lower right to upper left. It is seen that operating line 1 passes through the flutter region at part speed. A redesign, either by stiffening the blades (increasing $b\omega$) or reducing the incidence, has the effect of moving the operating line down and/or to the left relative to the flutter region and operating line 2 is seen to be flutter free.

Choking flutter in the bending mode is a phenomenon that has received relatively little attention in the published literature. It is mentioned briefly in Refs. 6 and 7 and its position on the performance map is also shown schematically in Fig. 7.2. It usually occurs at negative incidence angles at a part speed condition. The difference in the flow conditions for stall and choke are shown schematically in Fig. 7.4. It is seen that in the stall condition the inlet flow can expand through a larger in-passage stream area, but in the choked condition the inlet flow is constrained to pass through a smaller in-passage stream area. Thus, it is possible to produce higher in-passage Mach numbers, sometimes exceeding local critical values, than the external velocities would indicate. When this happens, in-passage shocks can form which can either cause flow separation or can directly couple adjacent blades. In either case, large amplitude oscillations are then possible.

In Ref. 6, Carter states that choking flutter in compressors is not likely to be a problem because the operating line will not approach the choke region. This statement was written in the early 1950's, prior to the use of variable geometry (i.e., movable stators) in modern compressors. Thus, although one would not deliberately run an engine into the choke region, it is possible that an inadvertent malfunction of the stator control might cause such an encounter. It is also possible that choking flutter might be experienced during prototype testing, when the effects of various stator positions were being explored.

In view of the meager information available on this subject, it will not be discussed at length in any subsequent section of this chapter.

It will be shown below that a single compressor blade with well-separated frequencies in the fundamental torsion and bending modes will not be susceptible to the classical coupled flutter sometimes encountered on aircraft wings. This has always been true for practical, unshrouded compressor blades mounted on reasonably stiff disks. However, with the advent of fan stages and necessarily longer blades, frequencies in the two fundamental modes were decreased. This had the effect of increasing the reduced velocity parameter $V/b\omega$, which is equivalent to shifting from operating line 2 to operating line 1 in Fig. 7.3. Part-span shrouds were then employed in the United States (snubbers or clappers in Great Britain) to introduce additional bending and torsion restraints, which increased the single-degree-of-freedom mode frequencies. Thus, the system was stiffened and the operating line was again shifted from position 1 to position 2, away from single-degree-of-freedom flutter.

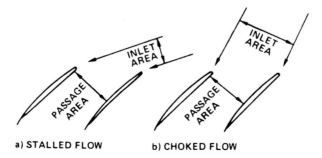

a) STALLED FLOW   b) CHOKED FLOW

Fig. 7.4 Differences between stalled flow and choked flow through a cascade (schematic).

Unfortunately, the additional part-span constraint also provided an additional degree of freedom to the structure and a pseudodisk mode was introduced that carried the blades in a combined bending-torsion mode, automatically coupled, which could extract energy from the airstream and sustain a self-excited flutter. This phenomenon was first described in Ref. 8 and a detailed analysis will be presented in a later section of this chapter. A description of the structural dynamics of the blade-disk-shroud modes will be found in Ref. 9. Fundamental information on the structural dynamics problem can be found in the pioneering work of Campbell or in the later analyses of Ehrich et al., and Ewins (cf. Bibliography). The region of occurrence of this phenomenon cannot be precisely defined on the compressor map and, hence, is not indicated on Fig. 7.2.

In recent years supersonic unstalled flutter has become an increasingly important factor in the design of high-performance turbomachinery fan blades. This flutter occurs at a supersonic relative flow, but at a subsonic axial flow; i.e., the vector sum of the subsonic axial velocity and the rotational velocity is a supersonic relative velocity (see Fig. 7.5). Only a small portion of the blade row has been schematically shown in Fig. 7.5 to illustrate the complexity of the flow geometry. It is seen that because the axial flow Mach number is less than 1, Mach waves can propagate forward of the plane of rotation (referred to here as the leading-edge locus). If the blades are fluttering, the Mach wave positions and strengths will become

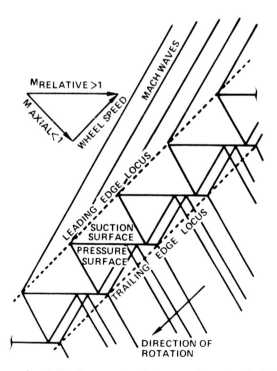

Fig. 7.5 Supersonic relative flow past a blade row with subsonic leading-edge locus.

variable and constitute an unsteady energy path ahead of the adjacent blades in the rotor. It is seen that the internal flow geometry is also very complicated, involving several internal Mach wave reflections and also the influence of a portion of the unsteady wake of any blade on the upper adjacent blade near the trailing edge.

The problem has manifested itself largely in thin fan blades operating at supersonic relative flow velocities at the tip. It generally occurs in the vicinity of the operating line or below, so it is not generally categorized as a stall flutter. Furthermore, its intensity, which can be catastrophically high, diminishes rapidly as one traverses a constant-speed line toward higher pressure ratio, as shown in the boundary of Fig. 7.2. A more comprehensive description of this phenomenon can be found in Ref. 10. Here, it is further noted that the flutter boundary sets a real limit on high-speed operation and that, at high stress levels, all blades tend to flutter at a common frequency with a constant interblade phase angle between blades. Thus, the phenomenon is amenable to analysis, using the coupled flutter energy method of Ref. 8 and the more advanced supersonic aerodynamic analysis of Refs. 11 and/or 12. This has been successfully implemented as part of the design system and, as stated in Ref. 10, has provided an accurate flutter prediction of the unstalled supersonic flutter boundary for high-speed rotor blades.

It was stated earlier that supersonic unstalled flutter can be diminished and even eliminated by increasing the pressure ratio while traversing a constant-speed line. However, this relief has its limitations because high-speed operation near the surge line can lead to a supersonic stalled flutter, as shown by the boundary in Fig. 7.2. Because this flutter is associated with high pressure ratios, there can be strong shocks present within the blade passages and, in some circumstances, ahead of the blade row. Hence, supersonic stalled flutter is not amenable to analysis, and no headway has been made to provide a reliable prediction technique for use in a compressor design system. To date, only limited data are available for empirical predictions.

7.4 ELEMENTARY CONSIDERATIONS OF AIRCRAFT WING FLUTTER

In pursuing our objective here of dealing with turbomachinery flutter, it is advantageous to discuss first the elementary flutter theory for aircraft wings developed by Theodorsen.[13] Once these elementary considerations are understood, it will be possible to point out the similarities where they exist and to identify the origins of the differences between aircraft wing flutter and the various types of turbomachinery flutter to be discussed below.

Equations of Motion

Before launching into the discussion of the equations of motion of an oscillating airfoil, it is useful to review briefly the method of approach to be used in this section. We shall begin with the generalized equations of motion, using unsteady aerodynamic forcing functions on the right-hand side that have not yet been derived (but will be derived in a subsequent section). The utility of this approach will lie in the use of specific aerodynamic functions to show the general form of the right-hand side and to use this general form to investigate the underlying physics of the problem at a relatively early point in the development. Later, in subsequent sections, the several relevant unsteady aerodynamic theories will be derived and their use in solving turbomachinery-related flutter problems will be discussed.

The equations of motion of a classical, two-dimensional flat-plate airfoil that is executing a combined plunging and pitching motion in a moving airstream (cf. Fig. 7.6) are given by

$$mb\ddot{h} + S_\alpha\ddot{\alpha} + K_h bh = Q_h \tag{7.4}$$

$$S_\alpha b\ddot{h} + I_\alpha\ddot{\alpha} + K_\alpha\alpha = Q_\alpha \tag{7.5}$$

where h and α are the dimensionless displacements in plunge and pitch. (See the Nomenclature for the definitions of the quantities used herein and Ref. 1 or the texts listed in the Bibliography for further details of the concepts discussed here.)

There are many possible ways to express the quantities Q_h and Q_α on the right-hand sides of Eqs. (7.4) and (7.5). For the simple case being considered here, it is easily shown[13] that the generalized forces Q_h and Q_α are simply the two-dimensional unsteady aerodynamic lift and moment and are expressed as

$$Q_h = L = -\pi\rho b^3 \left\{ -\ddot{h} - \frac{2V}{b} C(k)\dot{h} + a\ddot{\alpha} \right.$$

$$\left. - \left[1 - 2\left(a - \frac{1}{2}\right)C(k)\right]\frac{V}{b}\dot{\alpha} - \frac{2V^2}{b^2}C(k)\alpha \right\} \tag{7.6}$$

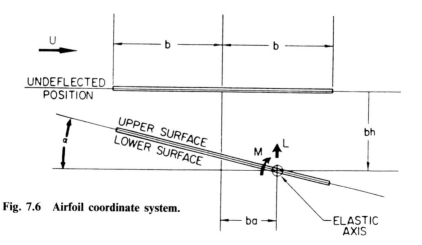

Fig. 7.6 Airfoil coordinate system.

$$Q_\alpha = M = \pi \rho b^4 \left\{ a\ddot{h} + 2 \left(\frac{1}{2} + a \right) C(k) \frac{V}{b} \dot{h} \right.$$

$$\left. - \left(\frac{1}{8} + a^2 \right) \ddot{\alpha} + \left[a - \frac{1}{2} + 2 \left(\frac{1}{4} - a^2 \right) C(k) \right] \frac{V}{b} \dot{\alpha} + 2 \left(\frac{1}{2} + a \right) C(k) \frac{V^2}{b^2} \alpha \right\}$$

$$(7.7)$$

Thus, it is seen that the right-hand sides of Eqs. (7.4) and (7.5) are also dependent on the displacements and their derivatives. As a first approximation, it is possible to identify the coefficients of the second derivatives as aerodynamic inertia or mass terms, the coefficients of the first derivatives as aerodynamic damping terms, and the coefficient of the angular displacement as an aerodynamic stiffness term. Actually, this is not strictly true because the complex Theodorsen function

$$C(k) = F(k) + iG(k) \qquad (7.8)$$

enters into certain terms of the right-hand side and complicates the situation. Nevertheless, the unsteady aerodynamic terms directly affect virtually all of the terms in the equations of motion. It will be shown below that this is accomplished through changes in the frequencies of the natural system modes that permit the two modes to couple, producing flutter.

Theodorsen Theory

In 1935, Theodorsen[13] published his now famous theory that laid the foundations for modern flutter theory. The key to this theoretical development lay in the way in which vorticity was shed into the moving wake to

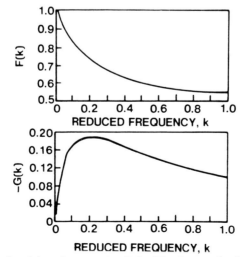

Fig. 7.7   Real and imaginary parts of the Theodorsen circulation function.

account for changes in the bound vorticity caused by a sinusoidal variation in either plunging or pitching displacement. In the course of his mathematical development (independently verified by Schwarz[14]), Theodorsen derived a complex circulation function, $C(k)$, which is defined above in Eq. (7.8) in terms of its real and imaginary parts and expressible as a ratio of Hankel functions,

$$C(k) = \frac{H_1^{(2)}(k)}{H_1^{(2)}(k) + iH_0^{(2)}(k)} \tag{7.9}$$

The real and imaginary parts, $F(k)$ and $G(k)$, are plotted vs k in Fig. 7.7 and a phase plane diagram of the entire function is shown in Fig. 7.8.

Theodorsen's method of solution was to assume a coupled sinusoidal motion of the system at a common (but unknown) frequency, say ω, and then seek the values of V and ω that satisfied the equations of motion [Eqs. (7.4) and (7.5)]. Specifically, he assumed that

$$h = \bar{h}e^{i\omega t} \tag{7.10}$$

and

$$\alpha = \bar{\alpha}e^{i(\omega t + \theta)} \tag{7.11}$$

where θ is the phase angle between the motions. When these quantities were inserted into the left-hand sides of Eqs. (7.4) and (7.5), he obtained

$$-\omega^2 mbh - \omega^2 S_\alpha \alpha + \omega_h^2 mbh = Q_h = L \tag{7.12}$$

$$-\omega^2 S_\alpha bh - \omega^2 I_\alpha \alpha + \omega_\alpha^2 I_\alpha \alpha = Q_\alpha = M \tag{7.13}$$

after previously relating the stiffness to the natural frequencies of the separate modes,

$$K_h = \omega_h^2 m \tag{7.14}$$

$$K_\alpha = \omega_\alpha^2 I_\alpha \tag{7.15}$$

A similar substitution of Eqs. (7.10) and (7.11) into the right-hand sides of Eqs. (7.6) and (7.7) and a great deal of algebraic manipulation yields the following results:

$$L = -\pi\rho b^3 \omega^2 \left\{ L_h h + \left[L_\alpha - \left(\frac{1}{2} + a\right) L_h \right] \alpha \right\} \tag{7.16}$$

$$M = \pi \rho b^4 \omega^2 \left\{ \left[M_h - \left(\frac{1}{2} + a\right) L_h \right] h \right.$$

$$\left. + \left[M_\alpha - \left(\frac{1}{2} + a\right)(M_h + L_\alpha) + \left(\frac{1}{2} + a\right)^2 L_h \right] \alpha \right\} \tag{7.17}$$

where

$$L_h = 1 - \frac{2i}{k} C(k), \qquad L_\alpha = \frac{1}{2} - \frac{i}{k}[1 + 2C(k)] - \frac{2}{k^2} C(k)$$

$$M_h = \tfrac{1}{2}, \qquad M_\alpha = \tfrac{3}{8} - (i/k) \tag{7.18}$$

The quantities L_h, L_α, M_h, and M_α are taken from the report by Smilg and Wasserman[15] in which the unsteady aerodynamic quantities are referred to the aerodynamic center of the airfoil at the quarter chord. When the pivot axis is also at this point, than $a = -\frac{1}{2}$ and Eqs. (7.16) and (7.17) are greatly simplified. Note that in Ref. 15 the positive lift and vertical translation are both directed downward, whereas in the present instance only the vertical displacement is positive downward. This accounts for the negative algebraic sign on the right-hand side of Eq. (7.16).

For simplicity in the present application, rewrite Eqs. (7.16) and (7.17) in influence coefficient form,

$$L = \pi \rho b^3 \omega^2 (A_h h + A_\alpha \alpha) \tag{7.19}$$

$$M = \pi \rho b^4 \omega^2 (B_h h + B_\alpha \alpha) \tag{7.20}$$

where all of the unsteady quantities have been gathered together into four quantities,

$$A_h = -L_h, \qquad A_\alpha = -L_\alpha + (\tfrac{1}{2} + a)L_h$$

$$B_h = M_h - (\tfrac{1}{2} + a)L_h$$

$$B_\alpha = M_\alpha - (\tfrac{1}{2} + a)(M_h + L_\alpha) + (\tfrac{1}{2} + a)^2 L_h \tag{7.21}$$

Equations (7.19) and (7.20) are substituted into Eqs. (7.12) and (7.13), and after considerable algebraic manipulation we obtain

$$\left[\frac{m}{\pi \rho b^2} \left(1 - \frac{\omega_h^2}{\omega^2}\right) + A_h \right] h + \left(\frac{S_\alpha}{\pi \rho b^3} + A_\alpha \right) \alpha = 0 \tag{7.22}$$

$$\left(\frac{S_\alpha}{\pi \rho b^3} + B_h \right) h + \left[\frac{I_\alpha}{\pi \rho b^4} \left(1 - \frac{\omega_\alpha^2}{\omega^2}\right) + B_\alpha \right] \alpha = 0 \tag{7.23}$$

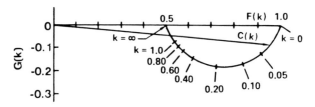

Fig. 7.8   Complex phase plane diagram of the Theodorsen circulation function.

or, more simply, using the notation of Ref. 16,

$$\bar{A}h + \bar{B}\alpha = 0 \quad \text{and} \quad \bar{D}h + \bar{E}\alpha = 0 \tag{7.24}$$

where

$$\bar{A} = \mu \left[1 - \left(\frac{\omega_h}{\omega_\alpha}\right)^2 \left(\frac{\omega_\alpha}{\omega}\right)^2 \right] + A_h$$

$$\bar{B} = \mu X_\alpha + A_\alpha$$

$$\bar{D} = \mu X_\alpha + B_h$$

$$\bar{E} = \mu r_\alpha^2 \left[1 - \left(\frac{\omega_\alpha}{\omega}\right)^2 \right] + B_\alpha \tag{7.25}$$

and where μ is the mass parameter,

$$\mu = m/\pi\rho b^2 \tag{7.26}$$

X_α is the center-of-gravity (c.g.) position aft of the elastic axis,

$$X_\alpha = S_\alpha/mb \tag{7.27}$$

and r_α is the radius of gyration of the airfoil about the elastic axis,

$$r_\alpha^2 = I_\alpha/mb^2 \tag{7.28}$$

Coupled flutter occurs when the determinant of the complex coefficients in Eq. (7.24) vanishes, as

$$\begin{vmatrix} \bar{A} & \bar{B} \\ \bar{D} & \bar{E} \end{vmatrix} = 0 \tag{7.29}$$

This equation must be satisfied separately by the vanishing of the real and imaginary parts. These two equations are all that are needed to form a

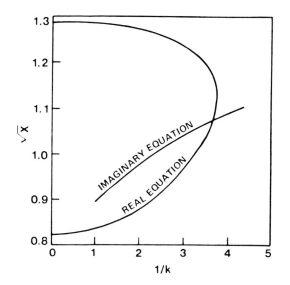

Fig. 7.9   Typical flutter solution using Theodorsen's method.

determinate set, because the only unknowns in the system are the reduced frequency $k = b\omega/V$ and the frequency ratio $\sqrt{X} = \omega_\alpha/\omega$. The procedure used by Theodorsen was to select a series of values of k, substitute them into the real and imaginary solutions of Eq. (7.29), and solve for the values of $\sqrt{X}$ that constituted the real and imaginary roots of the determinant. The intersection of the real and imaginary branches of the curves was the flutter point. This is schematically shown in Fig. 7.9.

Simplified Approach to Flutter

Theodorsen's theory has been in use now for more than 50 years and it has proved to be a very durable theory that has spawned many modifications and extensions. Within the limits of the assumptions made by Theodorsen, it has proved to be reasonably accurate. It has been found to be amenable to solution with a hand calculator, although moderately time-consuming, and it has been possible to automate it without difficulty for solution on high-speed digital computers. However, one over-riding fact becomes apparent to anyone who uses this theory. It is virtually impossible to derive any physical insight into the flutter mechanism from its direct application.

Fortunately, it is possible to simplify the unsteady aerodynamic quantities to varying degrees and obtain flutter solutions that, although somewhat inaccurate, possess sufficient simplicity to reveal much of the underlying physics of the instability. A particularly lucid survey of a number of simple applications is found on pp. 258–277 of Ref. 17. In the present discussion, we shall take the simplest possible approach in an effort to focus on the specific items that constitute the major differences between aircraft wing flutter and compressor blade flutter. In this section, we shall concentrate on developing the necessary formulas and facts, while the comparison between wing and compressor flutter will be deferred to a subsequent section.

In this very simple approach, consider a single-degree-of-freedom system capable of torsional motion only. Actually, this is equivalent to a study of the torsional mode of a binary system with well separated modes. To this end, combine Eqs. (7.5) and (7.7) and suppress the bending mode. The resulting equation is given by

$$I_\alpha \ddot{\alpha} + K_\alpha \alpha = M_1 \ddot{\alpha} + M_2 \dot{\alpha} + M_3 \alpha \tag{7.30}$$

where

$$M_1 = -\pi \rho b^4 (\tfrac{1}{8} + a^2) \tag{7.31}$$

$$M_2 = \pi \rho b^4 \left[a - \frac{1}{2} + 2 \left(\frac{1}{4} - a^2 \right) C(k) \right] \frac{V}{b} \tag{7.32}$$

$$M_3 = 2\pi \rho b^4 \left(\frac{1}{2} + a \right) C(k) \frac{V^2}{b^2} \tag{7.33}$$

We shall continue to simplify by assuming that k is very small and therefore the system can be represented by the quasisteady approximation in which the normally complex circulation function goes asymptotically to

$$\lim_{k \to 0} C(k) = 1 + i \times 0 \tag{7.34}$$

With this additional simplification, M_1 remains the same, but M_2 and M_3 are approximated by

$$M_2 \cong \pi \rho b^4 a (1 - 2a) \frac{V}{b} \tag{7.35}$$

$$M_3 \cong 2\pi \rho b^4 \left(\frac{1}{2} + a \right) \frac{V^2}{b^2} \tag{7.36}$$

which are both real. Thus, Eq. (7.30) becomes a simple, second-order differential equation with constant coefficients,

$$(I_\alpha - M_1)\ddot{\alpha} - M_2 \dot{\alpha} + (K_\alpha - M_3)\alpha = 0 \tag{7.37}$$

whose solution is

$$\alpha = \bar{\alpha} e^{\lambda t} \tag{7.38}$$

where

$$\lambda = \frac{M_2}{2(I_\alpha - M_1)} \pm i \sqrt{\frac{K_\alpha - M_3}{I_\alpha - M_1} - \left[\frac{M_2}{2(I_\alpha - M_1)}\right]^2} \qquad (7.39)$$

First consider the zero velocity limit. (Strictly speaking, this is an invalid limit for this equation since we have already assumed that $k = b\omega/V$ is vanishingly small. It is obvious that this assumption is incompatible with the zero velocity limit. However, our physical understanding will not be altered in this limit because the same terms will vanish here as would vanish in the case of nonvanishing k.) When $V \to 0$, both M_2 and M_3 vanish, so Eq. (7.39) reduces to

$$\lambda_0 = \pm i \sqrt{\frac{K_\alpha}{I_\alpha - M_1}} = \pm i\omega_{\alpha_0} \qquad (7.40)$$

where ω_{α_0} is the natural torsional frequency of the system in still air. When $V \neq 0$, Eq. (7.39) can be rewritten as

$$\lambda = -\frac{C_\alpha}{2(I_\alpha - M_1)} \pm i \sqrt{\omega_\alpha^2 - \left[\frac{C_\alpha}{2(I_\alpha - M_1)}\right]^2} \qquad (7.41)$$

where

$$C_\alpha = -M_2 = \pi\rho b^4 a(2a - 1)\frac{V}{b} \qquad (7.42)$$

is the aerodynamic damping and where

$$\omega_\alpha^2 = \frac{K_\alpha - M_3}{I_\alpha - M_1} = \omega_{\alpha_0}^2 - \frac{M_3}{I_\alpha - M_1} \qquad (7.43)$$

is the square of the natural wind-on frequency of the system. In the second form of Eq. (7.43), ω_{α_0} has been taken from Eq. (7.40) and in Eq. (7.42) the new definition of aerodynamic damping C_α has been introduced as a positive quantity for convenience. Actually, the value of C_α is unimportant here in our attempt to gain physical understanding. The more exact value of the aerodynamic damping in torsion would be a strong function of the circulation function $C(k)$, which has been neglected in the present analysis. Furthermore, the system damping for the coupled modes would be all-important in determining the instability point and this is beyond the scope of the present simplified analysis.

In the equations just derived, $I_\alpha - M_1$ is the effective inertia in pitch, including virtual mass effects. It is customary (cf., Ref. 16, p. 69) to define critical damping in terms of system mass or inertia as

$$C_c = 2(I_\alpha - M_1)\omega_\alpha \qquad (7.44)$$

and to define the damping ratio as

$$\gamma = \frac{C_\alpha}{C_c} = \frac{C_\alpha}{2(I_\alpha - M_1)\omega_\alpha} \qquad (7.45)$$

Then, Eq. (7.41) becomes

$$\lambda = -\gamma\omega_\alpha \pm i\omega_\alpha \sqrt{1 - \gamma^2} \qquad (7.46)$$

which, when substituted into Eq. (7.38), represents a damped, oscillatory pitching motion.

The most important thing here is the definition of ω_α given above in Eq. (7.43). When M_1 and M_3 are inserted from Eqs. (7.31) and (7.36), this becomes

$$\omega_\alpha^2 = \omega_{\alpha_0}^2 - \frac{2\pi\rho b^4(\frac{1}{2} + a)V^2/b^2}{I_\alpha + \pi\rho b^4(\frac{1}{8} + a^2)} \qquad (7.47)$$

For our purposes, it will be convenient to select a midchord pivot axis, $a = 0$, and when Eqs. (7.26) and (7.28) are used, this equation simplifies to

$$\omega_\alpha = \omega_{\alpha_0} \sqrt{1 - \frac{(V/b\omega_{\alpha_0})^2}{\mu r_\alpha^2 + 0.125}} \qquad (7.48)$$

Thus, it is seen that as the velocity increases from zero, the square of the frequency decreases monotonically as the square of the velocity. To put this in simpler terms, the presence of a moving airstream introduces an aerodynamic spring term M_3 into Eq. (7.43) that is proportional to the pitch displacement and that reduces the torsional stiffness of the system. Thus, the torsional frequency initially decreases as the velocity increases.

A similar but slightly more elaborate analysis of the bending mode will now be performed. A combination of Eqs. (7.4) and (7.6) with the torsion mode suppressed yields the result

$$mb\ddot{h} + K_h bh = L_1\dot{h} + L_2 h \qquad (7.49)$$

where

$$L_1 = -\pi\rho b^3 \qquad (7.50)$$

$$L_2 = -2\pi\rho b^3 \frac{V}{b} C(k) \qquad (7.51)$$

Here there is no direct aerodynamic stiffness term proportional to h in Eq. (7.6). Instead, an aerodynamic stiffness effect will be obtained by retaining the complex $C(k)$ function in the analysis. Rewrite Eq. (7.49) as

$$(mb - L_1)\ddot{h} - L_2\dot{h} + K_h bh = 0 \qquad (7.52)$$

assume a solution having the form

$$h = \bar{h}e^{i\omega_h t} \tag{7.53}$$

and upon introducing Eqs. (7.8), (7.10), (7.50), and (7.51), the real part of Eq. (7.52) becomes

$$-\omega_h^2 (m + \pi\rho b^2) - 2\pi\rho b^2 \omega_h \frac{V}{b} G + K_h = 0 \tag{7.54}$$

Note that for the no-flow condition, V and G both vanish and $\omega_h \rightarrow \omega_{h_0}$, where

$$\omega_{h_0} = \sqrt{\frac{K_h}{m + \pi\rho b^2}} \tag{7.55}$$

is the natural bending frequency of the system in still air. After dividing through by $\pi\rho b^2$ and using Eq. (7.26), these equations yield the result

$$\omega_h = \omega_{h_0} \left[\sqrt{1 + \left(\frac{(V/b\omega_{h_0})G}{\mu + 1} \right)^2} - \frac{(V/b\omega_{h_0})G}{\mu + 1} \right] \tag{7.56}$$

In this equation, G is the imaginary part of the Theodorsen function, which is negative for all values of k, and it is seen that the bending frequency tends to increase with increasing velocity from its still air value. Thus, as the velocity increases, the originally well-separated torsion and bending frequencies approach one another, as shown in Fig. 7.10, in which the frequency is plotted vs velocity. A comparable plot of the frequency vs the modal damping, taken from the Pines analysis of Ref. 18 (see also pp.

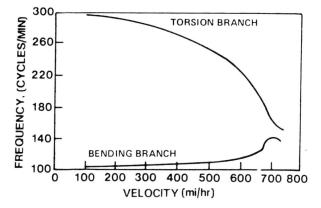

Fig. 7.10   Example of frequency coalescence at the flutter point.

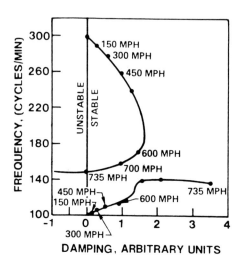

Fig. 7.11 **Effect of frequency coalescence on modal damping.**

269–274 of Ref. 17), is shown in Fig. 7.11. Here it is seen that as the frequencies first depart from their no-flow values the system damping in each mode initially increases. However, when the frequencies become close enough, the damping in one of the modes suddenly decreases and passes through zero and the system becomes unstable. Another interpretation is that when the two frequencies are sufficiently close an energy exchange between the two modes can take place and if the conditions are favorable, a coupled flutter will occur.

Before going on to the discussion of turbomachinery flutter, it should be re-emphasized that this section has used extremely simple concepts and does not purport to be accurate. Instead, its primary purpose has been to attempt to introduce physical meaning into an otherwise complicated and sometimes bewildering mathematical exercise.

7.5 FUNDAMENTAL DIFFERENCES BETWEEN TURBOMACHINERY FLUTTER AND WING FLUTTER

In this section, we shall attempt to point out the differences between aircraft wings and turbomachinery blades, particularly as these differences affect the flutter experienced by these configurations. Certain of these differences will be related to the fundamental discussions in the previous section, while other differences will be newly developed. To begin with, the discussion will center on some pertinent characteristics of the vibrations of rotating systems. Following this, the concept of the frequency shifts in both the bending and torsion modes will be amplified and extended.

Elementary Vibration Concepts for Rotating Systems

The object of this discussion is not to expound on vibration theory. It is assumed that the reader already has an adequate grasp of the fundamentals of vibration theory or that he can refer to one or more of the excellent texts on the subject (e.g., Refs. 19 or 20). Instead, the object here is to concentrate on one or two specific concepts that are unique to rotating machinery and hence are germane to the turbomachinery flutter problem.

As in many other flexible systems involving continuously distributed mass and stiffness, a turbomachine blade can vibrate in an infinity of discrete, natural modes, although, practically, only the lowest several modes are of any interest. However, even here there will be an added complication relative to a fixed-base vibrating system because the rotation of the turbomachine causes variations in both mode shape and frequency. Only the latter will be considered here and the discussion will be further restricted to the relatively simple concept of centrifugal stiffening. To understand this concept as it applies to turbomachinery blades, consider first the rotation of a chain possessing no inherent (or at-rest) stiffness. At zero rotational speed, the chain will hang limply from its support, but under rotation (say, about a vertical axis) the chain will be raised from its limp position and will approach the plane of rotation as the rotational speed increases (cf. Fig. 7.12).

This centrifugal stiffening effect will also cause changes in the frequency of vibration of the chain. If the system is displaced from equilibrium, a restoring force, equivalent to a spring force, will be imposed on each element of the chain and will tend to return the chain to its equilibrium position. This force is caused by the lateral displacement of the radial tension vector and is proportional to the square of the rotation speed Ω (cf. Ref. 19). It is easily shown in Ref. 19 that the frequency of such a chain is equal to Ω. (See curve labeled $\omega_1 = 0$ in Fig. 7.13.)

For a cantilever blade having inherent stiffness and hence having an at-rest frequency of ω_1, the result is similar, although somewhat more complicated. A simplification is afforded by Southwell's theorem (cf. p. 270 of Ref. 19), which states that the natural frequency of the system is approximately equal to

$$\omega = \sqrt{\omega_1^2 + \omega_2^2} \qquad (7.57)$$

where ω_1 is the at-rest frequency and ω_2 the frequency due solely to rotation. For the general case of a rotating, flexible beam, this is equivalent

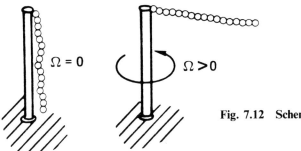

$\Omega = 0$ $\Omega > 0$

Fig. 7.12   Schematic of rotating chain.

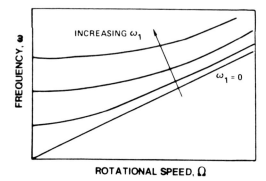

Fig. 7.13   Effect of rotation on frequency of a flexible cantilever beam (Southwell's theorem).

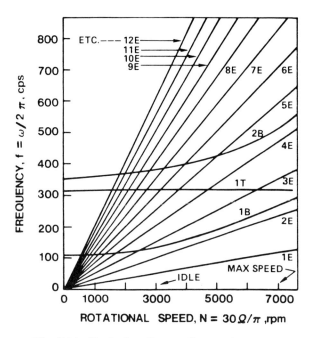

Fig. 7.14   Excitation diagram for rotating system.

to

$$\omega = \sqrt{\omega_1^2 + \beta\Omega^2} \qquad (7.58)$$

which yields a hyperbolic relationship between ω and Ω for $\omega_1 \neq 0$, as shown in the family of curves in Fig. 7.13. In Eq. (7.58), β is of order 1 for the lowest mode and increases in magnitude as the mode number increases. It should be noted that this phenomenon manifests itself primarily in the bending (or flatwise) modes and has virtually no effect on the torsion mode frequency under rotational conditions.

The importance of this type of behavior lies in the multiple sources of excitation for any rotating system. This is illustrated in Fig. 7.14, in which frequency of excitation or vibration is plotted vs rotational speed. First, consider the radial lines from the origin labeled 1E, 2E, 3E, ... These lines represent the loci of available excitation energy at any rotational speed for 1 excitation/revolution, 2 excitations/revolution, 3 excitations/revolution, etc. (Note the coordinate intersection of the 1E line at 6000 rpm and 100 cycle/s.) Superimposed on this diagram are three lines labeled 1B, 1T and 2B, which are hypothetical plots of the first bending, first torsion, and second bending frequencies, respectively. In this hypothetical plot, the designer has managed to keep the first bending frequency high enough so that it never intersects the 2E line, which is usually the source of most forced vibration energy. If such an intersection had occurred, the rotor blades could be subjected to a severe 2/rev bending vibration at the rpm of intersection, sufficiently strong to cause fatigue failures of the blades. Also in the diagram, the designer has permitted an intersection of the first bending mode line with the 3E line at approximately 2600 rpm. If this rotor is designed to idle at 3000 rpm and has a maximum speed of 7000 rpm, then the designer has succeeded in avoiding all integral-order vibration of the first bending mode over the entire operating range.

Note further in this diagram that the first torsion mode, which is unaffected by Southwell stiffening and is therefore a horizontal line, will intersect the 6E, 5E, 4E, and 3E lines as rotor rpm is increased from idle to full speed. This could conceivably cause trouble for a rotor blade operating at high aerodynamic load at part speed (4E) or near full speed (3E) rpm. Under such conditions, a stall flutter could occur (cf. Sec. 7.9), aggravated by an integral-order excitation in the vicinity of the intersection with the integral-order line. Thus, the designer must be aware of the existence of these problems and must be in a position to redesign his blading to eliminate the problem.

Finally, the second bending mode, 2B, will encounter excitations at all integral orders from 8/rev to 5/rev as rpm is increased. However, the 4/rev excitation will be avoided over the operating range of the rotor.

The reader should be aware that only a simple example has been demonstrated here. An actual rotor blade system, particularly one with a flexible disk or a part-span shroud, will also experience coupled blade/disk or blade/shroud modes that are not characterized here and that are beyond the scope of the present discussion. However, these concepts are covered adequately in the advanced literature dealing with such coupled motions.

Mass Ratio, Stiffness, and Frequency Differences

In Sec. 7.4, we described the coupled flutter phenomenon in terms of frequency coalescence of the two fundamental bending and torsion modes. It was shown that the torsional frequency would decrease with increasing velocity [Eq. (7.48)] and that the bending frequency would increase with increasing velocity [Eq. (7.56)]. However, the extent of the frequency change over the operational range was not discussed. This is of extreme importance in deciding whether or not a given system will be susceptible to classical coupled flutter. (Note that these frequency changes are aerodynamically induced and are not caused by rpm changes.)

Refer now to Eqs. (7.48) and (7.56). It is seen here that the structural and geometric parameters governing the frequency changes for the two modes are the mass and inertia ratios, μ and μr_α^2, and the stiffness parameters, $b\omega_{0h}$ and $b\omega_{0\alpha}$. Before we can examine the orders of magnitudes of these parameters, we must first consider the differences in the structural makeup of aircraft wings and compressor blades.

The traditional subsonic aircraft wing has a built-up section consisting of a box beam or spar that supports a stressed skin and stringer shell. It is usually made of aluminum. Hence, much of the wing consists of empty space and what little metal there is, is relatively lightweight. Typical values of the mass and inertia ratios for aircraft wings lie in the ranges of $5 < \mu < 20$ and $1 < \mu r_\alpha^2 < 15$. (Of course, these values would change if the empty space was replaced by fuel.)

The traditional compressor blade is a solid forged and machined blade made of titanium or steel. It may have a cambered, double circular arc profile, and an average thickness-to-chord ratio of 5%. If it is made of titanium, the blade described above will have mass and inertia ratios of approximately $\mu \cong 190$ and $\mu r_\alpha^2 \cong 3$. If it is made of steel, this same blade will have mass and inertia ratios 1.7 times larger than the values just cited.

To a first approximation, then, the inertia parameter μr_α^2 for a compressor blade is of the same order of magnitude as that for an aircraft wing. However, the mass parameter μ for a compressor blade is an order of magnitude (or more) greater than that for an aircraft wing. Thus, with everything else held the same in Eqs. (7.48) and (7.56), we could expect roughly the same change in ω_α with V, but a much smaller change in ω_h with V. Under these circumstances, frequency coalescence and coupled flutter would take place at a higher value of V for a compressor blade than it would for an aircraft wing.

Another factor that enters into the situation here is the inherent stiffness of compressor blades vs the inherent flexibility of aircraft wings. Although the wing chord is many times the compressor blade chord, the compressor blade frequencies are often many more times the wing frequencies. Thus, the stiffness parameters $b\omega_{0\alpha}$ and $b\omega_{0h}$ are at least the same for blade and wing and quite often are larger for the blade than for the wing. This is another way in which the blade frequency change with velocity is smaller than that of the wing.

There are many additional reasons for the observed differences between wing flutter and blade flutter. Most of these are subtle and contribute to the differences in minor ways. However, one additional difference exists which makes a major contribution. This is related to the Pines criteria (cf. Fig. 7.15) that (1) *no* flutter exists for center-of-gravity (c.g.) positions forward of the elastic axis (e.a.) and (2) if the c.g. is aft of the e.a. and if the aerodynamic center (a.c.) is forward of the e.a., flutter is possible only if the

PINES CRITERIA:

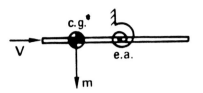

NO FLUTTER IF c.g. IS
FORWARD OF e.a.

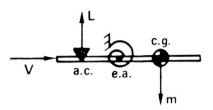

FLUTTER POSSIBLE IF c.g.
IS AFT OF e.a., IF a.c. IS
FORWARD OF e.a. AND IF
$$\frac{\omega h}{\omega \alpha} < f \, (GEOM, r_\alpha)$$

APPLIED TO:

COMPRESSOR BLADE

AIRCRAFT WING

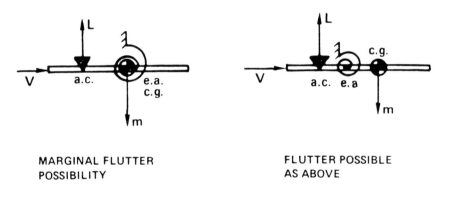

MARGINAL FLUTTER
POSSIBILITY

FLUTTER POSSIBLE
AS ABOVE

Fig. 7.15   Pines criteria applied to compressor blade and aircraft wing.

frequency ratio ω_h/ω_α is less than some specified amount involving these locations and the radius of gyration. In most aircraft wings, the e.a. is quite often in the neighborhood of the spar, between 25 and 40% of the chord, the c.g. is in the neighborhood of the 50% chord location, and the a.c. is at the 25% chord location for subsonic flows. Hence, the c.g. is aft of the e.a., the a.c. is forward of the e.a., and flutter is possible if the frequency criterion is met. In contrast to this, the solid compressor blade usually has a circular arc profile that is symmetric fore and aft. Hence, the c.g. and the e.a. are coincident or very nearly so at the 50% chord location. According to Pines,[18] no coupled flutter is possible if the c.g. is forward of the e.a., no matter how small the distance. Thus, the compressor blade with coincident c.g. and e.a. would constitute a borderline case between flutter and no flutter; thus, we would expect much less susceptibility to coupled flutter in such a configuration than we would in an aircraft wing with conventional separation between the e.a. and c.g.

7.6 FUNDAMENTALS OF UNSTEADY AERODYNAMIC THEORY FOR ISOLATED AIRFOILS

Before we examine the unsteady aerodynamics of multiblade systems, we shall look briefly into the theory for isolated airfoils. We shall proceed from the simpler, more physical view into the more complicated analysis and thereby establish some groundwork for further study. As stated earlier, the scope of this single chapter limits us to an exposition of concepts with a minimum of mathematical derivations; the reader will be directed to the appropriate sources for further study. The notation and the dimensionality of the variables will usually remain consistent with the primary source used for the derivations.

Incompressible Flow

Although Theodorsen[13] is generally credited with the first comprehensive analysis of airfoil flutter, the simpler and more rigorous version by Schwarz[14] will be discussed here. In addition to the basic reference, a detailed analysis of this version of the theory is available in Ref. 1 (pp. 272–278), so only the highlights will be of concern to us.

In the analysis of the isolated airfoil the coordinate system of Fig. 7.6 will be used, together with the physical picture shown below in Fig. 7.16. All lengths in this derivation have been made dimensionless with respect to the semichord b and the coordinate system has been located at the midchord of the airfoil. It is assumed that the airfoil is executing small timewise displacements relative to its mean position ($z = 0$) and that its vertical velocity is given by

$$w_a(x,t) = b \frac{\partial za}{\partial t} + U \frac{\partial za}{\partial x}$$

$$z = 0, \qquad -1 \leqslant x \leqslant 1 \tag{7.59}$$

(where the subscript a denotes the region over the airfoil). It should be noted that this equation is used in all unsteady aerodynamic derivations to define the interfacial condition between the fluid and the body in motion. For a prescribed mode of oscillation Eq. (7.59) defines the known vertical velocity (or downwash) over the extent of the airfoil chord.

The quantities γ_a and γ_w in Fig. 7.16 are the vorticity per unit chord in the vortex sheet over the airfoil chord and the wake, respectively. Use of the Biot-Savart law yields the relationship

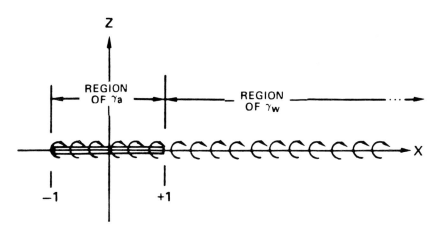

Fig. 7.16 Vortex sheet representation of isolated airfoil.

$$w_a(x,t) = -\frac{1}{2\pi} \oint_{-1}^{1} \frac{\gamma_a(\xi,t)}{x-\xi}\, \mathrm{d}\xi - \frac{1}{2\pi} \int_{1}^{\infty} \frac{\gamma_w(\xi,t)}{x-\xi}\, \mathrm{d}\xi \qquad (7.60)$$

in which the known downwash is equal to an integral of the unknown chordwise vorticity distribution. However, from potential flow theory, it can be shown that the chordwise pressure difference distribution is given by

$$\frac{\Delta p(x,t)}{\rho U} = -\gamma_a(x,t) - ik \int_{-1}^{x} \gamma_a(\xi,t)\, \mathrm{d}\xi \qquad (7.61)$$

Hence, the determination of the unsteady loads on an oscillating airfoil requires the successful inversion of the integral equation (7.60).

The assumption will now be made that the airfoil is oscillating in simple harmonic motion, with frequency ω such that all time-dependent quantities in the foregoing equations can be rewritten in the typical form

$$w_a(x,t) = \bar{w}_a(x)e^{i\omega t} \qquad (7.62)$$

whereupon Eq. (7.60) becomes

$$\bar{w}_a(x) = -\frac{1}{2\pi} \oint_{-1}^{1} \frac{\bar{\gamma}_a(\xi)}{x-\xi}\, \mathrm{d}\xi - \frac{1}{2\pi} \int_{1}^{\infty} \frac{\bar{\gamma}_w(\xi)}{x-\xi}\, \mathrm{d}\xi \qquad (7.63)$$

At this point in the analysis, Schwarz makes the key assumption that any circulation change on the airfoil appears as shed vorticity in the wake having equal magnitude and opposite sign, or

$$d\Gamma(t) = -b\gamma_w(1,t)\,dx \qquad (7.64)$$

where

$$\Gamma(t) \equiv b \int_{-1}^{1} \gamma_a(x,t)\,dx \qquad (7.65)$$

Equation (7.64) can be rewritten as

$$\frac{d\Gamma(t)}{dt} = -b\gamma_w(1,t)\frac{dx}{dt} = -U\gamma_w(1,t) \qquad (7.66)$$

Schwarz now makes the further assumption that this shed velocity is convected downstream at freestream velocity U and after a series of logical manipulations, arrives at the formula

$$U\gamma_w\left[\xi,t + \frac{b(\xi-1)}{U}\right] = -\frac{d\Gamma(t)}{dt} \qquad (7.67)$$

With the assumption of simple harmonic motion, all variables are proportional to $e^{i\omega t}$, so this becomes

$$U\bar{\gamma}_w(\xi)\exp\left[i\omega\left(t + \frac{b(\xi-1)}{U}\right)\right] = -i\omega\bar{\Gamma}e^{i\omega t} \qquad (7.68)$$

which further reduces to

$$\bar{\gamma}_w(\xi) = -\frac{i\omega}{U}\bar{\Gamma}e^{ik}e^{-ik\xi} \qquad (7.69)$$

Following Reissner,[21] we introduce a reduced circulation function (which is unknown but constant and can be evaluated at the conclusion of the analysis)

$$\bar{\Omega} = \frac{\bar{\Gamma}}{b}e^{ik} \qquad (7.70)$$

so Eq. (7.69) becomes

$$\bar{\gamma}_w(\xi) = -ik\bar{\Omega}e^{-ik\xi} \qquad (7.71)$$

and when this is substituted into Eq. (7.63), the result is

$$\bar{w}_a(x) = -\frac{1}{2\pi}\oint_{-1}^{1}\frac{\bar{\gamma}_a(\xi)}{x-\xi}\,d\xi + \frac{ik\bar{\Omega}}{2\pi}\int_{1}^{\infty}\frac{e^{-ik\xi}}{x-\xi}\,d\xi \qquad (7.72)$$

Now only one unknown $\bar{\gamma}_a(\xi)$ is under the integral and the remaining unknown $\bar{\Omega}$ is a constant. Thus, the integral equation can be inverted by Söhngen's[22] method and the resulting equation now yields the unknown $\bar{\gamma}_a(x)$ in terms of an integral of the known downwash $\bar{w}_a(\xi)$,

$$
\bar{\gamma}_a(x) = -\frac{2}{\pi}\sqrt{\frac{1-x}{1+x}}\left[-\oint_{-1}^{1}\sqrt{\frac{1+\xi}{1-\xi}}\frac{\bar{w}_a(\xi)}{x-\xi}\,\mathrm{d}\xi\right.
$$
$$
\left.+\frac{ik\bar{\Omega}}{2\pi}\oint_{-1}^{1}\sqrt{\frac{1+\xi}{1-\xi}}\left(\int_{1}^{\infty}\frac{e^{-ik\lambda}}{\xi-\lambda}\,\mathrm{d}\lambda\right)\frac{\mathrm{d}\xi}{x-\xi}\right] \tag{7.73}
$$

Once this crucial step has been performed, the remainder of the problem, although formidable in appearance, involves nothing more than a sequence of complicated algebraic manipulations and integrations to yield the final desired results. [Note that this solution for $\bar{\gamma}_a(x)$ yields $\bar{\Omega}$ by integration, via Eqs. (7.65) and (7.70).] Specifically, substitution of Eq. (7.73) into the time-independent form of Eq. (7.61) yields the pressure difference distribution

$$
-\frac{\Delta\bar{p}(x)}{\rho U} = \frac{2}{\pi}[1-C(k)]\sqrt{\frac{1-x}{1+x}}\int_{-1}^{1}\sqrt{\frac{1+\xi}{1-\xi}}\,\bar{w}_a(\xi)\,\mathrm{d}\xi
$$
$$
+\frac{2}{\pi}\oint_{-1}^{1}\left[\sqrt{\frac{1-x}{1+x}}\sqrt{\frac{1+\xi}{1-\xi}}\frac{1}{x-\xi}-ik\Lambda_1(x,\xi)\right]\bar{w}_a(\xi)\,\mathrm{d}\xi \tag{7.74}
$$

where

$$
\Lambda_1(x,\xi) = \frac{1}{2}\log\left(\frac{1-x\xi+\sqrt{1-\xi^2}\sqrt{1-x^2}}{1-x\xi-\sqrt{1-\xi^2}\sqrt{1-x^2}}\right) \tag{7.75}
$$

Further integration of Eq. (7.74) yields the unsteady lift and moment per unit span,

$$
L(t) = -b\int_{-1}^{1}\Delta p(x,t)\,\mathrm{d}x \tag{7.76}
$$

$$
M(t) = b^2\int_{-1}^{1}(x-a)\,\Delta p(x,t)\,\mathrm{d}x \tag{7.77}
$$

Actual values for these integrals can be obtained by introducing the generalized dimensionless motion

$$
z = -h - \alpha(x-a) \tag{7.78}
$$

into Eq. (7.59) for the downwash

$$w_a(x,t) = -b\dot{h} - b\dot{\alpha}(x - a) - U\alpha \qquad (7.79)$$

and thence into the equations given above. It can be shown (cf. Ref. 1) that this procedure will yield the time-independent versions of Eqs. (7.6) and (7.7). Furthermore, if the bars are removed from the quantities in Eq. (7.74), the resulting lift and moment formulas are identical to Eqs. (7.6) and (7.7), although it should be clearly understood that these are to be regarded as the time-dependent loads due to simple harmonic motions. (The reader should be cautioned that the symbol h in the present document is equivalent to h/b in Ref. 1.)

Although this chapter is devoted primarily to the unsteady aerodynamics of cascaded airfoils, it would appear that an undue degree of attention has been placed on the isolated airfoil solution. However, it will be seen in Sec. 7.7 that this method is partially extendable to the study of cascaded airfoils in incompressible flow. Although not all of the mathematical techniques (such as the Söhngen inversion of the integral equation) can be used, the fundamental approach is still based on the vortex sheet concept, which makes the Schwarz method particularly valuable as a basic reference.

Compressible Subsonic Flow

There are two basic approaches to the solution of the unsteady aerodynamic problem of an isolated airfoil oscillating in a subsonic compressible flow. The first, historically, involves a solution of an integral equation for a distribution of acceleration potential doublets. It was solved by Possio[23] with subsequent refinements and computations by Dietze[24] and Fettis.[25] The second approach by Timman et al.[26] involves a transformation of the differential equation for the acceleration potential into elliptical coordinates and expanding the solution as an infinite series of Mathieu functions. Although the Mathieu function approach leads to a mathematically exact solution, it is necessary to compute several infinite sums, even for uncoupled, simple harmonic motions. Furthermore, these sums converge slowly at rates that depend on both reduced frequency and Mach number. Finally, no serious effort has yet been made to extend this method to a solution of the subsonic oscillating cascade problem. Hence, it will not be considered here.

Conversely, Possio's method, although tedious, is relatively straightforward and has been the basis for several solutions to the unsteady cascade problem. It will be described briefly below. The reader should be aware that the notation of this section is dimensional and the equations outlined below are a condensation of the discussion in Ref. 1, pp. 318–325.

The starting point for this analysis is the differential equation for the acceleration potential ψ,

$$\nabla^2\psi - \left(\frac{1}{a_\infty^2}\frac{\partial^2\psi}{\partial t^2} + \frac{2M}{a_\infty}\frac{\partial^2\psi}{\partial x\,\partial t} + M^2\frac{\partial^2\psi}{\partial x^2}\right) = 0 \qquad (7.80)$$

where ψ is defined in terms of the disturbance pressure

$$p - p_\infty = -\rho_\infty \psi \qquad (7.81)$$

It is shown in Ref. 1 that the substantial derivative of the downwash w is given by

$$Dw = \frac{\partial \psi}{\partial z} \, dt \qquad (7.82)$$

and that integration along the trajectory of a fluid particle leads to the useful result

$$w(x,z,t) = \int_{-\infty}^{x} \frac{\partial \psi \left(\xi',z,t - \dfrac{x - \xi'}{U} \right)}{\partial z} \, \frac{d\xi'}{U} \qquad (7.83)$$

where U is the freestream velocity. As in the previous section, w on the airfoil surface is regarded as the prescribed quantity

$$w_a(x,t) = \frac{\partial z_a}{\partial t} + U \frac{\partial z_a}{\partial x} \qquad \text{for } z = 0, \quad -b \leqslant x \leqslant b \qquad (7.84)$$

(Note that the length variables used herein are dimensional quantities.) We shall assume simple harmonic motion such that a common time-dependence factor $e^{i\omega t}$ can be removed and we shall prescribe the additional boundary conditions: (1) there is zero pressure discontinuity across the x axis both ahead of and in the wake of the blade or $\psi = 0$ for $|x| > b$; (2) all signals must propagate outward from the blade without reflection; and (3) the Kutta condition must be satisfied, so ψ is not discontinuous at $x = b$.

The method used in Ref. 1 to obtain Possio's integral equation is to represent the airfoil with a sheet of acceleration potential doublets along the projection of the airfoil. Each elementary doublet is the mathematically derived superposition of a source-sink pair and each source (or sink) is a solution of the differential equation (7.80) and is referred to as a source pulse,

$$\psi_{\text{SP}} = \frac{A(\xi,\zeta,T)}{R} \qquad (7.85)$$

where

$$R^2 = a_\infty^2 (t - T)^2 - [(x - \xi) - U(t - T)]^2 - (z - \zeta)^2 \qquad (7.86)$$

and where ψ_{SP} is a two-dimensional pressure disturbance originating at ξ, ζ at time T that propagates outward in a cylindrical fluid region while being carried downstream at the freestream velocity U. $A(\xi,\zeta,T)$ is a time-dependent amplitude function of the disturbance that will be evaluated later by means of the boundary conditions. Both ψ_{SP} and A in Eq. (7.85) can be regarded as infinitesimal quantities since they represent a single source pulse out of a continuous sheet; in Ref. 1, it is stated that for simple harmonic motion we can write

$$A(\xi,\zeta,T) = -\frac{\bar{A}(\xi,\zeta)U^2 a_\infty e^{i\omega T}\,\mathrm{d}T}{2\pi} \tag{7.87}$$

which allows us to integrate Eq. (7.85) over T and obtain

$$\psi_s(x,z,t) = -\frac{\bar{A}(\xi,\zeta)U^2 a_\infty}{2\pi} \int_{-\infty}^{g} \frac{e^{i\omega T}\,\mathrm{d}T}{R} \tag{7.88}$$

In this equation, ψ_s is related to the pressure disturbance at x,z due to a disturbance with complex amplitude $\bar{A}$ located at ξ,ζ. The upper limit of integration

$$g = t - \frac{\sqrt{(x-\xi)^2 + (z-\zeta)^2(1-M^2)} - M(x-\xi)}{a_\infty(1-M^2)} \tag{7.89}$$

is the most recent value of T for which a disturbance at ξ,ζ can be felt at x,z at time t. After some manipulations of the integrand, Eq. (7.88) may be integrated to yield

$$\psi_s(x,z,t) = \frac{\bar{A}(\xi,\zeta)U^2 e^{i(\omega t + \bar{k})}}{4\sqrt{1-M^2}}\,iH_0^{(2)}(w') \tag{7.90}$$

where

$$w' = \frac{kM}{1-M^2}\sqrt{\left(\frac{x-\xi}{b}\right)^2 + (1-M^2)\left(\frac{z-\zeta}{b}\right)^2}$$

$$\bar{k} = \frac{kM^2(x-\xi)}{(1-M^2)b} \tag{7.91}$$

A sinusoidally pulsating doublet is now obtained by differentiating the source solution with respect to the vertical coordinate, with the result

$$\psi_D(x,z,t) = \left(\frac{\partial \psi_s}{\partial \zeta}\right)_{\zeta=0} = \frac{\bar{A}(\xi)\omega^2 M^2 z}{4(1-M^2)^{\frac{3}{2}}}\,i\,\frac{H_1^{(2)}(w')}{w'}\,e^{i(\omega t + \bar{k})} \tag{7.92}$$

Finally, the acceleration potential of the entire sheet of doublets is obtained by integrating over the chord,

$$\psi(x,z,t) = \int_{-b}^{b} \psi_D\,\mathrm{d}\xi = \frac{\omega^2 M^2 e^{i\omega t}}{4(1-M^2)^{\frac{3}{2}}}\,iz \int_{-b}^{b} \bar{A}(\xi)\frac{H_1^{(2)}(w')}{w'}\,e^{i\bar{k}}\,\mathrm{d}\xi \tag{7.93}$$

Use is now made of the singular behavior of the integrand. Note that the integral is multiplied by z. Hence, the imposition of the limit $z \to 0$ as the airfoil surface is approached causes the right-hand side to vanish everywhere except in the neighborhood of $\zeta = x$, where the integrand becomes large without bound in the limit. Use of appropriate limiting techniques as z is allowed to vanish from above (approaching the suction surface) and from below (approaching the pressure surface) yields the results

$$\psi(x,0^+,t) = -\frac{U^2 \bar{A}(x)}{2} e^{i\omega t}$$

$$\psi(x,0^-,t) = \frac{U^2 \bar{A}(x)}{2} e^{i\omega t} \qquad (7.94)$$

and from Eq. (7.81), we can identify the quantity $\bar{A}(x)$ as

$$\bar{A}(x) = \frac{\Delta \bar{p}_a(x)}{\rho_\infty U^2} \qquad (7.95)$$

where $\Delta \bar{p}_a(x)$ is the desired but as yet unknown pressure difference across the airfoil. When this is substituted into Eq. (7.93), the result is

$$\psi(x,z,t) = \frac{\omega^2 M^2 i z e^{i\omega t}}{4\rho_\infty U^2 (1 - M^2)^{\frac{3}{2}}} \int_{-b}^{b} \frac{H_1^{(2)}(w')}{w'} e^{i\bar{K}} \Delta \bar{p}_a(\xi) \, d\xi$$

$$= \frac{-i e^{i\omega t}}{4\rho_\infty \sqrt{1 - M^2}} \int_{-b}^{b} \left[\frac{\partial H_0^{(2)}(w')}{\partial z} \right]_{\zeta = 0} e^{i\bar{K}} \Delta \bar{p}_a(\xi) \, d\xi \qquad (7.96)$$

Use is now made of Eq. (7.83) relating downwash to the disturbance potential. When this is evaluated along the x axis for a simple harmonic motion, we obtain

$$w(x,0,t) = \frac{1}{U} e^{-i\omega x/U} \int_{-\infty}^{x} \left[\frac{\partial \psi}{\partial z} \right]_{z \to 0^+} e^{i\omega \xi'/U} \, d\xi' \qquad (7.97)$$

and substitution of Eq. (7.96) yields the result

$$\bar{w}_a(x) = \frac{-i e^{-ikx/b}}{4\rho_\infty U \sqrt{1 - M^2}} \int_{-b}^{b} \Delta \bar{p}_a(\xi) e^{ik\xi/b} \int_{-\infty}^{x} \left[\frac{\partial^2 H_0^{(2)}(w')}{\partial z^2} \right]_{\substack{z \to 0^+ \\ \zeta \to 0}} e^{i\bar{K}} \, d\xi' \, d\xi \qquad (7.98)$$

This is still an inconvenient form and considerably more manipulation must be performed (as shown in Ref. 1) before the final solution is obtained,

$$\bar{w}_a(x) = -\frac{\omega}{\rho_\infty U^2} \oint_{-b}^{b} \Delta \bar{p}_a(\xi) K \left[M, \frac{k(x - \xi)}{b} \right] d\xi \qquad \text{for } -b \leqslant x \leqslant b \qquad (7.99)$$

where the kernel function is given by

$$
K\left[M, \frac{k(x-\xi)}{b}\right] = \frac{1}{4\sqrt{1-M^2}} \left\{ e^{ik}\left[iM\frac{|x-\xi|}{x-\xi} H_1^{(2)}(M\lambda) - H_0^{(2)}(M\lambda) \right] \right.
$$

$$
+ i(1-M^2)e^{-ik(x-\xi)/b}\left[\frac{2}{\pi\sqrt{1-M^2}} \log\frac{1+\sqrt{1-M^2}}{M} \right.
$$

$$
\left. \left. + \int_0^{k/M^2} e^{iu}H_0^{(2)}(M|u|)\, du \right] \right\} \tag{7.100}
$$

and where

$$
\lambda = \frac{k|x-\xi|}{(1-M^2)b}, \qquad u = \frac{k(\xi-\xi')}{(1-M^2)b} \tag{7.101}
$$

Equation (7.99) is Possio's integral equation in which the known down-wash is written in terms of an integral of the unknown pressure difference distribution. Unlike the Schwarz integral equation (7.72), there is no convenient inversion formula available and it has been necessary for the several authors working on this problem to resort to approximate iterative techniques. Tabulations of the unsteady aerodynamic coefficients are found in Refs. 24, 27, and 28.

As in the case of incompressible flow, the form of this solution will be important in our discussion of compressible subsonic flow past cascades and of acoustical resonance (Sec. 7.7).

Compressible Supersonic Flow

Although several researchers attacked the problem of compressible supersonic flow past an oscillating airfoil, the first systematic and straightforward approach was that of Garrick and Rubinow[29] who extended some of Possio's early work and used the method of distributed source pulses. However, this solution has little or no relevance to the cascade problem except for its extensive tabulated values, which are used later to provide isolated airfoil reference values. It is mentioned here primarily for its historical significance and will not be studied in any detail.

More relevant to the work on cascades that follows is the use of the Laplace transform (cf. Refs. 1, pp. 364–367, and 30, pp. 37, 38, 50). Our starting point is the wave equation (7.80) with ψ replaced by the disturbance potential ϕ and with the assumption of simple harmonic motion

$$
\phi(x,z,t) = \bar{\phi}(x,z)e^{i\omega t} \tag{7.102}
$$

such that

$$
\nabla^2\bar{\phi} = M^2\frac{\partial^2\bar{\phi}}{\partial x^2} + \frac{2Mi\omega}{a_\infty}\frac{\partial\bar{\phi}}{\partial x} - \frac{\omega^2}{a_\infty^2}\bar{\phi} \tag{7.103}
$$

with boundary condition

$$\frac{\partial \bar{\phi}}{\partial z} = \bar{w}_a(x) \qquad \begin{array}{l} \text{for } z = 0 \\ \text{over } 0 \leqslant x \leqslant 2b \end{array} \qquad (7.104)$$

where the origin of the dimensional coordinates is now at the leading edge. A straightforward use of Laplace transform technique, defined by

$$\bar{\phi}^*(s,z) = \mathscr{L}[\bar{\phi}(x,z)] = \int_0^\infty e^{-sx}\bar{\phi}(x,z)\,\mathrm{d}x \qquad (7.105)$$

leads ultimately to the solution

$$\bar{\phi}(x^*,0^+) = -\frac{b}{\sqrt{M^2-1}} \int_0^{x^*} \bar{w}_a(\xi^*)e^{-(i\bar{\omega}/2)(x^*-\xi^*)} J_0\left[\frac{\bar{\omega}}{2M}(x^*-\xi^*)\right] \mathrm{d}\xi^*$$
$$(7.106)$$

where

$$\bar{\omega} = \frac{2kM^2}{M^2-1} \qquad (7.107)$$

and where the starred coordinates are dimensionless with respect to b. Equation (7.106) can be substituted into the pressure relationship (cf. Ref. 1, p. 357)

$$\Delta \bar{p}_a(x^*) = -2\rho_\infty \frac{U}{b}\left(ik\bar{\phi} + \frac{\partial \bar{\phi}}{\partial x^*}\right)\bigg|_{z^*=0^+} \qquad (7.108)$$

and use of Eqs. (7.76) and (7.77) yields the desired values of the unsteady lift and pitching moment.

7.7 UNSTEADY AERODYNAMIC THEORY FOR CASCADED AIRFOILS

Incompressible Flow

Several attempts were made in the early 1950's to extend isolated unsteady theories to solve the unsteady cascade problem, most notably those of Lilley,[31], Sisto,[32] and Lane and Wang.[33] As with any early work on a difficult technical problem, these pioneering efforts were hampered by lack of precedent to serve as a mathematical guide and by a lack of adequate and efficient computational equipment to cope with the need to account for a doubly-infinite array of airfoils in the flowfield. In contrast to the isolated airfoil in potential flow, in which the reduced frequency was the only explicit parameter (for incompressible flow), the unsteady cascade analysis had the additional parametric influence of blade gap, leading-edge locus stagger angle, and interblade phase angle. Nevertheless, the solutions obtained in the cited works represented valid approaches and the few available tabulated values clearly indicated the differences between cascades and isolated airfoils and pointed up the need for additional work.

In the early 1960's, Whitehead published his theoretical analysis of the incompressible unsteady cascade problem,[34] based largely on his earlier dissertation.[35] In it, he used a standard vortex sheet approach similar to that described earlier (Sec. 7.6) in connection with the theory of Schwarz. However, the integral inversion technique of Söhngen has no counterpart in the cascade problem and Whitehead resorted to a matrix formulation and a subsequent matrix inversion technique. By virtue of an ingenious use of the Multhopp cosine transformation, he showed that a trapezoidal integration would yield exact results for certain key integrals and an analytical removal of the leading-edge logarithmic singularity produced a tractable result that was relatively easy to tabulate. In the following summary of Ref. 34, only a few of the key steps in Whitehead's derivation can be included and an effort will be made to bridge the gap between these

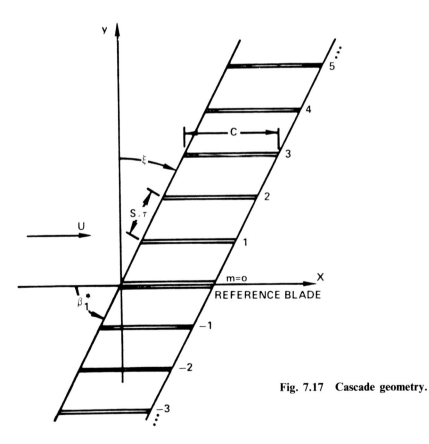

Fig. 7.17 Cascade geometry.

steps with references to his report. Use will be made of the notation of the report throughout.

The cascade geometry is shown in Fig. 7.17 in which c is the chord, s the slant gap, and ξ the stagger angle of the leading-edge locus relative to the vertical coordinate axis, which passes through the leading edge of the reference blade. In addition to the usual assumptions of two-dimensional, incompressible, inviscid potential flow past flat-plate airfoils at zero incidence angle, it is further assumed that all blades vibrate with the same amplitude and with constant phase angle between the motions of any pair of adjacent blades.

To begin with, Whitehead assumes a vortex sheet distribution on the blades and defines the strength of one element of bound vorticity located at a distance x behind the leading edge of the reference blade to be $\gamma(x)e^{i\omega t}$ to account for periodic motions. If the motion of the upper adjacent blade ($m = 1$) leads the motion of the reference blade, then blade 1 has a corresponding element of vorticity of strength $\gamma(x + s \sin\xi, s \cos\xi)e^{i(\omega t + \beta)}$ located at the point $(x + s \sin\xi, s \cos\xi)$ and, in general, for the mth blade in the array, we have $\gamma e^{i(\omega t + m\beta)}$ at $(x + ms \sin\xi, ms \cos\xi)$, which establishes the periodicity condition over the array of blades. Because the motion is assumed to be sinusoidal, any change in γ appears in the wake as free vorticity. As in the Schwarz theory, we can relate the elementary free vorticity $\epsilon(x_1,x)$ at x_1 in the wake to a bound vortex element at x by the formula

$$\epsilon(x_1,x) = -\frac{i\omega}{U}\gamma(x)e^{i\omega/U(x-x_1)} \qquad (7.109)$$

The total free vorticity at x_1 in the wake due to the entire bound vorticity sheet is obtained from an integration of Eq. (7.109),

$$\epsilon(x_1)e^{i\omega x_1/U} = -\frac{i\omega}{U}\int_0^{x_1}\gamma(x)e^{i\omega x/U}\,\mathrm{d}x \qquad (7.110)$$

After differentiating this with respect to x_1 and rearranging, the result is

$$\frac{\mathrm{d}\epsilon}{\mathrm{d}x} + \frac{i\omega}{U}(\epsilon + \gamma) = 0 \qquad (7.111)$$

which will be useful below.

Next, the bound vorticity is related to the pressure difference across the blade. Since the theory is to be used in situations in which the whole flow contains vorticity (i.e., the wake vorticity of adjacent blades), the velocity potential concept cannot be used. Instead we use the equation of motion in the x direction, linearized to read

$$\left(\frac{\partial}{\partial t} + U\frac{\partial}{\partial x}\right)(ue^{i\omega t}) = -\frac{1}{\rho}\frac{\partial p}{\partial x} \qquad (7.112)$$

where u is the horizontal perturbation velocity. When this equation is written for points just above $(+)$ and just below $(-)$ the plane of the airfoil and the difference is taken, we have

$$\left(\frac{\partial}{\partial t} + U\frac{\partial}{\partial x}\right)[(u_- - u_+)e^{i\omega t}] = -\frac{1}{\rho}\frac{\partial}{\partial x}(p_- - p_+) \qquad (7.113)$$

but the difference between the perturbation velocity yields the total vorticity of the system,

$$u_- - u_+ = \gamma + \epsilon \qquad (7.114)$$

When this is substituted into Eq. (7.113) and manipulated, the result is

$$-\frac{1}{\rho}\frac{\partial}{\partial x}(p_- - p_+) = U\left\{\left[\frac{i\omega}{U}(\gamma + \epsilon) + \frac{d\epsilon}{dx}\right] + \frac{d\gamma}{dx}\right\}e^{i\omega t} \qquad (7.115)$$

and substitution of Eq. (7.111) simplifies this to

$$-\frac{1}{\rho}\frac{\partial}{\partial x}(p_- - p_+) = U\frac{d\gamma}{dx}e^{i\omega t} \qquad (7.116)$$

This can be directly integrated to yield

$$p_- - p_+ = -\rho U\gamma e^{i\omega t} \qquad (7.117)$$

whereupon the lift and pitching moment (about the leading edge) are obtained by the further integration of Eq. (7.117),

$$F = -\rho U \int_0^c \gamma\, dx \qquad (7.118)$$

$$M = -\rho U \int_0^c \gamma x\, dx \qquad (7.119)$$

Note, however, that up to this point in the analysis the blade motion has not yet been prescribed and γ is an unknown function of x.

The next step in the analysis is to relate the vorticity distribution γ to the motion-induced vertical velocity on the reference blade. We begin by postulating a single, staggered row of vortices, each being located on a separate blade of the cascade and each at the same distance behind the leading edge of its respective blade. If the vortex on the reference blade at $(x,0)$ has strength Γ_0, then the vortex on the mth blade will have a strength

$$\Gamma_m = \Gamma_0 e^{im\beta} \qquad (7.120)$$

(assuming the existence of an interblade phase angle β) and will be located at

$$x_m = x + ms \sin\xi$$

$$y_m = ms \cos\xi \qquad (7.121)$$

For this one vortex Γ_m at (x_m, y_m), the velocity induced on the reference blade at $(\eta, 0)$ is, by the Biot-Savart law,

$$v = \frac{\Gamma_m}{2\pi} \frac{\eta - x_m}{(\eta - x_m)^2 + y_m^2} \qquad (7.122)$$

Hence, the velocity for the entire array of vortices is obtained from the doubly infinite sum

$$v = \frac{\Gamma_0}{2\pi} \sum_{m=-\infty}^{\infty} \frac{e^{im\beta}(\eta - x - ms \sin\xi)}{(\eta - x - ms \sin\xi)^2 + (ms \cos\xi)^2} \qquad (7.123)$$

or, symbolically,

$$v = \frac{\Gamma_0}{c} V(z) \qquad (7.124)$$

where $z = (\eta - x)/c$ and where

$$V(z) = \frac{1}{2\pi} \sum_{m=-\infty}^{\infty} \frac{e^{im\beta}[z - m(s/c) \sin\xi]}{[z - m(s/c) \sin\xi]^2 + [m(s/c) \cos\xi]^2} \qquad (7.125)$$

This can be summed, using the methods described on pp. 364-370 of Bromwich,[36] to yield the closed-form result

$$V(z) = \frac{(a + ib)e^{-(\pi - \beta)z(a + ib)}}{4 \sinh[\pi z(a + ib)]} + \frac{(a - ib)e^{(\pi - \beta)z(a - ib)}}{4 \sinh[\pi z(a - ib)]} \qquad (7.126)$$

where

$$a = \frac{c}{s} \cos\xi, \qquad b = \frac{c}{s} \sin\xi \qquad (7.127)$$

There are some subtleties associated with the values of V for the endpoint values of β, but Whitehead disposes of this problem by showing that the function $V(z)$ will appear only as a difference in the subsequent steps.

Once the form of Eq. (7.124) has been established, we can rewrite it for a similar row of elements of bound vorticity on the oscillating blades. Specifically, the normal velocity induced at a point $(\eta, 0)$ by a set of bound vorticity elements $\gamma(x)e^{i(\omega t + m\beta)}$ is obtained by writing v'_γ in place of v and $\gamma(x)$ in place of Γ_0/c in Eq. (7.124),

$$v'_\gamma(\eta,x) = \gamma(x)V(\eta - x) \qquad (7.128)$$

(At this point in the analysis, Whitehead assumes that x and y are now dimensionless quantities with respect to c.) The strength of an element of free vorticity at some point $(x_1,0)$ due to the element of bound vorticity at $(x,0)$ is given by Eq. (7.109) as

$$\epsilon(x_1,x) = -i\lambda\gamma(x)e^{i\lambda(x - x_1)} \qquad (7.129)$$

where λ is the reduced frequency relative to the full blade chord,

$$\lambda = \frac{c\omega}{U}\left(= 2k = \frac{2b\omega}{U} \right) \qquad (7.130)$$

The velocity induced at $(\eta,0)$ by a single element of this free vorticity is obtained, as in Eq. (7.128), by writing

$$v'_{\delta\epsilon}(\eta,x,x_1) = \epsilon(x_1,x)V(\eta - x_1) \qquad (7.131)$$

Now, for each element of bound vorticity $\gamma(x)$ at $(x,0)$, there is a sheet of free vorticity represented by $\epsilon(x_1,0)$, where the elements are located at points $(x_1,0)$ extending over the range $x \leqslant x_1 < \infty$ and each element of ε is caused by the element γ. Thus, the velocity induced at $(\eta,0)$ by all the free vorticity associated with $\gamma(x)$ is obtained by integrating Eq. (7.131) over all x_1 in the range $x \leqslant x_1 < \infty$,

$$v'_\epsilon(\eta,x) = \int_x^\infty \epsilon(x_1,x)V(\eta - x_1)\,\mathrm{d}x_1 \qquad (7.132)$$

The sum of Eqs. (7.128) and (7.132) yields the velocity induced at $(\eta,0)$ as a result of a single element of bound vorticity and its associated free vortex sheet. However, as Whitehead points out, when $\beta = 0$, the integral of Eq. (7.132) remains finite at the upper limit and the integral fails to converge. If the assumption is made that the system started from rest, then the total vorticity on each blade and its wake is zero, or

$$\int_0^\infty (\gamma + \epsilon)\,\mathrm{d}x = 0 \qquad (7.133)$$

and if this integral is rewritten for a single bound vortex element and is multiplied by the limiting value $V(-\infty)$ and subtracted from the sum of Eqs. (7.128) and (7.132), a convergent result is obtained. Ultimately, this can be written in the form

$$v'(\eta,x) = \gamma(x)K(x - \eta) \tag{7.134}$$

where K is the kernel function

$$K(z) = V(-z) - V(-\infty) - i\lambda e^{i\lambda z} \int_z^\infty [V(-z_1) - V(-\infty)]e^{-i\lambda z_1} \, dz_1 \tag{7.135}$$

Finally, the velocity at η due to all vorticity elements is given by the integral of Eq. (7.134) over the blade chord,

$$v(\eta) = \int_0^1 v'(\eta,x) \, dx = \int_0^1 K(x - \eta)\gamma(x) \, dx \tag{7.136}$$

As in previous analyses of this sort, $v(\eta)$ is prescribed by the motion and is known, while $\gamma(x)$, which is needed for load calculations, is unknown.

This is the essence of the physical portion of Whitehead's analysis. Three possible vertical velocities, associated with vertical translation, pitch about the blade leading edge, and an imposed sinusoidal vertical gust velocity imbedded in the freestream, are postulated as

$$v = q + \alpha U(1 + i\lambda\eta) - we^{-i\lambda\eta} \tag{7.137}$$

An influence function approach, in which each component of vorticity is defined as the vorticity per unit vertical velocity amplitude, yields the total strength of the bound vorticity sheet as

$$\gamma = q\gamma_q + \alpha U\gamma_\alpha - w\gamma_w \tag{7.138}$$

where the γ are solutions of the following three integral equations, expressed in a single matrix equation as

$$\int_0^1 [\gamma_q,\gamma_\alpha,\gamma_w]K(x - \eta) \, dx = [1,(1 + i\lambda\eta),e^{-i\lambda\eta}] \tag{7.139}$$

where the notation $[\cdots]$ represents a 1×3 row matrix.

Space does not permit a detailed account of Whitehead's solution procedure. In essence, he used a Multhopp transformation for the two variables, $x = \frac{1}{2}(1 - \cos\theta)$ and $\eta = \frac{1}{2}(1 - \cos\phi)$ over n stations along the blade chord and found that the trapezoidal rule yielded exact results for several of the integral equations in the transformed notation. Although the process was tedious and painstaking, the inversion of the integral equations written in matrix form was relatively straightforward, the only exception being the leading-edge logarithmic singularity, which was analytically accounted for. Specifically, Eq. (7.139) was written in matrix form as

$$A\Gamma = B \tag{7.140}$$

where A is an $n \times n$ matrix of kernel function elements, Γ an $n \times 3$ matrix of bound vorticity elements, and B an $n \times 3$ matrix representing the right-hand side of Eq. (7.139). Here, it is seen that A and B are known and Γ is unknown and Whitehead's report details the steps necessary to yield the formal solution

$$\Gamma = A^{-1}B \tag{7.141}$$

Substitution of this solution for each component of Γ into Eqs. (7.118) and (7.119) rewritten in matrix form as

$$C = -\frac{1}{\pi} X\Gamma = -\frac{1}{\pi} XA^{-1}B \tag{7.142}$$

yields the desired values of normal force coefficient and moment coefficient for each component of imposed vertical velocity. The $2 \times n$ matrix X in Eq. (7.142) contains a factor 1 in its first row and a sequence of moment arm weighting functions in its second row, such that the C matrix consists of normal force components in its first row and moment components in its second row, as

$$C = \begin{bmatrix} C_{F_q} & C_{F_\alpha} & C_{F_w} \\ C_{M_q} & C_{M_\alpha} & C_{M_w} \end{bmatrix} \tag{7.143}$$

(It should be noted that the force and moment for vertical translation are defined per unit vertical velocity rather than the customary vertical displacement. This is discussed further by Whitehead in Ref. 34.)

Whitehead's report contains tabulated values of all of his coefficients for a variety of parameter values. Although it is not the purpose of this document to examine these tabulations in detail, it is instructive to note the effect of cascading on two of the unsteady coefficients: the unsteady force due to plunging C_{F_q} and the unsteady moment due to pitch C_{M_α}. These two coefficients are important determinants of the flow of energy in single-degree-of-motion oscillations, as will be seen presently. In Figs. 7.18 and 7.19 are found phase plane plots of C_{F_q} and C_{M_α}, respectively; i.e., the real part of each quantity is plotted along the abscissa and the imaginary part along the ordinate. Both plots are for $s/c = 1$ and $\zeta = 60$ deg and each plot contains three connected curves for $\lambda = 0.5$, 1.0, and 2.0 ($k = 0.25$, 0.5, and 1.0). On each curve, the point $\beta = 0$ is noted, and subsequent points, in the direction of the arrows, are for incremental increases of $\beta/2\pi = 0.1$. A dominant effect in each plot is that of reduced frequency, but this is not particularly surprising since comparable changes occur for isolated airfoils. What is of importance here is the large change that occurs from point to

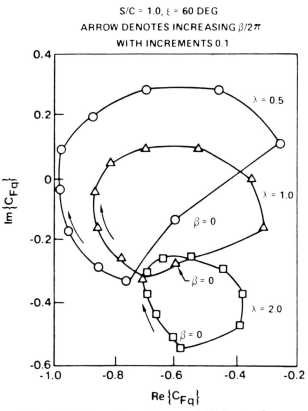

S/C = 1.0, ξ = 60 DEG

ARROW DENOTES INCREASING $\beta/2\pi$

WITH INCREMENTS 0.1

Fig. 7.18   Phase plane diagram of lift due to plunge.

point as the interblade phase angle is varied. This is uniquely associated with the presence of neighboring blades and is indicative of the importance of these cascade parameters on the ultimate determination of blade stability.

Compressible Subsonic Flow

Several early attempts to solve the problem of cascaded flat-plate airfoils oscillating in a compressible stream employed an extension of the Possio acceleration potential method to account for the presence of neighboring blades (Refs. 37–41 to name a few). In Ref. 37, the present author extended the method to the point of writing the formal solution in integral equation form, but was unable to obtain numerical values because of the limitations of the available computing equipment in the mid-1950's. Similarly, Lane and Friedman[38] employed a Fourier transform approach, but they too were limited by inadequate computer facilities to a few calculations for an unstaggered cascade oscillating at an interblade phase angle of $\sigma = 180$ deg. This case was specifically chosen by Lane and Friedman[38] because it

represented the cascade equivalent of compressible wind-tunnel wall interference (Ref. 42), and they were able to show good agreement with
previous theory for the few cases they calculated.

The three later analyses concentrate primarily on the acoustical properties of the oscillating cascade in a compressible flow. This is especially true
of the Kaji-Okazaki[40] analysis in which the results are confined to an
equation for the acceleration potential function and to plots of acoustical
transmission coefficients. More to the point are the analyses of Whitehead[39]
and Smith,[41] who both seek primarily acoustical answers, but who also
discuss the results for unsteady lift and moment. Of the two, it appears that
the Smith calculation procedure is somewhat more straightforward and
easier to implement. Unfortunately, although the analysis is completely
general, the only coefficients tabulated are either for incompressible flow or
for zero stagger with antiphase motion, in both instances for comparison of
Smith's limiting cases with the work of previous authors.

For several years, the study of unsteady subsonic compressible cascade
flow remained dormant and the theoreticians concentrated their efforts on
the more tractable (and critical) problem of supersonic relative flow flutter
(cf. next section). It was clear that a direct solution of the flat-plate cascade

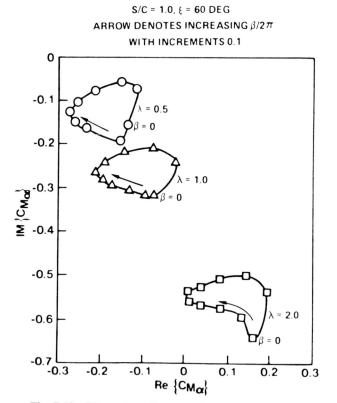

Fig. 7.19 Phase plane diagram of moment due to pitch.

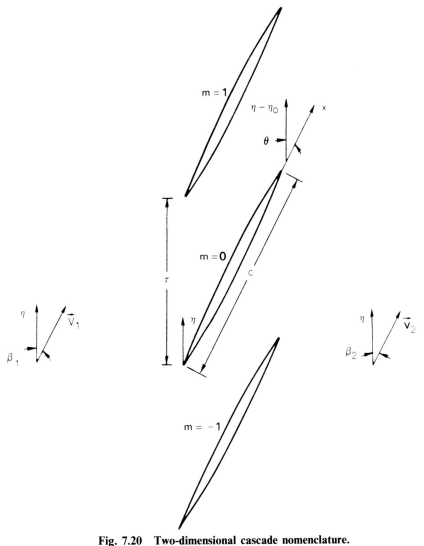

Fig. 7.20   Two-dimensional cascade nomenclature.

problem was beyond the scope of the current state-of-the-art in the 1960's and early 1970's and it was becoming apparent that the observed subsonic instabilities were invariably associated with an increase in the loading of "real" blades. In other words, a useful solution would require the inclusion of nonzero flow deflection (or turning) through a cascade of thick, cambered airfoils. These requirements precluded the use of classical linear aerodynamic theory in which the unsteady flow is treated as a small disturbance about uniform steady flow. Such solutions do not apply if velocity gradients due to incidence, blade shape, or operation in the transonic Mach number regime are significant.

In the early 1970's, an aerodynamic model was formulated to overcome these limitations. It permitted subsonic inlet and exit conditions and included the effects of blade geometry and flow turning on unsteady response.[43,44] The unsteady equations are derived from the assumption that unsteady disturbances are of small amplitude and are harmonic in time relative to a fully nonuniform irrotational mean or steady background flow. The resulting set of equations are linear, time independent, and contain variable coefficients that depend on the underlying mean flow. These equations are solved using an implicit least-squares finite-difference approximation that is applicable on arbitrary grids—an important feature for turbomachinery applications. In a sequence of reports and papers, numerical solutions, based on this linearized analysis, have been applied to subsonic flows through vibrating cascades of double-circular-arc (DCA) airfoils and NACA 0012 airfoils[45,46] and for subsonic and transonic flows through vibrating cascades of flat-bottomed DCA airfoils.[47,48] The analytical predictions based on this theory are generally in very good agreement with measured results and will be dealt with in the subsequent section on experimental results. The discussion of the analysis, which follows below, is borrowed liberally from the survey report of Ref. 49.

Assume an isentropic and irrotational flow of a perfect gas through a two-dimensional cascade (Fig. 7.20) of vibrating airfoils. The blades are undergoing identical harmonic motions at frequency ω, but with a constant phase angle σ between the motions of adjacent blades. It is assumed that the flow remains attached to the blade surfaces and that the blade motion is the only source of unsteady excitation.

As a result of the foregoing assumptions, the flow through the cascade is governed by the field equations

$$\frac{\partial \tilde{\rho}}{\partial t} + \nabla \cdot (\tilde{\rho} \nabla \tilde{\Phi}) = 0 \qquad (7.144)$$

and

$$\frac{\tilde{\rho}}{\bar{\rho}_1} = \frac{1 - (\gamma - 1)\bar{\rho}_1 \{\tilde{\Phi}_t + [(\nabla\tilde{\Phi})^2 - V_1^2]/2\}}{\gamma P_1} \qquad (7.145)$$

where $\tilde{\Phi}(X,t)$ and $\tilde{\rho}(X,t)$ are the time-dependent velocity potential and fluid density; $\bar{\rho}_1$, P_1, and V_1 are the upstream freestream density, pressure, and velocity, respectively; γ is the specific heat ratio of the fluid; X a position vector; and t the time. In addition to Eqs. (7.144) and (7.145), the flow must be tangential to the moving blade surfaces and the acoustic energy must either attenuate or propagate away from or parallel to the blade row in the far field. Finally, mass and tangential momentum must be conserved across shocks and the pressure and the normal component of the fluid velocity must be continuous across the vortex-sheet unsteady wakes that emanate from the blade trailing edges and extend downstream.

Equations (7.144) and (7.145) along with equations based on the foregoing conditions at the blade, shock, and wake surfaces and in the far field are sufficient to determine the unsteady flow. However, the computing resources required by this nonlinear time-dependent unsteady aerodynamic formulation limits its usefulness for turbomachinery aeroelastic investigations. Instead, an assumption of a small unsteady disturbance is usually invoked. This assumption permits an efficient approximate description of the unsteady flow that is suitable for aeroelastic calculations.

Thus, following the approach used in Refs. 45–48, assume that the blades are undergoing small-amplitude (i.e., of $\mathcal{O}(\epsilon) \ll 1$) unsteady motions and expand the flow variables in asymptotic series in ϵ; e.g.,

$$\tilde{\Phi}(X,t) = \Phi(X) + \tilde{\phi}(X,t) + \cdots \tag{7.146}$$

where, $\Phi(X)$ is the zeroth-order or steady-flow potential, $\tilde{\phi}(X,t) = \mathrm{Re}\{\phi(X)e^{i\omega t}\}$ is the first-order (in ϵ) unsteady perturbation potential produced by harmonic blade motions, the dots refer to the higher-order terms and $\mathrm{Re}\{\ \}$ denotes the real part of $\{\ \}$. In addition to Eq. (7.146), Taylor series expansions are used to refer information on the moving blade, shock, and wake surfaces to the respective mean positions of these surfaces. After substituting these expansions into the full governing equations, equating terms of like power in ϵ, and neglecting terms of higher than first order in ϵ, time-independent nonlinear and linear variable-coefficient boundary value problems are obtained, respectively, for the zeroth- and first-order flows.

The field equations governing the steady flow follow from Eqs. (7.144) and (7.145) after replacing the time-dependent variables $\tilde{\Phi}(X,t)$ and $\tilde{\rho}(X,t)$ by their zeroth-order or steady-flow counterparts $\Phi(X)$ and $\bar{\rho}(X)$ and setting temporal derivative terms equal to zero. The resulting equations, when combined with the associated zeroth-order boundary condition of flow tangency at the mean blade surfaces, prescribed uniform flow conditions at the inflow boundary, and a Kutta condition at blade trailing edges, describe the steady background flow through the stationary cascade.

The differential equation governing the first-order or linearized unsteady flow, i.e.,

$$A^2\nabla^2\phi = \frac{D_S^2}{Dt^2}\phi + (\gamma - 1)\nabla^2\Phi\frac{D_S}{Dt}\phi + \nabla(\nabla\Phi)^2 \cdot \frac{\nabla\phi}{2} \tag{7.147}$$

follows from the mass conservation law [Eq. (7.144)], Bernoulli's equation (7.145), the isentropic relations, and the asymptotic expansions for the flow variables. Here $A = \gamma P/\bar{\rho}$ is the speed of sound propagation in the steady background flow, $D_S/Dt = i\omega + \nabla\Phi \cdot \nabla$ is a mean flow convection derivative operator, and ϕ is the complex amplitude of the linearized unsteady potential. Solutions to Eq. (7.147) are subject to both boundary conditions at the mean positions of the blade, shock, and wake surfaces, and the requirements on the behavior of unsteady disturbances far upstream and downstream from the blade row. Shock and wake (i.e., the steady downstream stagnation streamlines) mean positions are determined from the steady solution. The unsteady surface and far-field conditions are given explicitly in Refs. 45–48 and will not be repeated here.

The preceding aerodynamic formulation for determining the unsteady flow through a cascade of airfoils undergoing small-amplitude harmonic oscillations requires the solution of a nonlinear boundary value problem for the zeroth-order or steady flow, followed by the solution of a linear variable-coefficient boundary value problem for the first-order or linearized unsteady flow. Both of these problems are time independent. Moreover, because of the cascade geometry and the assumed form of the blade motion, the steady and linearized unsteady flows must exhibit blade-to-blade periodicity. Thus, for example,

$$P(X + m\tau e_n) = P(X) \tag{7.148}$$

and

$$p(X + m\tau e_n) = p(X)e^{im\sigma} \tag{7.149}$$

where $P(X)$ is the steady pressure, $p(X)$ the complex amplitude of the linearized unsteady pressure, $m = 0, \pm 1, \pm 2,...$ a blade number index, τ the blade spacing, and e_n a unit vector in the "circumferential" or η direction. (See Fig. 7.20.)

Equations (7.148) and (7.149) allow a numerical resolution of the steady and the linearized unsteady flow equations to be restricted to a single extended blade passage region of the cascade. Although the unsteady solution is dependent on the steady solution, the numerical procedures used to solve the two equation sets can be independent of each other.

In view of the stringent and often conflicting requirements placed on the construction of a computational mesh suitable for the resolution of cascade flowfields, numerical approximations to the steady[50,51] and the linearized unsteady[45–48] problems have been based on a two-step solution procedure. First, large-scale phenomena are determined on a sheared H-type cascade mesh of moderate density. Then, for blades with rounded leading edges or flows containing shocks, the second step is to determine detailed local solutions on body-fitted polar-type meshes of high density. The local mesh domains are chosen to cover and extend well beyond limited regions of high mean velocity gradient. The final solutions to the steady and unsteady boundary value problems are taken to be composites of the corresponding cascade and local mesh solutions.

The steady or mean flow solutions of Ref. 49 were determined using the two-dimensional finite-area approximation described in Refs. 50 and 51. In this method, the integral form of the continuity equation is approximated over polygons (constructed by a triangularization of the computational mesh) in the physical plane. For transonic flows, artificial compressibility is introduced in supersonic regions to stabilize the solution scheme and to capture shocks.

The unsteady solutions were determined using the finite-difference approximation described in Refs. 45–48. Discrete approximations to the linear unsteady equations are obtained using an implicit least-squares interpolation procedure. Thus, an algebraic approximation $L\phi$ to the linear differential operator $\mathscr{L}\phi$ at the mesh point Q_0, is written in terms of the values of ϕ at Q_0 and at certain neighboring points $Q_1,...,Q_N$ as

$$(\mathscr{L}\phi)_0 = (L\phi) = q^0\phi_0 + \sum_{n=1}^{N} \delta_n(\phi_n - \phi_0) \qquad (7.150)$$

where q^0 is a multiplicative constant. The difference coefficients δ_n in Eq. (7.150) are evaluated in terms of a prescribed set of interpolating functions and a set of interpolating coefficients. The latter are determined by a weighted least-squares procedure.

The points Q_0 through Q_N, termed a neighbor set, are defined in a "centered" fashion for interior or field points and in a one-sided fashion for boundary points. For transonic applications one must distinguish between regions of subsonic flow, where the unsteady differential equation is elliptic, and supersonic flow, where it is hyperbolic. This change in equation type depends on the local mean flow Mach number and is accommodated through the use of local type-dependent differencing approximations (see Refs. 47 and 48). With proper ordering, the discrete unsteady equations can be assembled into a single block-pentadiagonal system, which can be solved conveniently using Gaussian elimination.

In a typical case, the unsteady excitation is caused by blades undergoing prescribed single-degree-of-freedom torsional (pitching) motions about an axis at or near midchord. Furthermore, the blades are assumed to vibrate at a frequency ω (or a reduced frequency k based on blade semichord) with a constant phase angle σ between the motions of adjacent blades. Thus, the angular displacement $\tilde{\alpha}_m(t)$ of the mth blade is given by

$$\tilde{\alpha}_m(t) = \text{Re}\{\alpha_m e^{i\omega t}\} = \text{Re}\{\alpha_0 e^{i(\omega t + m\sigma)}\}, \qquad m = 0, \pm 1, \pm 2,... \qquad (7.151)$$

where α_m is the complex amplitude of the mth blade angular displacement.

The linearized unsteady response parameters obey relations similar to Eq. (7.151) and thus it is necessary to consider only the response parameters associated with the reference blade. For convenience, the subscript $m = 0$, which refers to the reference blade, will be omitted in the following discussion. The unsteady aerodynamic response quantities of interest include the perturbation unsteady pressure and pressure-difference coefficients, $\tilde{C}_p$ and $\Delta\tilde{C}_p$, respectively; the unsteady aerodynamic moment coefficient $\tilde{C}_m$, acting on the moving reference blade surface; the aerodynamic work done by the airstream on this blade over one cycle of its motion C_w; and the aerodynamic damping coefficient Ξ. These quantities are defined below.

The perturbation unsteady pressure coefficient at the moving reference blade surface is given by

$$\tilde{C}_p(x,t) = \text{Re}\{C_p(x)e^{i\omega t}\} = \text{Re}\{|C_p|e^{i(\omega t + \phi_p)}\} = \frac{p_\mathscr{B}(x,t)}{\frac{1}{2}\bar{\rho}_1 V_1^2|\alpha|}, \qquad 0 \leqslant x \leqslant c \qquad (7.152)$$

where $C_p(x)$ is the complex amplitude of this pressure coefficient, x a coordinate measuring distance along the mean blade-chord line, $|\ \ |$ the magnitude of a complex quantity, and ϕ_p the phase angle by which the complex vector $C_p(x)$ leads the complex-displacement vector α. Furthermore, $p_{\mathscr{B}}(x,t)$ is the unsteady pressure perturbation at the instantaneous position of the reference blade surface $\mathscr{B}$, which is determined theoretically from the solutions for the steady and linearized unsteady velocity potentials; ρ_1 and V_1 are the uniform density and flow speed, respectively, far upstream of the cascade (see Fig. 7.20); and c is the blade chord length. The unsteady pressure-difference coefficient is defined by

$$\Delta \tilde{C}_p(x,t) = \text{Re}\{\Delta C_p(x)e^{i\omega t}\} = \text{Re}\{|\Delta C_p|e^{i(\omega t + \phi_{\Delta p})}\}$$

$$= \tilde{C}_p(x_-,t) - \tilde{C}_p(x_+,t), \qquad 0 \leqslant x \leqslant c \qquad (7.153)$$

where the subscripts $-$ and $+$ refer to the lower (pressure) and upper (suction) surfaces of the blade, respectively, and $\phi_{\Delta p}$ is the phase angle by which the complex pressure difference ΔC_p leads the complex angular displacement α.

The perturbation unsteady moment coefficient is defined by

$$\tilde{C}_m(t) = \text{Re}\{C_m e^{i\omega t}\} = \text{Re}\{|C_m|e^{i(\omega t + \phi_m)}\} = -c^{-2}\oint_{\mathscr{B}} \tilde{C}_p(x,t)(\mathbf{R}_p \cdot d\mathbf{s})$$

$$(7.154)$$

where ϕ_m is the phase angle by which the moment leads the angular displacement, $\mathbf{R}_p$ a position vector extending from the reference blade axis of rotation to a point on the moving reference blade surface, and $d\mathbf{s}$ a differential vector tangent to this blade surface and directed counterclockwise. Both the angular displacement and the moment are regarded here as being positive in the clockwise direction. It should be noted that, if shock discontinuities are present, additional terms must be added to the right-hand sides of Eqs. (7.152) and (7.154) to account for the concentrated loads produced by the motions of the shocks along the blade surface (see Refs. 47 and 48).

Finally, the aerodynamic work per cycle and the aerodynamic damping coefficient for the pure torsional vibrations are given by

$$C_w = -\pi|\alpha|\Xi = \pi|\alpha|\,|C_m|\sin\phi_m \qquad (7.155)$$

(A further discussion and a rigorous derivation of these work-related quantities are found in a subsequent section on two-dimensional work per cycle and aerodynamic damping.)

When α is prescribed as a real quantity, the right-hand side of Eq. (7.155) can be replaced by $\pi\alpha\,\text{Im}\{C_m\}$, where $\text{Im}\{\ \ \}$ denotes the imaginary part of $\{\ \ \}$. The stability of the torsional blade motion (according to linearized

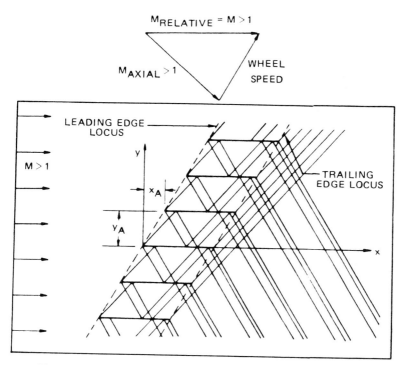

Fig. 7.21   Supersonic cascade with supersonic leading-edge locus.

theory) depends upon whether $C_w \lessgtr 0$ (or $\Xi \gtrless 0$). Thus, if $C_w < 0$, the airstream removes energy from the blade motion and this motion is stable; if $C_w = 0$, there is no net transfer of energy and the blade motion is neutrally stable; and finally, if $C_w > 0$, the airstream supplies energy to the blade motion and this motion is unstable.

The preceding discussion, taken largely from the survey report by Verdon and Usab,[49] provides a framework from which the reader, with the help of the cited references, can reconstruct the analytical procedures used in the prediction of unsteady subsonic cascade aerodynamics. Further discussion of the results of these calculations will be deferred to a later section on experimental results, where a direct comparison between theory and experiment will be made.

The reader should note that a further exposition of these and other related topics in turbomachinery aeroelasticity can be found in the two-volume AGARD Manual,[52,53] which contains contributions by Verdon, Whitehead, and several other prominent authors in this field. Space limitations preclude the inclusion of additional material by other authors in this chapter and the reader is urged to refer to the bibliography attached to this chapter.

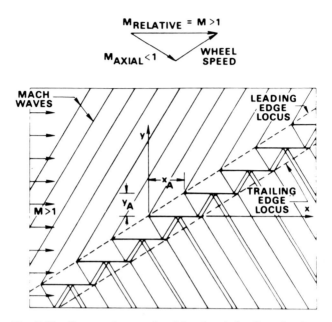

Fig. 7.22   Supersonic cascade with subsonic leading-edge locus.

Compressible Supersonic Flow

Supersonic unsteady aerodynamics of multiblade systems was first discussed by Lane[54] in 1957, who analyzed the problem in which the axial velocity into the blade row is supersonic. As seen in Fig. 7.21, all velocities relative to the plane of the leading edge are supersonic and, accordingly, this is referred to as the supersonic leading-edge locus condition. Although such a configuration has several reflected waves within any blade passage, no disturbances exist upstream of the blade passages and, further, the unsteady wakes cannot influence the flow adjacent to neighboring blade surfaces. Hence, this problem was amenable to Laplace transform techniques, as demonstrated by Lane.[54] However, it soon became evident that this solution was of academic interest only because no practical high-speed design, either within the current state-of-the-art at that time or for the foreseeable future, would have a supersonic axial throughflow. Instead, all high-speed engines were to be configured with a subsonic axial flow and a supersonic relative flow, as pictured in Fig. 7.5 and repeated in Fig. 7.22 for use in the analytical description that follows. In this case, referred to as the subsonic leading-edge locus condition, unsteady disturbances exist infinitely far upstream of each blade passage and the unsteady wakes influence the flowfield adjacent to the lower surfaces of the blades above them.

The first practical solution to this problem was obtained by Verdon in the late 1960's and subsequently published as Ref. 11 in 1973. This was a numerical solution for a cascade consisting of a finite number of blades. It was found that, in such a finite cascade approximation, the number of blades in the cascade had to be chosen sufficiently large so that a limiting behavior for the aerodynamic forces and moments could be estimated. It was then assumed that these limit values were representative of the blades of an infinite array. However, there were no mathematical proofs to validate this reasoning and, further, it was recognized (by virtue of the slow convergence of the series in the solution) that the numerical process was inefficient computationally. In addition, its use did not reveal the physics of the phenomena as well as an analytic solution would. Nevertheless, it was shown by Mikolajczak et al.[10] and by Snyder and Commerford[55] that the results of this finite cascade approximation yielded predictions that are in agreement with rotor fan experience.

Despite the apparent success of this method, it was seen that, in addition to the shortcomings cited above, the necessity to build a solution by successively adding blades to the cascade yielded a procedure that lacked computational efficiency. Consequently, an improved "closed-form" solution was developed by Verdon and McCune[12] that eliminated most of the objections raised by the previous solution, but largely confirmed the approximations to be valid. This method is described in summary form below.

In this discussion, all quantities are dimensionless. The fluid is assumed to be an inviscid, nonconducting, ideal gas with constant specific heats and the flow is assumed to be irrotational and isentropic. Disturbances in the supersonic stream are caused by an infinite array of thin, slightly cambered, lifting surfaces or blades that are performing rapid harmonic motions of small amplitude. These motions are generally normal to the blade chord lines, which are aligned parallel to the freestream direction. Based on these flowfield assumptions, the differential equation governing the disturbed flowfield is

$$\frac{\partial^2 \psi}{\partial y^2} - \mu^2 \frac{\partial^2 \psi}{\partial x^2} - \mu^2 k^2 \psi = 0 \qquad (7.156)$$

where

$$\mu^2 = M^2 - 1$$

$$k = \omega M / \mu^2 \qquad (7.157)$$

and

$$\psi(x,y) = \phi(x,y,t)e^{i(kMx - \omega t)} \qquad (7.158)$$

Here M is the freestream Mach number, ω the reduced frequency of the blade motion (based on full chord), ψ the modified velocity potential of the unsteady flowfield, and ϕ the velocity potential defined in the usual manner. The pressure p at a point in the flowfield is given by

$$(p - p_\infty)e^{i(kMx - \omega t)} = P(x,y)$$

$$= -2\left(\frac{\partial}{\partial x} - \frac{i\omega}{\mu^2}\right)\psi(x,y) \qquad (7.159)$$

where p_∞ is the freestream pressure and $P(x,y)$ the modified relative pressure. In addition, a blade-to-blade periodicity condition

$$\psi(x + nx_A, ny_A)e^{-in\Omega} = \psi(x,y), \qquad n = 0, \pm 1, \pm 2,\dots \qquad (7.160)$$

must be satisfied, where

$$\Omega = \sigma + kMx_A \qquad (7.161)$$

and where σ is the interblade phase angle. With this periodicity condition, it is sufficient to determine ψ in a single extended blade passage region, defined in Ref. 12 by $|x| < \infty$, $0 < y < y_A$. The boundary conditions applied to the upper and lower boundaries of this region are: (1) modified potential and normal velocity components are continuous in y along the upstream extensions of the blade chordlines; (2) normal velocity component is continuous across blade and wake surfaces; (3) flow must be tangent to blade surfaces; and (4) pressure must be continuous across wake surfaces. Furthermore, there can be no upstream propagation of disturbances in supersonic flow, unsteady disturbances must be bounded at an infinite distance from their origin, disturbance waves impinging on blade surfaces must be reflected, and disturbance waves impinging on wake surfaces must be transmitted through the wake.

In the analysis of Ref. 12, the expression for the modified potential in the reference region is developed in terms of component potentials that account for disturbances originating from different sources. Component potential functions are then superposed to obtain the complete expression for the modified potential that satisfies the conditions listed above. There then follows in Ref. 12 a long and rather complicated analysis in which the isolated airfoil solution of Miles[30] is generalized to represent the potential due to the reference blade and wake and in which the supersonic leading-edge locus solution of Lane[54] is generalized to account for interference effects of neighboring blades. Although the complete solution is too involved to be repeated here, its salient features will now be reviewed.

To begin with, two cascade geometries are considered, as shown in Fig. 7.23. In the upper part, referred to as cascade A, a parameter D is defined as

$$D = x_A + \mu y_A > 1 \qquad (7.162)$$

and the lower leading-edge Mach wave from any blade passes behind the lower blades. Here, there is a single Mach wave intersection on each blade, denoted by point 1 on the pressure surface of the reference blade. In the lower part of the figure, referred to as cascade B, we have

$$D = x_A + \mu y_A < 1 < x_A + 3\mu y_A \qquad (7.163)$$

and the leading-edge Mach waves are reflected once by the adjacent blades below. Here, there are three Mach wave intersections on each blade,

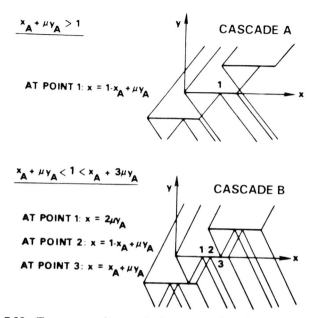

Fig. 7.23   Two supersonic cascade flow geometries of current interest.

denoted by points 1 and 2 on the pressure surface and by point 3 on the suction surface of the reference blade. Discontinuities occur in the reference blade pressure distributions at these points.

Verdon chooses to begin his analysis with the construction of the modified potential for cascade B and shows that the result for the simpler flow geometry of cascade A is readily obtained by neglecting the terms for the additional reflections. Without going into mathematical detail, the analysis proceeds as follows. The Miles modified normal velocity distribution $V(x, ny_A)$ is replaced by the disturbance function distributions $A_n(x)$, which are not known a priori, but are determined from relations derived from the boundary conditions on the unsteady flow. Five different potential functions, $\psi_i(x,y)$, $i = 1,2,...,5$, can be written to represent the various disturbance sources, including those originating on the reference blade and its wake, and the adjacent upper blade and its wake. These potential functions also account for those disturbance sources that cause a propagation into the reference passage from both upstream and downstream locations. Each of the five is an integral expression, typified by that for the fourth component potential,

$$\mu\psi_4(x,y) = - \sum_{n=-\infty}^{-1} \int_{1+\mu ny_A}^{\infty} A_n(\xi)I_{0,n}^{-}(x-\xi,y)\,\mathrm{d}\xi$$

$$- \sum_{n=-\infty}^{-1} \int_{1+\mu ny_A}^{1+B+\mu ny_A} A_n(\xi)I_{0,2-n}^{+}(x-\xi,y)\,\mathrm{d}\xi \qquad (0 < y < y_A) \qquad (7.164)$$

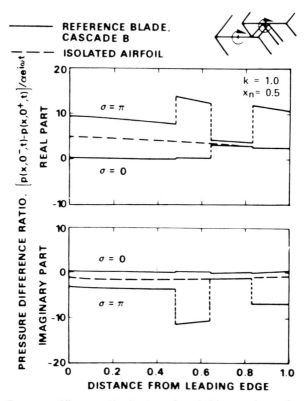

Fig. 7.24 **Pressure difference distributions for pitching motions of cascade B and isolated airfoil.**

where the influence functions $I_{0,n}^{\pm}$ are algebraic combinations of Bessel functions and unit step functions. (A complete definition of these quantities is found in Ref. 12.) Then, the modified potential distribution in the reference passage region is obtained by a summation of the component terms,

$$\psi(x,y) = \sum_{i=1}^{5} \psi_i(x,y), \qquad 0 < y < y_A \qquad (7.165)$$

which satisfies all of the boundary conditions cited above. Verdon then uses the flow tangency and pressure continuity conditions and specifies an unknown disturbance function distribution $A(x)$, which now appears in the several integrals on the right-hand side of Eq. (7.165). {This function for the reference blade is obtained from each of the more general terms, $A_n(x)$, on the nth blade by virtue of the blade motion periodicity condition [Eq. (7.160)], which involves the interblade phase angle σ.}

    Ultimately, all of this analysis must be directed toward computing the unsteady load on the reference blade, and to accomplish this Eq. (7.165) is substituted into Eq. (7.159), whereupon the reference blade pressure distribution can be expressed in terms of several integrals and summations of the unknown disturbance function $A(x)$. This is determined by using the boundary conditions of flow tangency at the upper surface of the reference blade,

$$A(x) = V(x,0) - \int_0^{x_A - \mu y_A} A(\xi)K(x - \xi)\,d\xi \qquad (0 \leqslant x \leqslant 1) \qquad (7.166)$$

and of continuity of pressure across the reference wake,

$$A(x) - \int_1^x A(\xi)K_{0,0}(x - \xi)\,d\xi = F(x) - G[A(x),x] \qquad (x > 1) \qquad (7.167)$$

where $K(x)$ is a semi-infinite series involving Bessel functions, $V(x,0)$ is the prescribed vertical velocity on the airfoil surface due to blade motion and where F and G are functionals defined by Verdon in Ref. 12. For $0 \leqslant x \leqslant x_A - \mu y_A$ Eq. (7.167) is a Fredholm integral equation with unknown function $A(x)$. The solutions of these equations were obtained numerically. Substitution into the potential function equations and thence

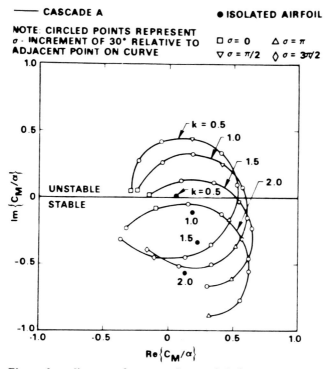

Fig. 7.25   Phase plane diagram of moment due to pitch for cascade A and isolated airfoil ($M = 1.345$).

into Eq. (7.166) yields the pressure and subsequent integrals of $P(x,y)$ yield the unsteady normal force and moment on the reference blade.

It is beyond both the scope and the purpose of this chapter to go any deeper into the details of this theory, which is adequately covered in Ref. 12. However, it is instructive to look at some of the results of the theory and to examine the implications of these results. To this end, consider Fig. 7.24 in which the real and imaginary parts of the unsteady chordwise pressure difference distribution due to pitch are plotted for cascade B (cf. Fig. 7.23). The results for two interblade phase angles, $\sigma = 0$ and π, representing in-phase and antiphase motions of adjacent blades, are presented here and are compared to the results from isolated airfoil theory (e.g., Ref. 29 or 30). The several discontinuities that occur in the cascade results are directly related to the various primary and reflected waves impinging on the reference blade. This is discussed in detail in Ref. 12. It is clear that significant differences exist between isolated airfoil theory and cascade theory; this is borne out in an examination of the predicted stability of the blade row.

It will be shown presently that a convenient measure of blade stability can be obtained from the exchange of energy between the blade and the surrounding airstream (Ref. 8). For single-degree-of-freedom pitching motions of an airfoil (either isolated or within a cascade), the work done by the airstream over one cycle of blade motion will be shown to be

$$W_{\text{per cycle}} = \pi\alpha \operatorname{Im}\{C_M\} \tag{7.168}$$

where the moment coefficient is obtained by direct integration of the chordwise pressure difference distribution,

$$C_M e^{i\omega t} = - \int_0^1 (x - x_n)[p(x,0^-) - p(x,0^+)]\, dx \tag{7.169}$$

For positive work per cycle, the flow of energy is from the airstream into the airfoil. Hence, from Eq. (7.168), $\operatorname{Im}\{C_m\} > 0$ implies instability. The usefulness of this concept is illustrated in Figs. 7.25 and 7.26 in which moment phase plane diagrams are presented for cascades A and B, respectively. Each continuous curve is for a specific value of k and the points along each curve denote variations in interblade phase angle. Also included in each diagram is a set of comparable values of C_m for an isolated airfoil.

It is seen from both figures that an increase in k is beneficial in stabilizing the system. However, it is also seen that use of isolated airfoil theory is significantly unconservative, that a dominant stability parameter is σ, and that changes in blade-to-blade interference strongly affect the stability of the system (compare cascade A with cascade B). In general, unstable cascade pitching motions are possible over a broad range of frequencies and interblade phase angles and experience has shown that these parameter ranges are relevant to current state-of-the-art turbomachines.[10] This aerodynamic theory, together with the coupled blade-disk-shroud stability

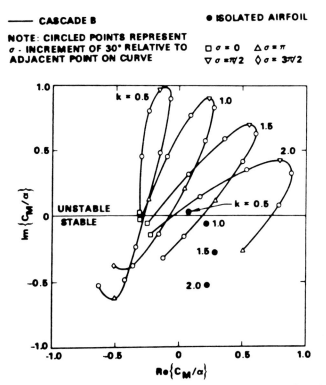

Fig. 7.26   Phase plane diagram of moment due to pitch for cascade B and isolated airfoil ($M = 1.281$).

theory of Ref. 8 (to be discussed in Sec. 7.9), has provided the designer with an essential tool to avoid serious rotor blade instabilities within the flight envelope of modern aircraft.

It should be noted that the phase plane diagrams in Figs. 7.25 and 7.26 appear to be incomplete. As noted in Ref. 12, this was related to the failure of the wake iteration procedure to converge for certain combinations of the cascade parameters. The condition for this convergence failure is referred to as "acoustical resonance" and is discussed in the next section. These convergence difficulties were ultimately overcome and solutions were obtained over the missing values of σ,[56] excluding the immediate neighborhood of the resonance points. This is further amplified in Ref. 57 where the concepts of subresonance and super-resonance are discussed. (It is noted that, at supersonic freestream Mach numbers, the far-field acoustic waves are primarily of the propagating type. Blade motions are subresonant when all waves propagate away from the blade row, resonant when at least one wave travels parallel to the blade row, and super-resonant when at least one wave attenuates with increasing axial distance from the blade row.)

In closing this section, it should be noted that, commensurate with the importance of this problem, a large number of investigators have worked

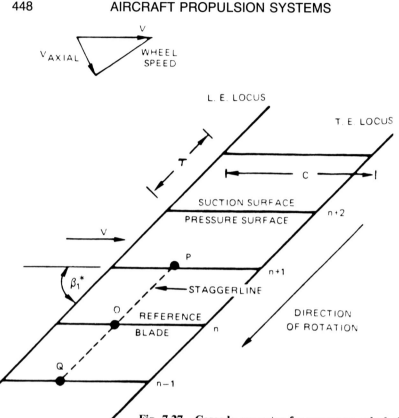

Fig. 7.27 Cascade geometry for resonance calculations.

on it, using a variety of techniques and with various constraints on their parameter ranges (e.g., Kurosaka[58] and Brix and Platzer[59]). Subsequently, the work by Goldstein et al.[60] treated the presence of a strong normal shock in the blade passage and, more recently, Surampudi and Adamczyk[61] have bridged the gap for transonic flow conditions with results on either side of $M = 1.0$ that agree with the respective classical subsonic and supersonic conditions.

Acoustical Resonance

The acoustical resonance condition will now be examined from two points of view: its implication for blade aerodynamic damping and its depiction in terms of a geometric construction of the phenomenon. First, the aerodynamic damping effects can be studied by referring to the form of the flat-plate cascade integral equation. One version of this solution, taken from Ref. 37, is

$$\bar{w}(x) = -\frac{1}{4\rho_\infty V\beta} \int_{-b}^{b} \Delta\bar{p}(x_0)[K_0(M,x-x_0) + K_1(M,x-x_0,\tau)] \, dx_0 \quad (7.170)$$

where $\beta = \sqrt{1 - M^2}$, τ is the slant gap between blades, $\bar{w}(x)$ the known prescribed downwash on the blade at x, and $\Delta \bar{p}(x_0)$ the unknown pressure difference at x_0. In this equation,

$$K_0(M, x - x_0) = -e^{-(i\omega/V)(x - x_0)} \lim_{\eta \to 0} \int_{-\infty}^{x - x_0} e^{i\omega\xi/V\beta^2}$$

$$\times \left(\frac{\partial^2}{\partial \xi^2} + \frac{\omega^2}{a^2 \beta^2} \right) H_0^{(2)} \left(\frac{\omega r_0}{a \beta^2} \right) d\xi \qquad (7.171)$$

is the Possio kernel function for an isolated airfoil [cf. a different form of this function in Eq. (7.100)] and where

$$K_1(M, x - x_0, \tau) = -e^{-(i\omega/V)(x - x_0)} \lim_{\eta \to 0} \int_{-\infty}^{x - x_0} e^{i\omega\xi/V\beta^2}$$

$$\sum_{\substack{n = -\infty \\ n \neq 0}}^{\infty} e^{2i\pi Rn} \left(\frac{\partial^2}{\partial \xi^2} + \frac{\omega^2}{a^2 \beta^2} \right) H_0^{(2)} \left(\frac{\omega r}{a \beta^2} \right) d\xi \qquad (7.172)$$

is the kernel function associated with the cascaded blades. In these equations,

$$r = \sqrt{(\xi - n\tau \sin\theta)^2 + \beta^2(\eta - n\tau \cos\theta)^2} \qquad (7.173)$$

$$r_0 = \sqrt{\xi^2 + \beta^2 \eta^2} \qquad (7.174)$$

$$R = \frac{\sigma}{\pi} - \frac{\omega M}{2\pi a \beta^2} \tau \sin\theta \qquad (7.175)$$

where, in addition, θ is the stagger angle and σ the interblade phase angle.

It was found in Ref. 37 that for certain values of the cascade parameters (to be discussed below) the cascade kernel function K_1 became unbounded. This was determined upon integrating Eq. (7.172) by parts, applying some well-known Hankel function identities, and taking the limit as $\eta \to 0$. (In this formulation η is the coordinate normal to the blade chord. The limit as $\eta \to 0$ in Eq. (7.172) is required to satisfy a downwash boundary condition in the plane of the reference airfoil, which is centered on the origin of the coordinate system.) The resulting series of Hankel functions were examined for convergence and were found to diverge at these resonance points. The implications of this series divergence is quite profound, as can be seen from an examination of Eq. (7.170). By virtue of the assumption of small displacements, $\bar{w}(x)$ must represent a bounded amplitude, and experience has shown that K_0 is always bounded. If K_1 is unbounded for certain values of the cascade parameters, then the only way this equation can be satisfied is for $\Delta \bar{p}$ to vanish identically. However, by Eqs. (7.76) and (7.77), both the unsteady lift and moment must also vanish and, hence, the naturally

occurring aerodynamic damping associated with the quadrature components of these quantities will also vanish. Under these conditions, if an
external oscillatory energy source imparts energy to the blades at a
frequency at or near the natural blade frequency, there is no aerodynamic
dissipation of this energy. Consequently, the amplitude of the blade motion
is limited only by the structural damping of the system and, under
conditions of extreme excitation, such as an integral order encounter (cf.
Fig. 7.14), large and damaging blade response can occur.

The remainder of this section will concentrate on the evaluation of the
specific flow and configuration parameter values that lead to resonance and
will be based solely on the geometric relationships between adjacent blades,
the propagation vectors, and the transit time of acoustic waves from one
blade to the next.

First, consider the cascade geometry of Fig. 7.27, which has an approach
velocity V at zero incidence in the relative frame of reference. The object
will be to determine the conditions under which a signal emanating from
point O on the reference blade will travel to either point P or Q on the
adjacent blades in time to be in phase with the production of the same
signal on the neighboring blade. The line connecting points, Q, O, and P
must be parallel to the leading-edge locus and will be referred to as a
stagger line. The signal will be assumed to be propagated outward as a
cylindrical wave from O at the speed of sound a and returned to either
point P or Q as the vector sum of a and V. The resonance condition will
be satisfied when the time required to propagate the signal along the
stagger line to a neighboring blade is equal to the time lag between the
motions of adjacent blades. The latter is concerned with the interblade
phase angle σ, which will be defined here to be positive for blade n leading
blade $n - 1$ (forward traveling wave).

Assume all blades to be executing an unsteady motion at frequency ω
and assume further that blade n performs this motion before blade $n - 1$ so
σ is a positive number. The time lag between the motions of blades n and
$n - 1$ will be any of the following quantities:

$$t_{L(+)} = \frac{\sigma}{\omega}, \frac{2\pi + \sigma}{\omega}, \dots, \frac{2\pi v + \sigma}{\omega} \qquad (v = 0,1,2,\dots) \qquad (7.176)$$

depending on the harmonic being considered. The time lag between the
motions of blades n and $n + 1$ will be

$$t_{L(-)} = \frac{2\pi - \sigma}{\omega}, \frac{4\pi - \sigma}{\omega}, \dots, \frac{2\pi v - \sigma}{\omega} \qquad (v = 1,2,3\dots) \qquad (7.177)$$

(The plus and minus subscripts have no bearing on the algebraic signs and
are merely used to designate the apparent direction of propagation, with +
associated with propagation in the direction of rotation.)

To determine the time of propagation t_P, first consider Fig. 7.28. This
diagram in velocity coordinates shows a sonic circle centered about point O
on the reference blade and shows several possible vector constructs, all of

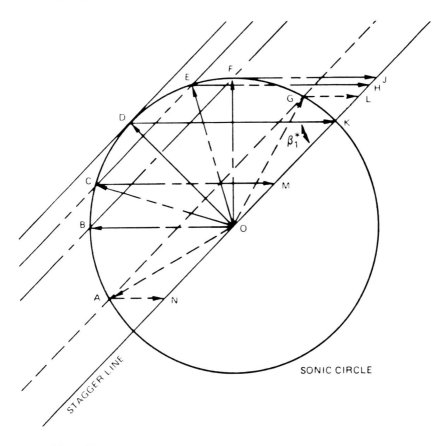

Fig. 7.28   Velocity field for several possible resonance conditions.

which terminate on the stagger line. Thus, for example, propagation from point O to point P in Fig. 7.27 could take place by virtue of the vector sum of the sonic velocity $\overline{OC} = a$ and the freestream velocity $\overline{CM} = V$. In this case, a propagation velocity $\overline{OM} = V_P$ would carry the signal from point O on the reference blade to point P on the upper adjacent blade and would be associated with the time lag of Eq. (7.177). To reduce the apparent complexity of this situation, two specific cases will be extracted from Fig. 7.28 and will be treated separately.

First, consider the case of subsonic relative flow, shown in Fig. 7.29, in which $V < a$. The two possible propagation velocities have been subscripted in accordance with the direction of propagation and are consistent with the time lag notation used earlier. (Note that these are not physical velocities so the propagation velocity V_P can be greater than a.) Use of the law of cosines on triangle OAN yields

$$a^2 = V_{P(+)}^2 + V^2 - 2VV_{P(+)} \cos(\pi - \beta_1^*) \qquad (7.178)$$

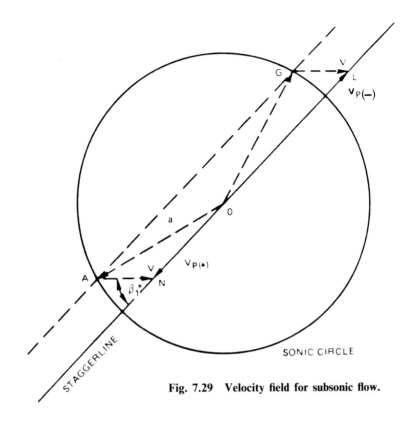

Fig. 7.29   Velocity field for subsonic flow.

and after some algebraic manipulations, it can be shown that

$$V_{P(+)} = \sqrt{a^2 - V^2 \sin^2\beta_1^*} - V \cos\beta_1^* \qquad (7.179)$$

for propagation in the direction of rotation. Similarly, for triangle OGL,

$$V_{P(-)} = \sqrt{a^2 - V^2 \sin^2\beta_1^*} + V \cos\beta_1^* \qquad (7.180)$$

The time required for the signal to propagate from point O to point Q wil$_P$ be

$$t_{P(+)} = \tau/V_{P(+)} \qquad (7.181)$$

and similarly the time of propagation from point O to point P will be

$$t_{P(-)} = \tau/V_{P(-)} \qquad (7.182)$$

The resonance condition occurs when propagation time equals lag time; after equating $t_{P(+)} = t_{L(+)}$ and $t_{P(-)} = t_{L(-)}$ and performing some inter-

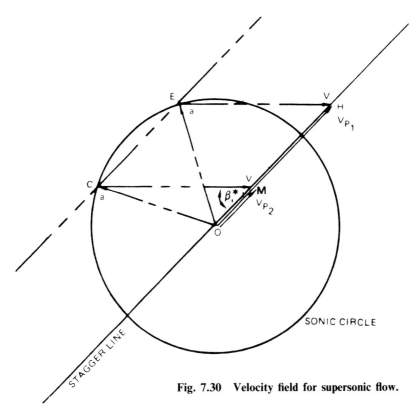

Fig. 7.30   Velocity field for supersonic flow.

mediate steps, a general formula for resonance can be written as

$$2\pi v \pm \sigma = \frac{2kM(\tau/c)}{\sqrt{1 - M^2 \sin^2\beta_1^* \mp M \cos\beta_1^*}} \tag{7.183}$$

forward, upper sign, $v = 0,1,2,...,M < 1$

backward, lower sign, $v = 1,2,3,...,M \leqslant 1$

where $M = V/a$ and $k = c\omega/2V$. The conditions under Eq. (7.183) indicate the direction of propagation and the valid ranges of parameter applicability. Note that there is a limiting case as $M \to 1$. As this case is approached, the appropriate vectors in Fig. 7.28 will be $\overline{OB}$ and $\overline{OF}$ and it is seen that the time required for forward propagation becomes arbitrarily large without bound. Hence, only backward propagation is possible for sonic and supersonic speeds.

Next consider the case of supersonic relative flow, shown in Fig. 7.30 in which $V > a$. This time only backward propagation is possible, so the velocities along the stagger line are defined simply as V_{P_1} and V_{P_2}. Once again, use of the law of cosines yields the two solutions

$$V_{P_{1,2}} = V\cos\beta_1^* \pm \sqrt{a^2 - V^2\sin^2\beta_1^*} \qquad (7.184)$$

where the subscripts 1,2 are associated with the upper and lower sign, respectively. When the time for propagation $t_{L(-)}$ from Eq. (7.177) is set equal to $t_{P(-)} = t_{P_{1,2}} = \tau/V_{P_{1,2}}$, the result is

$$2\pi v - \sigma = \frac{2kM(\tau/c)}{M\cos\beta_1^* \pm \sqrt{1 - M^2\sin^2\beta_1^*}} \qquad (7.185)$$

backward only, $v = 1,2,3,\ldots$

$$1 \leqslant M \leqslant \csc\beta_1^*$$

which can be shown to agree with the result obtained in Refs. 12 and 56.

The sonic Mach number limit has already been discussed. The upper limit represents the condition under which the axial Mach number becomes unity and is denoted in Fig. 7.28 by the single vector $\overline{OD}$. For any $M > \csc\beta_1^*$, the axial Mach number will be greater than unity and the Mach wave from point O on the reference blade (Fig. 7.27) will lie behind point P on blade $n + 1$. Hence, no resonance condition (as defined herein) can occur for a supersonic throughflow blade row. For further discussions of the resonance phenomenon, the reader is referred to the several articles listed in the bibliography.

7.8   DYNAMIC STALL—EMPIRICISM AND EXPERIMENT

Historical Background for Torsional Stall Flutter

Of all the flutter problems confronting the turbomachinery designer, the most persistent has been that associated with single-degree-of-freedom torsional motion under high load conditions. For many years, the designer had little choice in his approach to this problem. He either applied empirical "laws" obtained by the correlation and consolidation of previous flutter experiences on rotors, or he attempted to use predictions based on the growing body of isolated airfoil experience, which was then modified to fit cascade experience. Either choice was unsatisfactory and progress was ultimately made by encountering problems and overcoming them. This, at best, was a time-consuming and costly procedure.

Recently, considerable progress has been made (and continues to be made) in a direct attack on the multiblade flutter problem. In addition to the theoretical approaches cited in previous sections, the determination of empirical data for realistic multiblade configurations has also enhanced the ability of the designer to provide flutter-free operation over the flight envelope of the turbomachine. It is interesting to note, however, that current techniques still rely heavily on past practice and, in particular, on the considerable effort devoted to isolated airfoil stall flutter. Thus, it is instructive to review this field and to determine where its results are still applicable and where new discoveries indicate that large differences exist.

As early as 1938, Bratt and Scruton[62] were conducting tests to measure moment hysteresis in pitch and were using the concept of work per cycle around the moment loop to determine the stability derivatives. More will be said about this in a subsequent section. In 1943, Victory[63] published a definitive and comprehensive study of torsional stall flutter. In this experiment, she measured the quadrature, or out-of-phase, component of the moment to determine the aerodynamic damping derivative. This was found to decrease with increasing angle of attack. Victory used these empirical variations of aerodynamic damping in the classical theory and was able to calculate the critical flutter speed at stall with reasonable accuracy.

In the United States, Mendelson[64] attempted to derive an analysis for stall flutter. This work was only partially successful in that it required some rather arbitrary assumptions on the vector amplitudes of the unsteady forces and moments to make the analysis agree with the experiment. A critique of Mendelson's paper will be found in Ref. 65.

A significant contribution to stall flutter literature was published by Halfman and his colleagues in Ref. 66. This was an extensive study of the lift and moment responses of three symmetric profiles oscillating separately in both pure pitching and pure plunging motions. A balance was used to measure both the in-phase and out-of-phase components of lift and moment. The data are presented in tabular form and in hysteresis loop form and the aerodynamic stability of the airfoil is discussed using damping derivatives.

Baker[67,68] measured the stall flutter stress response as a function of velocity and incidence angle. A crossplot of these results at constant stress yielded the familiar stall flutter contour shown in Fig. 7.31. This figure is particularly interesting in that it shows the transition from classical flutter at low incidence and very high $V/b\omega$ to stall flutter at high incidence and moderate $V/b\omega$. In both of these reports, the lower limit of $V/b\omega$ for flutter seems to be $V/b\omega \cong 1.0$ for large values of incidence angle. Further work was done by Rainey[69] on a variety of planforms with different pitching axes. This report clearly showed the detrimental effect of a rearward movement of the pitching axis on stall flutter response. A crossplot of these results was published in Ref. 70 and is presented herein in Fig. 7.32.

In all of the work cited above, the measurements were made either with a dynamic balance or with strain gage beams or flexures. These devices permitted gross measurements of force, moment, or damping to be made, but the details of the unsteady flow over the blade could not be determined. In 1957, Rainey[71] published a report describing the use of miniature pressure transducers distributed chordwise over the airfoil to measure unsteady load distribution. Although Rainey did not discuss this detailed pressure data and confined his analysis to the integrated lift and moment responses, his technique paved the way for more elaborate pressure tests that have been and still are seeking to explain the underlying mechanism of stall flutter. A few of these subsequent investigations are reported in Refs. 72–74.

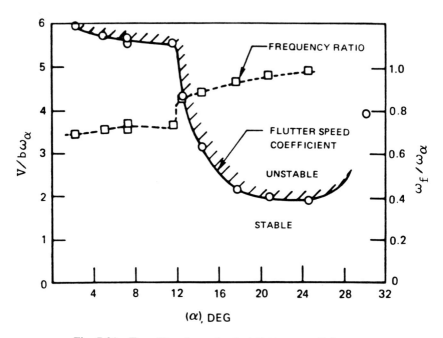

Fig. 7.31 Transition from classical flutter to stall flutter.

Consideration will now be given to these recent experiments and to some modern theoretical approaches.

Two-Dimensional Work Per Cycle and Aerodynamic Damping

In this section, the method of Ref. 75 will be followed, but the basic derivation will be equally applicable to both multiblade systems and isolated airfoils. To begin with, the work per cycle of torsional motion is given by the cyclic integral of the product of the real parts of the moment and differential pitch angle,

$$W = \oint M_R \, d\alpha_R \qquad (7.186)$$

With the assumption of sinusoidal torsional motion,

$$\alpha = \bar{\alpha} e^{i\omega t} \qquad (7.187)$$

where α is complex and $\bar{\alpha}$ a real amplitude. Similarly, the unsteady moment can be written as

$$M = \bar{M} e^{i\omega t} = (\bar{M}_R + i\bar{M}_I) e^{i\omega t} \qquad (7.188)$$

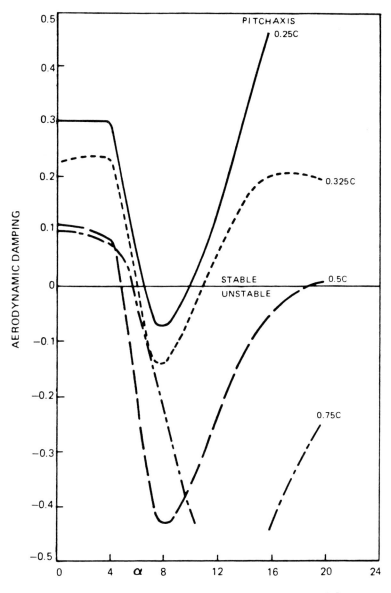

Fig. 7.32   Effect of rearward movement of pivot axis in stall flutter.

but here both M and $\bar{M}$ are complex. This is because $\bar{M}$ is generally not in phase with α. Note that the $\bar{M}_R$ in Eq. (7.188) is the real part of only the moment amplitude and is not the real part of the entire moment M_R that is needed for Eq. (7.186). To obtain this, expand the exponential in Eq. (7.188) and extract the real part,

$$M_R = \bar{M}_R \cos\omega t - \bar{M}_I \sin\omega t \qquad (7.189)$$

A similar expansion and differentiation of Eq. (7.187) yields the real part of the differential

$$d\alpha_R = -\bar{\alpha} \sin\omega t \; d(\omega t) \tag{7.190}$$

Substitute Eqs. (7.189) and (7.190) into Eq. (7.186)

$$W = \int_0^{2\pi} (\bar{M}_R \cos\omega t - \bar{M}_I \sin\omega t)(-\bar{\alpha} \sin\omega t) \; d(\omega t) \tag{7.191}$$

and by orthogonality, we have

$$W = \pi\bar{\alpha}\bar{M}_I \tag{7.192}$$

which says that the aerodynamic work per cycle is directly proportional to the quadrature, or out-of-phase, component of the moment. This is the work done by the air on the airfoil; hence, a positive value of W will indicate an unstable motion, since this implies a net energy exchange from the air to the airfoil, whereas a negative value of W will indicate a stable, or damped, motion. Equations (7.186) and (7.192) may be rewritten in coefficient form by dividing through by $\frac{1}{2}\rho V^2 (2b)^2$, with the result

$$C_w = \oint C_{M_R} \; d\alpha_R = \pi\bar{\alpha}\bar{C}_{M_I} \tag{7.193}$$

Next, we shall consider the aerodynamic damping of the airfoil, but first it is useful to review briefly the behavior of a linear, damped, torsional system such as the one described by the differential equation,

$$I\ddot{\alpha} + C\dot{\alpha} + K\alpha = 0 \tag{7.194}$$

where I is the inertia, C the damping, and K the stiffness. If the motion is essentially sinusoidal (i.e., only slightly damped and hence very nearly a constant-amplitude sinusoid), then Eq. (7.187) is a solution. The equation then becomes

$$(-\omega^2 I + i\omega C + K)\bar{\alpha} = 0 \tag{7.195}$$

It is seen that the damping coefficient is contained in the imaginary part of this expression, and we can assume that the equivalent damping terms for any similar linear system will also be contained in the imaginary part of the differential equation solution.

Equation (7.194) represents a system oscillating in torsion in a vacuum. If the same system were to oscillate in torsion in a moving airstream, the right-hand side would be replaced by the unsteady aerodynamic moment as in Eq. (7.30),

$$I\ddot{\alpha} + C\dot{\alpha} + K\alpha = M_1\ddot{\alpha} + M_2\dot{\alpha} + M_3\alpha \qquad (7.196)$$

From Eqs. (7.31), (7.32), and (7.33), we see that M_1 is a pure real quantity, whereas M_2 and M_3 are complex,

$$M_2 = M_{2R} + iM_{2I} \qquad (7.197)$$

$$M_3 = M_{3R} + iM_{3I} \qquad (7.198)$$

When these are substituted into Eq. (7.196) and the terms are rearranged, the result is

$$(I - M_1)\ddot{\alpha} + (C - M_{2R} - iM_{2I})\dot{\alpha} + (K - M_{3R} - iM_{3I})\alpha = 0 \qquad (7.199)$$

Once again, if the damping is sufficiently small, the motion will be nearly sinusoidal and Eq. (7.187) represents a solution, so Eq. (7.199) becomes

$$[-\omega^2(I - M_1) + i\omega(C - M_{2R} - iM_{2I}) + (K - M_{3R} - iM_{3I})]\bar{\alpha} = 0 \qquad (7.200)$$

After collecting real and imaginary parts,

$$\{[-\omega^2(I - M_1) + \omega M_{2I} + K - M_{3R}] + i[\omega(C - M_{2R}) - M_{3I}]\}\bar{\alpha} = 0 \quad (7.201)$$

As in the case of the system oscillating in a vacuum, the imaginary part of Eq. (7.201) will be the total damping of the system,

$$\text{Total damping} = \omega(C - M_{2R}) - M_{3I} \qquad (7.202)$$

The quantity ωC is the system damping in vacuum; therefore, the remainder of Eq. (7.202) must be the aerodynamic damping of the system, defined by

$$\xi = -\omega M_{2R} - M_{3I} \qquad (7.203)$$

The symbol ξ denotes the (dimensional) aerodynamic damping parameter of a system executing a single-degree-of-freedom torsional motion.

Now, the right-hand side of Eq. (7.196) is the unsteady aerodynamic moment acting on the airfoil, which is the same as the right-hand side of Eq. (7.188). If the two right-hand sides are equated, Eq. (7.187) is substituted for α and the exponential factor $e^{i\omega t}$ is canceled; the result is

$$\bar{M}_R + i\bar{M}_I = [(-\omega^2 M_1 - \omega M_{2I} + M_{3R}) + i(\omega M_{2R} + M_{3I})]\bar{\alpha} \qquad (7.204)$$

After equating real and imaginary parts, it is seen that

$$\bar{M}_I = [\omega M_{2R} + M_{3I}]\bar{\alpha} \tag{7.205}$$

A comparison of Eqs. (7.203) and (7.205) shows that the aerodynamic damping parameter ζ is equal to the negative of the derivative of the imaginary component of the unsteady moment with respect to the amplitude of motion,

$$\zeta = -\frac{d\bar{M}_I}{d\bar{\alpha}} = -(\omega M_{2R} + M_{3I}) \tag{7.206}$$

This may be rewritten in dimensionless form by dividing through by $\frac{1}{2}\rho V^2(2b)^2$,

$$\Xi \equiv \frac{\zeta}{\frac{1}{2}\rho V^2(2b)^2} = -\frac{\omega M_{2R} + M_{3I}}{\frac{1}{2}\rho V^2(2b)^2} \tag{7.207}$$

and hence

$$\Xi = -\frac{d\bar{C}_{MI}}{d\bar{\alpha}} \tag{7.208}$$

In the ultimate formulation to be obtained herein, it will be useful to express the aerodynamic damping parameter in terms of the work per cycle of motion, since the latter is a quantity that is usually easy to measure or to calculate. To accomplish this, a few manipulations are necessary since both $\bar{C}_{M_I}$ and C_w are implicit functions of $\bar{\alpha}$. First, return to the influence coefficient form of the moment equation in Eq. (7.20) and rewrite this for α motion only. If both M and α are put in amplitude-exponential form, the exponential can be canceled and the result is

$$\bar{M} = \pi\rho b^4\omega^2 B_\alpha\bar{\alpha} \tag{7.209}$$

After dividing through by $\frac{1}{2}\rho V^2(2b)^2$ and using the reduced frequency $k = b\omega/V$, this becomes

$$\bar{C}_M = \frac{\pi k^2}{2} B_\alpha\bar{\alpha} \tag{7.210}$$

which has an imaginary part written simply as

$$\bar{C}_{M_I} = \frac{\pi k^2}{2} B_{\alpha I}\bar{\alpha} \tag{7.211}$$

Note that the influence coefficient $B_{\alpha I}$ is independent of $\bar{\alpha}$, so Eq. (7.211) may be substituted into Eq. (7.208) to obtain

$$\Xi = -\frac{\pi k^2}{2} B_{\alpha I} \qquad (7.212)$$

Also, Eq. (7.211) can be substituted into Eq. (7.193) to yield

$$C_w = \pi \left(\frac{\pi k^2}{2} B_{\alpha I}\right) \bar{\alpha}^2 \qquad (7.213)$$

and a comparison of Eqs. (7.212) and (7.213) yields the useful formula

$$\Xi = -\frac{C_w}{\pi \bar{a}^2} \qquad (7.214)$$

To this point in the analysis, there is no restriction on the number of blades or the flow velocity.

Additional insight can be gained if we specialize the equations for the specific case of an isolated airfoil oscillating about its quarter chord $(a = -\frac{1}{2})$ in an incompressible potential flow. Although this is not the axis of rotation for compressor blades, a great deal of oscillatory data[72-74] were obtained with this pivot axis and, further, the equations for potential flow past an isolated airfoil reduce to an extremely simple form. With the substitution $a = -\frac{1}{2}$, Eq. (7.21) for B_α reduces to

$$B_\alpha = M_\alpha \qquad (7.215)$$

and from Eq. (7.18) the imaginary part of this is

$$B_{\alpha I} = M_{\alpha I} = -1/k \qquad (7.216)$$

When this is substituted into Eqs. (7.211), (7.212), and (7.213), we obtain

$$\bar{C}_{M_I} = -\frac{\pi k}{2} \bar{\alpha} \qquad (7.217)$$

$$\Xi = \pi k/2 \qquad (7.218)$$

$$C_w = -\frac{\pi^2 k}{2} \bar{\alpha}^2 \qquad (7.219)$$

Thus, we see from Eq. (7.219) that, for these ideal conditions of potential flow and pivot axis at the quarter chord, the work coefficient will always be negative, which means the net work flowing from the airstream into the airfoil is negative. In other words, the airfoil is dissipating energy and the

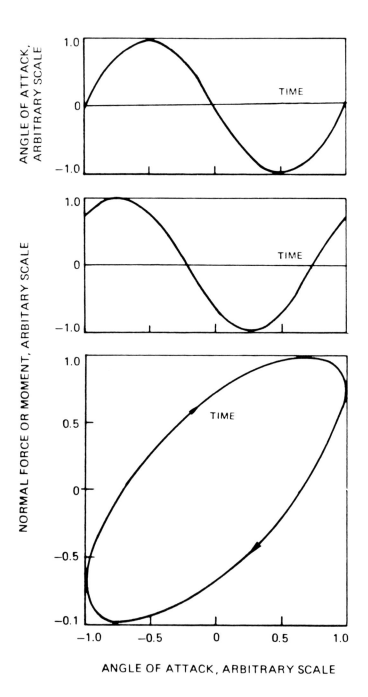

Fig. 7.33 Hypothetical normal force or moment and angle-of-attack hysteresis loops.

system is stable. Furthermore, the work coefficient is directly dependent on k and $\bar{\alpha}$. Similarly, Eq. (7.218) shows that the aerodynamic damping is always positive and is independent of amplitude.

Dynamic Stall and Stall Flutter
of a Pitching Isolated Airfoil

The process of dynamic stall is extremely complicated and has been treated phenomenologically by most investigators, as it will be in this section. To begin with, let us assume that an airfoil is oscillating in pitch and that a time history of the aerodynamic response shows that the normal force or moment is not in phase with the pitching motion, as shown in the two upper panels of Fig. 7.33. When these two time histories are combined and the time parameter is eliminated, the result will be a closed curve of normal force or moment vs angle of attack as shown in the bottom panel of Fig. 7.33. The arrows denote the direction of increasing time. The enclosed area, in the case of the moment, represents the energy absorbed or dissipated (i.e., the work per cycle) as discussed in the previous section. This is found to be the case in both classical and separated flows. In the former, the closed contour will be elliptical, whereas in the latter the contour will be distorted.

A few representative unsteady normal force coefficient loops for an isolated airfoil taken from Ref. 75 are presented in Fig. 7.34. The solid lines represent the unsteady data and the dashed lines the steady-state characteristics. Three of the inset figures are for a constant reduced frequency of $k = 0.075$ and show the effects of varying mean incidence angle from $\alpha_M = 6$ to 12 to 18 deg. The two right-hand inset figures are for $\alpha_M = 12$ deg and illustrate the effect of a change in frequency from $k = 0.075$ to 0.3. It is clear that the increase in α_M to values greater than the steady-state stall angle has a profound effect on the dynamic force response of the oscillating airfoil. It is also clear that an increase in k produces a radical change in the dynamic stalling behavior of the airfoil. Specifically, at low frequency, the dynamic force response reaches its peak value just before the maximum incidence angle is reached; it then drops precipitously to a value far below the steady-state stall value and remains there for almost the entire region of decreasing incidence. In contrast to this behavior, the effect of high frequency is to maintain a nearly elliptical response loop, even for incidence angles beyond stall, over the entire angle-of-attack range.

Until recently, very little work had been done to examine in detail the fluid dynamics of dynamic stall of isolated airfoils. Even with some of the fundamental studies now being conducted or newly completed,[76-85] a great deal of speculation remains on the nature of dynamic stall. We know that stall is produced in association with separation of flow from the airfoil surface and that separation is the disruption of the boundary layer on the airfoil. This disruption is invariably associated with an unfavorable pressure gradient (i.e., increasing pressure on the suction surface) that exceeds some unknown (for dynamic processes) critical amount. We would expect

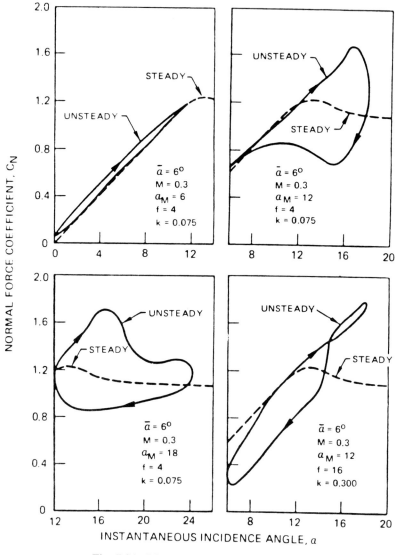

Fig. 7.34 Normal force hysteresis loops.

that two time-dependent effects would be present in the boundary layer: (1) a direct time lag associated with the inertia or mass of the fluid contained in the boundary layer and (2) a pressure time lag on the airfoil due to dynamic motion that changes the severity of the unfavorable pressure gradient. From these concepts, we can speculate that, at low frequency, there is ample time for the boundary layer to respond to the motion and separation can occur. Once the separation has taken place, recovery can occur only when the unfavorable conditions have been removed by a

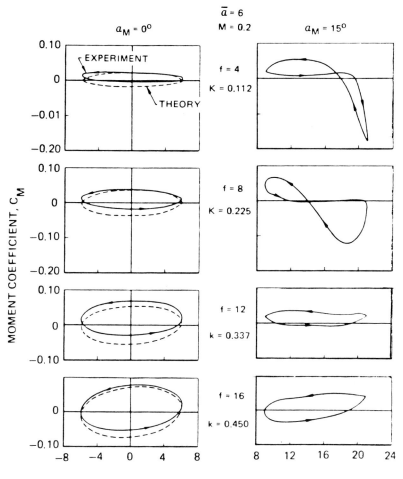

Fig. 7.35   Effect of frequency and mean incidence angle on moment hysteresis loops.

sufficiently large margin. Conversely, at high frequency, the time lag effects suppress the unfavorable effects and either no separation or only partial separation occurs. In other words, the separation does not have a chance to occur at the top of the motion before the airfoil has returned to lower angles of attack.

We now turn our attention to some representative unsteady moment loops taken from Ref. 75 and presented in Fig. 7.35. This figure shows the effects of increasing the reduced frequency from $k = 0.112$ to 0.450 for two mean angles of attack, $\alpha_M = 0$ and 15 deg. The solid curves are from the experiment and the potential flow theory predictions are shown as dashed curves for $\alpha_M = 0$ deg. At this low mean angle of attack, the theory and experiment are in good agreement and all of the loops, from low to high

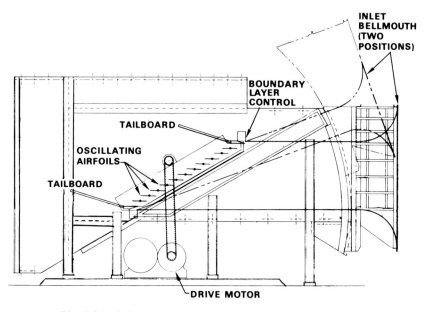

Fig. 7.36   Schematic of subsonic oscillating cascade wind tunnel.

frequency, exhibit the same general behavior. However, at $\alpha_M = 15$ deg, the low-frequency loops are badly distorted and exhibit a crossover behavior, while the high-frequency loops behave more like the theoretical loops. This crossover generally occurs at or above the static stall condition; it will now be shown that this behavior is the key to the stall flutter phenomenon for isolated airfoils.

It was stated in the previous section that system stability is related to work per cycle of motion and, from Eq. (7.186) or (7.193), this is given by the cyclic integral around the loop, which is equivalent to the area enclosed by the loop. From elementary considerations, a counterclockwise enclosure of the area represents negative work, which is equivalent to positive damping and hence a stable motion. Conversely, a clockwise enclosure of the area represents positive work, negative damping, and hence an unstable system. It is seen from Fig. 7.35 that all the $\alpha_M = 0$ deg loops as well as the two highest-frequency loops for $\alpha_M = 15$ deg have counterclockwise enclosures and hence are stable. However, the two lowest-frequency loops for $\alpha_M = 15$ deg have crossover or figure-eight characteristics in which the low-α portion is counterclockwise, implying stability, but the high-α portion is clockwise, implying instability. In other words, these crossover loops indicate an energy balance in which negative work done in the left-hand side of the loop is balanced by positive work done in the right-hand side of the loop. Actually, there is no actual balance of energy in the loops shown here because the airfoil was rigidly driven through its motion. If the airfoil had been free to oscillate, it would have reached an equilibrium amplitude, known as a limit cycle, in which a true energy balance could have been achieved.

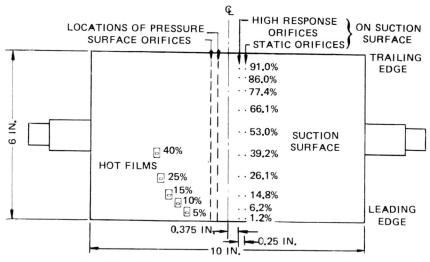

Fig. 7.37   Schematic plan view of instrumented airfoil.

It should be emphasized that this discussion is primarily applicable to isolated airfoils at high incidence angle. For many years, it was believed that these arguments could be applied to multiblade systems as well, provided that the right scaling factor or transformation variable could be found. To this end, a considerable amount of time and effort were expended in searching for these quantities. However, it has now been determined that other mechanisms involving strong interblade interactions appear to be dominant in the flutter of compressor blades under high aerodynamic loading. This will be discussed in the next section.

Unsteady Multiblade Aerodynamics at High Incidence

The very early efforts to employ empiricism in the prediction of rotor blade flutter have been described above and will not be referred to again. Instead, the purpose of this section is to document the recent and current efforts to develop a more rational approach to the problem. To date, these efforts still involve some degree of empiricism in their implementation, but they represent a significant improvement over the stress rise-flutter boundary approach of 30 years ago.

Unsteady aerodynamic tests have been conducted by the present author and co-workers in a linear, subsonic oscillating cascade wind tunnel (OCWT) shown schematically in Fig. 7.36. The general characteristics of this facility are fully discussed in Refs. 86–88 and only a brief description will be given here.

The test section of the OCWT is 10 in. wide and 25 in. high and is configured to have 11 shaft-mounted blades in cascade, equally placed along a line making a 30 deg angle with respect to the tunnel floor. Hence, the sidewall stagger of the OCWT is nominally 30 deg. Boundary-layer slots are located ahead of the test section on both sidewalls and on the tunnel ceiling and floor. The boundary-layer air is evacuated by means of an auxiliary vacuum pump. The main 10×25 in. test section receives its air from atmosphere through an upstream bellmouth and discharges downstream through two Allis-Chalmers centrifugal compressors. Inlet angle variations into the test section are obtained by rotating the floor and ceiling nozzle blocks about a pair of pivots whose centers lie along the locus of blade leading edges.

The initial cascade configuration under test consists of 11 blades, each of which has a chord of $c = 6$ in. and a span of $L = 10$ in. The airfoil has a NACA 65 series profile with 10 deg camber and a thickness-to-chord ratio of 0.06. The slant gap, measured along the blade-to-blade stagger line is $\tau = 4.5$ in. so the gap-to-chord ratio is $\tau/c = 0.75$. The blade stagger angle β_1^* is measured between the tangent to the blade mean camber line at the leading edge and the leading-edge locus line. Note that this angle is the complement of the NASA definition of blade angle (cf. Ref. 89, pp. 184, 185). Finally, the blade inlet angle β_1 is measured between the inlet velocity V and the leading-edge locus line.

The center airfoil of the cascade is heavily instrumented with both miniature high-response pressure transducers and with hot-film transducers. Figure 7.37 shows the permanent locations of all pressure orifices and the array of hot films at the leading-edge region used in the first series of tests. The entire set of airfoils is driven coherently in a sinusoidal pitching motion with an amplitude of $\alpha = \pm 2$ deg. The system is driven by an electric motor through a series of timing belts and pulleys. The direct drive to each blade culminates in a four-bar linkage that provides a sinusoidal motion with low harmonic distortion.

An initial experimental program was performed at a freestream velocity of 200 ft/s and an incidence angle of $\alpha_{MCL} = 8$ deg (relative to the mean camber line) at frequencies of 4.5, 11.0, and 17.1 cycles/s over a range in interblade phase angles from $\sigma = -60$ to $+60$ deg. The equivalent reduced frequencies were $k = 0.035, 0.086$, and 0.134. At this incidence angle, the blades were heavily loaded, but no actual stall was observed that was comparable to the stall customarily experienced on isolated airfoils. Indeed, the steady-state pressure distribution was devoid of any obvious signs of stall, as seen in Fig. 7.38. Nevertheless, a strong instability was found to exist on examination of the imaginary part of the moment coefficient [cf. Eq. (7.193)] or of the integrated aerodynamic damping coefficient [cf. Eq. (7.208)]. Typical phase plane plots for the three frequencies are shown in Fig. 7.39 and the aerodynamic damping variation with interblade phase angle is shown in Fig. 7.40. It is seen here that the dominant parameter affecting stability is σ, the interblade phase angle. This is reminiscent of the theoretical results for supersonic torsional flutter described in Figs. 7.25 and 7.26. In both cases, the comparable isolated airfoil results are found to be almost completely inadequate.

A more extensive experiment was undertaken in Ref. 90, with additional angles of attack, frequencies, and a comparably wide range of interblade phase angles. As before, the measured unsteady pressures were integrated to yield the aerodynamic damping Ξ. The results are shown in Fig. 7.41,

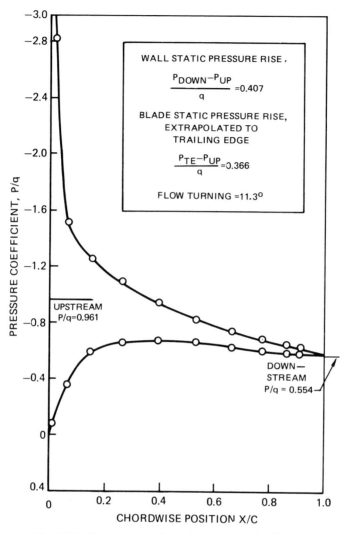

Fig. 7.38 Steady-state chordwise pressure distribution.

with $\Xi(\sigma)$ vs σ for each combination of k and α_{MCL}. In this array of plots, α_{MCL} increases from top to bottom and k increases from left to right. The staggered placement of the second row provides an approximate coordination of k values in the vertical column of plots.

At $\alpha_{MCL} = 6$ deg (top row) changes in k produce only minor changes in $\Xi(\sigma)$, which will subsequently be interpreted to mean that loading effects are minimal (but not absent). In contrast to this, at $\alpha_{MCL} = 10$ deg (bottom row), changes in k produce substantial changes in $\Xi(\sigma)$, which will be interpreted to mean that loading effects are significant. Similarly, a vertical

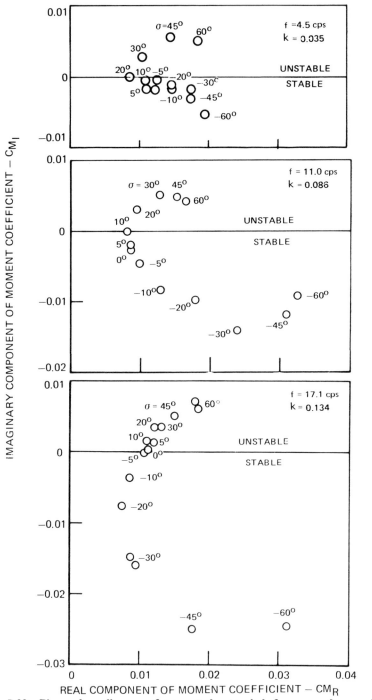

Fig. 7.39 Phase plane diagram of moment due to pitch from cascade experiment (wind-tunnel velocity $V = 200$ ft/s, mean incidence angle $\alpha = 8$ deg).

scan of any column for comparable values of k will show large changes at the smallest k and small changes at large k. (Theoretical results are also shown for $\alpha_{MCL} = 8$ deg, which will be discussed presently.)

To establish the basis for this reasoning the reader is referred to any number of reports and papers on the unsteady aerodynamics of dynamically stalled isolated airfoils (e.g., Refs. 75 and 91). In these and other research efforts, several basic principles have been established, as follows:

(1) At low load and hence in the absence of stall, increasing k has a minimal effect on changing the character of the unsteady isolated blade response. Specifically, the moment hysteresis loops generated by the blade response are qualitatively similar, while exhibiting regular and systematic changes with frequency.

(2) At high load and hence in the presence of stall, increasing the frequency has a profound effect on the character of the unsteady blade response. At low k, the dynamic stall phenomenon manifests itself in distorted hysteresis loops and higher harmonic reactions. At high k, the loops become similar to the potential flow loops, even when the static stall regime is deeply penetrated.

(3) Thus, at constant low k, an increase in load will produce strong dynamic stall effects, while at constant high k, increasing the load will not necessarily produce dynamic stall on the airfoil.

(4) If the isolated airfoil becomes unstable in single-degree-of-freedom torsional motion, it will do so at high load and low k. Increasing the frequency will tend to stabilize the motion.

Returning now to Fig. 7.41, it is seen that the cascade is always unstable for positive values of σ. This confirms the results of Ref. 86, shown earlier in Fig. 7.40. This was also predicted by the flat-plate cascade theory of Whitehead,[34] using the procedures developed by Smith.[41] (They were

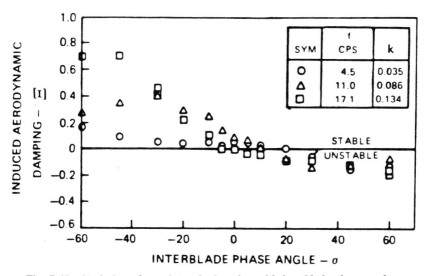

Fig. 7.40   Variation of aerodynamic damping with interblade phase angle.

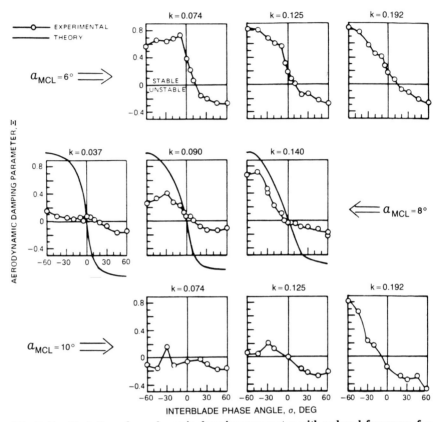

Fig. 7.41 Variation of aerodynamic damping parameter with reduced frequency for $\tau/c = 0.75$.

calculated only at $k = 0.035$, 0.086, and 0.134 and have been superimposed on the three panels for $\alpha_{MCL} = 8$ deg for convenience. The general trends for the theory are similar to the experimental results for low loading or for high frequency, although the magnitude of Ξ does not agree. This is hardly surprising because even at $\alpha_{MCL} = 6$ deg, the results in Ref. 90 show that significant steady turning has occurred and the theory is derived for a flat-plate cascade at zero incidence angle. Nevertheless, theory and experiment both agree in their prediction of instability for $\sigma > 0$, even under low-load conditions.)

It appears, then, that we can draw a diagonal line through the set of panels in Fig. 7.41, from upper left to lower right, to separate the region behaving like low loading from the region behaving like high loading, an effect experienced in isolated airfoil response, as described above. It is particularly significant that, in the upper-right diagonal region, the experimental curves are all similar to one another and strongly resemble the theoretical curves for the flat-plate cascade. We can presume, therefore, that the deterioration of stability for $\sigma < 0$ is probably associated with the

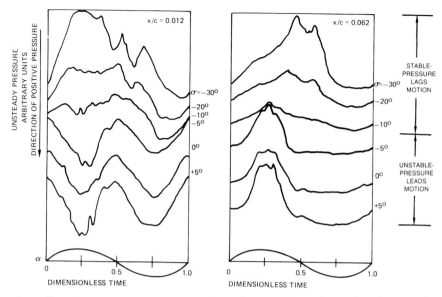

Fig. 7.42   Time-averaged pressure waveforms from airfoil suction surface for several interblade phase angles.

deterioration of flow quality under high load and that this effect is mitigated by the increase in frequency.

However, despite these apparent similarities between multiblade and isolated airfoil behavior, there is a profound difference that bears repeating here. The isolated airfoil has never been known to be unstable in subsonic single-degree-of-freedom torsional flutter except under a high-load condition, whereas the cascaded airfoil can be unstable for a variety of conditions including low loading and even potential flow. As noted several times in the preceding text, the interblade phase angle σ is consistently a key parameter in the stability of the system.

Some insight into the mechanism of this interaction effect may be gained by a brief study of the pressure time histories near the leading edge, which are shown in Fig. 7.42 over a range of interblade phase angles. The most revealing evidence here is found in the right half of this figure for the 6.2% chord location. A comparison with Fig. 7.39 shows that the system is stable for $\sigma < -5$ deg and unstable for $\sigma > -5$ deg and an examination of Fig. 7.42 shows that the pressure at $x/c = 0.062$ lags the motion for $\sigma < -5$ deg and leads the motion for $\sigma > -5$ deg. Thus, the pressure lead or lag, caused by interblade phase angle variations, plays a key role in the system stability. An even more interesting observation may be made of the left half of Fig. 7.42 for the 1.2% chord location. Here, the dominant effect of interblade phase angle variation appears to be the transition from a predominantly first harmonic response for $\sigma \simeq -30$ deg to a strong second harmonic response for $\sigma \geq -5$ deg. The pressure deficit in these latter cases occurs at peak incidence angle and has all of the outward appearances of the loss in suction peak associated with dynamic stall. Some additional details of these and other results will be found in Refs. 86, 88, and 90.

A second major experiment was conducted in the OCWT with the same 11 blade cascade and the same heavily instrumented center blade. In this case, additional pressure instrumentation was added to the leading-edge regions of five other neighboring blades in the central region of the blade row, with the object of examining unsteady blade-to-blade periodicity.[92,93] The impetus for this experimental objective was very simple and very important. The use of linear cascades to investigate phenomena related to turbomachinery blades had always been predicated on the ability of the rectilinear cascade to model blades in an annular array. To this end, steady-state experiments had customarily been devised with sufficient flow and geometric control to provide a uniform or periodic flow behavior over as much of the cascade center (i.e., the measurement region) as possible. Although this is desirable in dynamic testing as well, unsteady periodicity had not, as a rule, been verified in such tests. Virtually all of the unsteady cascade experiments reported in the open literature had generated data on one or two blades near the center of the cascade with no additional measurements away from the cascade center.

All tests previously conducted in the OCWT had been at the (relatively) large amplitude of $\bar{\alpha} = 2$ deg and the mean camber line incidence angle had been representative of modest to high loading ($\alpha_{MCL} \geqslant 6$ deg). Furthermore, all measurements had been made only on the center blade of the cascade with no opportunity to verify dynamic periodicity. Consequently, this experiment had a threefold objective, addressing the three limitations of previous experimental programs. The specific major tasks undertaken in this experiment were: (1) to examine the gapwise periodicity of the steady and unsteady blade loads under a variety of conditions; (2) to determine the effect of a smaller pitching amplitude on the unsteady response; and (3) to examine the effects of steady loading on the unsteady response by performing these tests at both low and modest incidence angles. In addition, comparisons with an advanced unsteady theory for thick, cambered airfoils were made and unsteady intergap pressure measurements were made along the leading-edge plane. Details of this study will be found in the references cited, so only the highlights will be discussed here.

As in previous tests, the center airfoil (blade 6) of the blade cascade was extensively instrumented to provide measurements of the unsteady pressure response. For this experiment, other blades were also instrumented with miniature transducers. The blades are located in the cascade as shown in the schematic diagram in Fig. 7.43. Blade 6 is the fully instrumented center blade. Partial instrumentation was placed on blades 3, 4, 5, 7, and 9. All but blade 4 had suction surface orifices at $x/c = \chi = 0.0120$ and 0.0622 and pressure surface orifices at $\chi = 0.0120$. Blade 4 also had suction surface orifices at $\chi = 0.0120$ and 0.0622 and had additional suction surface orifices at $\chi = 0.0050$ and 0.0350 with no orifice on the pressure surface. (This permitted a more detailed coverage of the pressure response near the leading edge of a centrally located blade without significantly interrupting the evaluation of cascade periodicity near the leading edge on the pressure surface.)

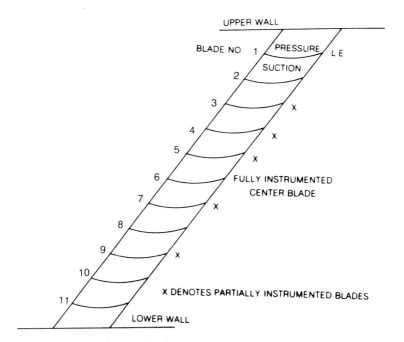

Fig. 7.43   Schematic of cascade showing instrumented blades.

Unsteady pressure data for each channel were Fourier analyzed, primarily to provide first, second, and third harmonic results for ease in analysis, but also to provide a compact means of data storage for subsequent use. (These data are completely tabulated in a data report.[94]) The measured pressure p was normalized with respect to the wind-tunnel freestream dynamic pressure q and by blade pitching amplitude in radians $\bar{\alpha}$ to yield the pressure coefficient,

$$C_p(\chi,t) = \frac{p(\chi,t)}{q\bar{\alpha}}$$ (7.220)

In general, the pressure coefficient in Eq. (7.220) can be expressed in terms of its amplitude and phase angle as

$$C_p(\chi,t) = \bar{C}_p(\chi)e^{i(\omega t + \phi_p(\chi))}$$ (7.221)

and in the plots that follow, the notation used, $\bar{C}_{p1}^{(n)}$ and $\phi_{p1}^{(n)}$, refers to the first harmonic (subscript 1) and blade number (superscript n).

A total of 96 test conditions were run. These were comprised of all possible combinations of two mean camber line incidence angles ($\alpha_{\mathrm{MCL}} = 2$, 6 deg), two pitching amplitudes ($\bar{\alpha} = 0.5$, 2 deg), three reduced frequencies

($k = 0.072$, 0.122, 0.151), and eight interblade phase angles ($\sigma = 0$, ± 45, ± 90, ± 135, 180 deg). In the discussion that follows, reference will be made to the two suction surface locations as 0.012U and 0.062U and to the pressure surface location as 0.012L. Data presented in this excerpt will be limited to a few examples.

Of the 12 possible combinations of parameters listed above, only 4 will be discussed in detail. A survey of all results has shown that the data vary only superficially with reduced frequency (for the range tested) and only the $k = 0.151$ conditions will be studied. Figures 7.44 and 7.45 contain gapwise distributions of pressure amplitudes for 2 ± 0.5 and 6 ± 0.5 deg and Figs. 7.46 and 7.47 contain gapwise distributions of pressure phase angles for 2 ± 2 and 6 ± 2 deg. Each figure has results for all interblade phase angles and, within each panel, the two suction surface measurements are depicted by solid symbols and the pressure surface measurement by the open symbol. In each case, only the first harmonic component is plotted.

Figure 7.44 shows the gapwise pressure amplitudes for 2 ± 0.5 deg to be relatively level for all three chord stations. This is particularly true for 0.062U and 0.012L. For the suction surface leading-edge station (0.012U), the measured results are generally level, but some deviation is evident at blade 4. No significant departures from these results are observed at the two lower frequencies (cf. Ref. 94). In general, these results show the cascade to be acceptably periodic in pressure amplitude response at $\alpha_{MCL} = 2$ deg.

The situation is somewhat altered for $\alpha_{MCL} = 6$ deg in Fig. 7.45. Here, the second suction surface station (0.062U) is still level, indicating good periodicity, and the pressure surface station (0.012L) is generally level with only mild deviations from completely periodic behavior. However, at 0.012U, there are strong gapwise gradients. This would suggest a significant loss in leading-edge periodicity at $\alpha_{MCL} = 6$ deg, but a recovery to periodic behavior within 5% of the chord aft of the leading edge. Once again, little change is evident for the other two reduced frequencies. It is probable that this reduced periodicity near the leading edge is associated with the increase in cascade loading.

Figure 7.46 contains plots of the gapwise distribution of first harmonic pressure phase angle measured at each of the three blade stations for 2 ± 2 deg. (Note that 0.012U and 0.062U are referred to the left scale and 0.012L to the right scale.) The phase angle for any blade is referenced to the motion of blade n by the procedure derived in Ref. 92. With the exception of the results for $\sigma = 0$ deg, the distributions are essentially flat, signifying good periodicity. The gapwise phase gradient for $\sigma = 0$ deg may be associated with the so-called acoustical resonance phenomenon that occurs near $\sigma = 0$ deg for the test conditions of this experiment. (This is discussed briefly below.) When these data are compared with data for the two lower frequencies (cf. Ref. 94), it is found that there is virtually no change for 2 ± 2 deg at all frequencies. There is also little change for 2 ± 0.5 deg at $k = 0.122$, but there is an appreciable increase in the scatter for $k = 0.072$. (This is associated with background noise and the fixed data acquisition rate, described in Ref. 92.)

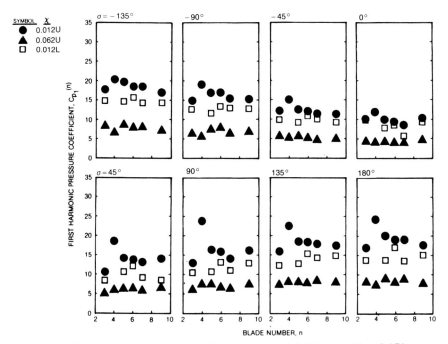

Fig. 7.44   Gapwise pressure amplitude for $\alpha = 2 \pm 0.5$ deg and $k = 0.151$.

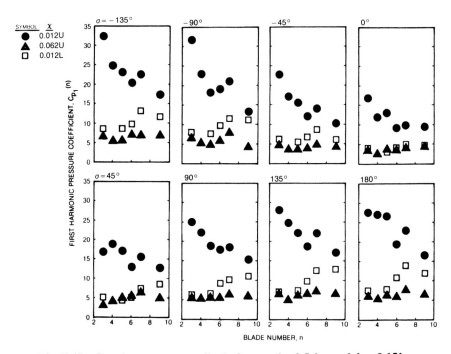

Fig. 7.45   Gapwise pressure amplitude for $\alpha = 6 \pm 0.5$ deg and $k = 0.151$.

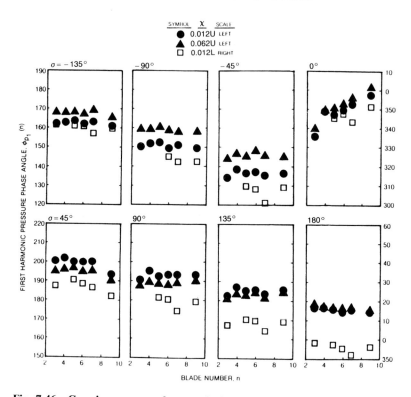

Fig. 7.46   Gapwise pressure phase angle for $\alpha = 2 \pm 2$ deg and $k = 0.151$.

Similar trends are observed for 6 ± 0.5 deg and 6 ± 2 deg in Fig. 7.47, with strong gapwise gradients for $\sigma = 0$ deg and a generally level behavior elsewhere. As before, recourse to the data in Ref. 94 shows that the scatter tends to increase at the lower frequencies at ± 0.5 deg, with less tendency to do so at ± 2 deg.

Overall, the experiment has shown that the cascade blade response in its present configuration is periodic at the lowest load condition ($\alpha_{MCL} = 2$ deg) for most parameter values tested, but has a gapwise gradient in phase angle at $\sigma = 0$ deg. Further, there is a significant gapwise gradient in magnitude at the airfoil leading edge over a wide range of σ at the modest load condition ($\alpha_{MCL} = 6$ deg), but within 5% of the chord aft of the leading edge, the amplitude response is again periodic. Phase periodicity for $\alpha_{MCL} = 6$ deg is comparable to that for the low-load condition. Thus, for these two load conditions, the measured data satisfy the periodicity condition over most of the operating ranges and over most of the blade leading-edge region, lending credence to the belief that the unsteady data obtained in this experiment are valid. This belief is considerably strengthened below when the data are compared with theory.

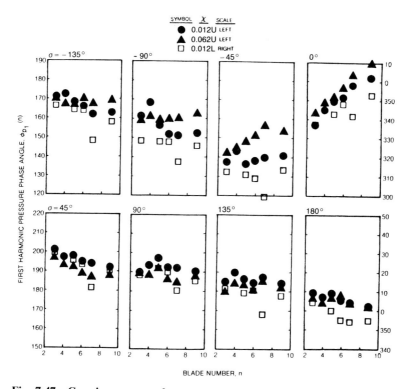

Fig. 7.47   Gapwise pressure phase angle for $\alpha = 6 \pm 2$ deg and $k = 0.151$.

Before proceeding further, the question of acoustical resonance at $\sigma = 0$ deg will be addressed. In an earlier section, Eq. (7.183) was derived (geometrically) for resonance in subsonic flow. This solution represents two possible modes: a forward wave with $v = 0$ and use of the upper sign, and a backward wave with $v = 1$ and the lower sign. Both of these modes are valid and both are computed. The parameters used here are: $M \cong 0.18$, $\tau/c = 0.75$, and $\beta_1^* = 30$ deg. A straightforward computation for the three test values of k leads to the results given in Table 7.1. With these values nested so closely about $\sigma = 0$, it is obvious that the opportunity for resonance exists.

Selected cases from these experimental results were chosen for comparison with the unsteady theory of Verdon and Caspar[44] for a blade having nonzero thickness and camber and operating in a subsonic, compressible flow. The basis of this theory is the unsteady perturbation of the potential function about the steady-state condition. Hence, it was necessary as an initial step to match the steady theory to the measured steady-state pressures. In the work cited, this is the only adjustment made to produce agreement between theory and experiment and will not be discussed in detail here since there is ample documentation elsewhere.[44,49]

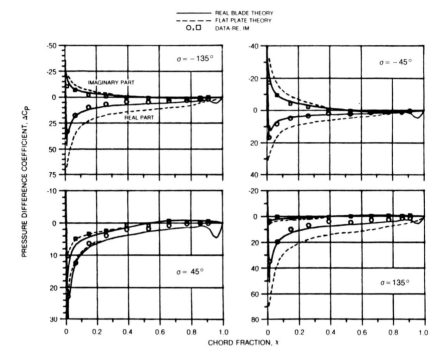

Fig. 7.48   Comparison of theory and experiment for $\alpha = 2 \pm 0.5$ deg and $k = 0.122$.

Table 7.1   Computed Acoustical Resonance
Conditions

k	Forward wave $v = 0$		Backward wave $v = 1$	
	σ rad	σ deg	σ rad	σ deg
0.072	0.023	1.3	6.266	-1.0
0.122	0.039	2.2	6.255	-1.6
0.151	0.049	2.8	6.248	-2.0

In Fig. 7.48, the real and imaginary parts (circles and squares) of the measured unsteady pressure difference coefficient for $\alpha_{MCL} = 2 + 0.5$ deg are compared with the Verdon-Caspar "real blade" theory (solid lines) and the flat-plate version of this theory (dashed lines) at $\sigma = \pm 45$ and ± 135 deg. In all cases except $\sigma = 45$ deg, the agreement of the data with the real blade

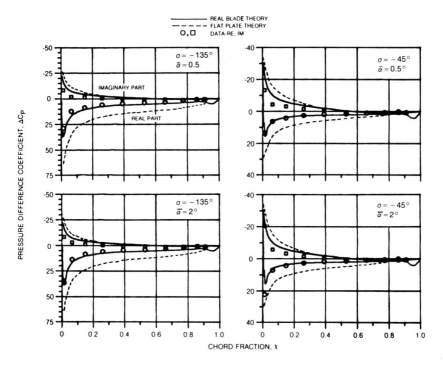

Fig. 7.49   Comparison of theory and experiment for $\alpha_{MCL} = 6$ deg and $k = 0.122$.

theory is better than with the flat-plate theory and, without exception, the agreement with the real blade theory is excellent. At this incidence angle, no distinction can be made between the data for the two separate amplitudes since they too are in nearly perfect agreement and, of course, the normalized theoretical values for the two amplitudes are identical.

In Fig. 7.49, the theory and experiment for $\alpha_{MCL} = 6$ deg are compared at $\sigma = -135$ and -45 deg in the left and right panels, respectively. Here, the experimental distributions have measurable differences at their leading edges, so the upper and lower panels are for $\bar{\alpha} = 0.5$ and 2 deg. As before, the real blade theory is independent of amplitude and is the same for upper and lower panels at each σ. Furthermore, the flat-plate theory is independent of incidence, so the plots in Fig. 7.49 are identical to those in Fig. 7.48 for $\sigma = -135$ and -45 deg. Once again, the agreement between real blade theory and the measured results is excellent and this more complete theory is shown to be superior to the flat-plate theory. (The small deviation in the real part of the Verdon-Caspar theory near the trailing edge appears to be caused by the difficulty of accurately capturing the singular behavior in unsteady pressure at a sharp trailing edge with a finite-difference approximation.)

In addition to this comparison of data with the subsonic real blade analysis of Verdon-Caspar, theoretical predictions for this profile were also

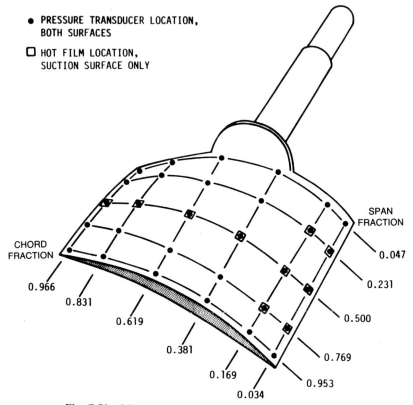

● PRESSURE TRANSDUCER LOCATION,
BOTH SURFACES

☐ HOT FILM LOCATION,
SUCTION SURFACE ONLY

Fig. 7.50 Measurement stations on equivalent blade.

made by Atassi, based on the incompressible real blade analysis described in Ref. 95. The results of this comparison are shown in Ref. 93 and the theoretical curves of Atassi are virtually the same as those of Verdon.

Shortly after these experiments and analyses were completed, a study was jointly initiated by the U.S. Air Force Office of Scientific Research (AFOSR) and the Swiss Federal Institute of Technology, Lausanne, in which data from a variety of unsteady multiblade experiments were compared with several theoretical methods.[96] Specifically, there were 9 "standard configurations" representing the experimental data and 19 aeroelastic prediction models. The experiment described above was "standard configuration 1" and the Verdon and Atassi theories were "methods 3 and 4." This study is an important contribution to the turbomachinery literature in that it covers a wide variety of configurations and methods and presents a self-consistent measure for the comparisons through the use of a (generally) uniform notation and geometric definitions. (A limited number of comparisons that were contributed to this study are contained in Ref. 49.)

One final example of the mutual corroboration of theory and experiment will be discussed in this section. A series of unsteady aerodynamic experi-

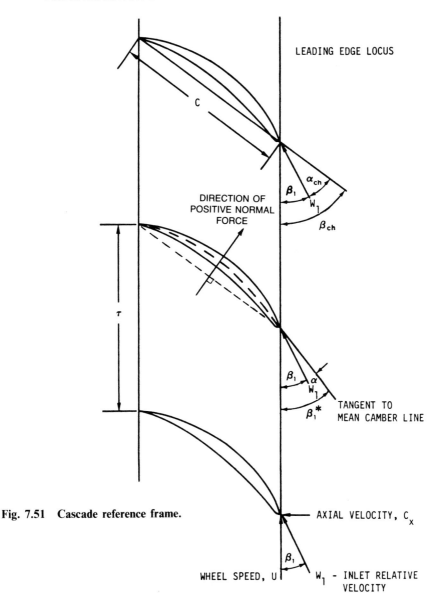

LEADING EDGE LOCUS

C

DIRECTION OF
POSITIVE NORMAL
FORCE

α_{ch}

β_1

W_1

β_{ch}

τ

β_1 α

W_1

β_1^* TANGENT TO
MEAN CAMBER LINE

Fig. 7.51   Cascade reference frame.

AXIAL VELOCITY, C_x

β_1

WHEEL SPEED, U W_1 - INLET RELATIVE
VELOCITY

ments were carried out on an isolated rotor blade row model in a
large-scale rotating rig designated LSRR2. In brief, air enters through a
12 ft diameter bellmouth, is contracted smoothly to a constant 5 ft outside
diameter of the test section, and exhausts beyond the model section into a
dump diffuser. The model centerbody is 4 ft in diameter, yielding a hub/tip
ratio of 0.8. The isolated rotor blade row consists of 28 blades, 6 in. in span
and nominally 6 in. in chord. The model geometry and the baseline

steady-state characteristics of this configuration are fully documented in Ref. 97. The model is driven by a variable-speed electric motor and the downstream flow is throttled by a variable-vortex valve prior to passing through a constant-speed centrifugal blower. In this manner, the model can mimic the speed line and pressure rise behavior of an engine compressor, although at a low absolute speed. However, the size of the model blades and the nominal operating condition of 510 rpm yields a Reynolds number, based on blade chord, of 5×10^5, which is typical of blades in a high compressor.

In this unsteady experiment, all blades were shaft mounted in a bearing support and each was individually connected to a servo motor through a four-bar linkage[93] to convert rotary motion of each motor to sinusoidal pitching motion of each blade. On-board electronic controls were used to set a common blade frequency and a variable interblade phase angle, as described in Ref. 98 and in related AFOSR Technical Reports (to be published). Additional on-board electronics provided gain and signal conditioning for the 60 miniature pressure transducers which measured both steady and unsteady pressures at six chordwise and five spanwise stations on the separate suction and pressure surfaces of an equivalent blade (cf. Fig. 7.50). (Actually, the transducers were arranged in chordwise arrays of 6 per blade on 10 separate blades and the signals were transformed by the data reduction procedures to represent the unsteady response of a single blade.)

The initial experiments on this model were run to compare the steady pressures measured with the miniature transducers with the original pneumatic data from Ref. 97. The results were in excellent agreement and are fully documented in Ref. 98. In the unsteady experiments, the blades were oscillated at a constant geometric amplitude of ± 2 deg. Primary variables of this part of the experiment were blade pitching frequency and flow coefficient. The results described in this document will relate to these two variables.

Flow coefficient is customarily defined in terms of the ratio of axial velocity C_x and wheel speed at the mean passage height U_m, or C_x/U_m. These and other parameters are depicted in Fig. 7.51.

The pitching frequency was set in terms of the number of oscillations per revolution N, which is related to the blade oscillatory frequency f by the formula

$$f = N[\text{rpm}]/60 \qquad (7.222)$$

This in turn was converted to a reduced frequency (based on full chord in Ref. 98) defined as

$$\omega = 2\pi f c/W_1 \qquad (7.223)$$

where W_1 is the inlet relative velocity into the rotor and is the vector sum of the wheel speed U and the axial velocity C_x. If the n blades of the rotor are oscillating in a standing wave pattern (in the fixed frame) and if N is the

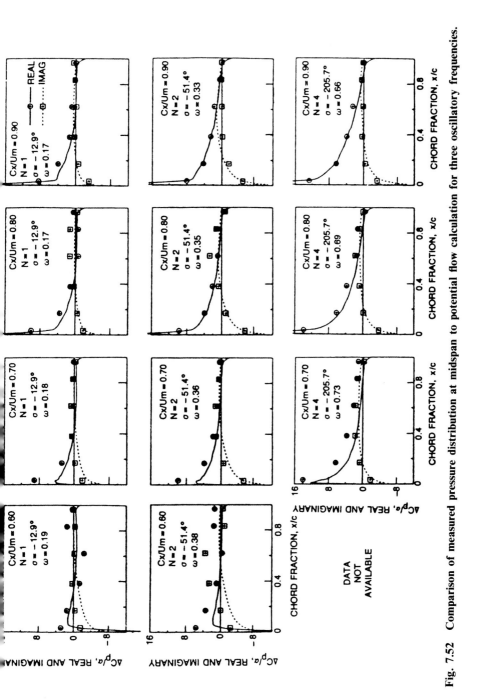

Fig. 7.52 Comparison of measured pressure distribution at midspan to potential flow calculation for three oscillatory frequencies.

number of oscillations per revolution, then the magnitude of the interblade phase angle in radians is the circumferential arc divided by the number of blades per wave, or

$$|\sigma| = \frac{2\pi}{(n/N)} = \frac{2\pi N}{n} \qquad (7.224)$$

The interblade phase angle is defined to be negative for a backward traveling wave in the rotating frame.

The original intent was to oscillate the blades at frequencies and interblade phase angles that would produce standing waves in the fixed frame of reference. However, a programming error in the blade oscillator microprocessor caused the interblade phase angle algorithm to be multiplied by 1, 2, and 4 for the 1, 2, and 4/rev conditions, so only the 1/rev wave was a standing wave. This affected only the direct comparison of the oscillatory results with the unsteady response to a sinusoidally distorted inlet flow and only for the two higher frequencies. In the oscillatory case, which is the subject of this discussion, the interblade phase angle was the only parameter affected, as described fully in Ref. 98.

Unsteady data were recorded at flow coefficients ranging 0.60–0.95 in 0.05 steps. Four flow coefficients (0.60, 0.70, 0.80, and 0.90) were selected from these as being representative of the entire set. As before, the Verdon-Caspar linearized potential flow analysis[44,48] for a cascade of blades having nonzero thickness and camber, operating in a subsonic, compressible flow, was exercised at each of these flow coefficients at the three reduced frequencies and interblade phase angles. The comparison between theory and experiment is shown in Fig. 7.52, in which the real and imaginary first harmonic components of the blade pressures at the midspan station, expressed in pressure difference coefficient form, is plotted vs chord station. The several panels in this figure show the results for three frequencies ($N = 1$, 2, 4 from top to bottom) and for four flow coefficients ($C_x/U_m = 0.6, 0.7, 0.8, 0.9$ from left to right).

The real and imaginary parts of the data are denoted by the circles and squares, and of the theory by the solid and dashed lines, respectively. Overall, the agreement is very good. The experimental data and the theory show essentially the same behavior for flow coefficients of 0.70–0.90, with the data having slightly higher amplitude at the high reduced frequency. The decrease in amplitude of the data in going from $C_x/U_m = 0.70$ to 0.60 appears to be caused by the formation of a separation bubble on the suction surface of the leading edge, which is discussed in detail in Ref. 98.

Space limitations preclude a detailed study of any of the cases cited in this document. The work reported in Ref. 98 and the several other cited references should be consulted for details of both the experiments and the analyses discussed here. Of importance to the reader (and to the turbomachinery community) are the ability to run experiments of realistic cascade or rotor configurations and the ability to obtain reasonable agreement with a theory that is tailored to the specific blade geometry under study. Both of these realizations are the culmination of a significant effort by many individuals over a considerable span of time. This does not imply that the problems are fully solved—it only indicates that many of the tools required for flutter-free design are now in place.

It is obvious that the work cited above represents only a small portion of the empirical investigations that were carried out over the past several years (see, for example, Refs. 96, 99–106, and the Bibliography). It was chosen for examination here primarily because of this author's intimate knowledge of this work, but also because it represents an approach that, while empirical, has proved to be of immediate use to the theoretician. The direct comparison between theory and experiment at the fundamental level of local surface measurements can (and does) provide the necessary insight to guide any modifications required to correct the theory or to clarify the experiment. Furthermore, the observations made in such experimental studies provide the necessary information for the development of new and advanced theoretical work.

7.9 COUPLED BLADE-DISK-SHROUD STABILITY THEORY

Genesis of the Problem

In Sec. 7.3, the design changes leading to the shrouded fan configuration were described. This additional part-span constraint dramatically raised both the torsional and bending frequencies of the blades without materially affecting the overall weight and, in some instances, permitted the use of thinner (and hence lighter) hardware. The need for this configuration was driven by the introduction of the fan engine, which required one or more stages of extra-long blades at the compressor inlet to provide an annulus of air to bypass the central core of the engine. This solved the stall flutter problem, but introduced a more insidious problem, initially termed "non-integral order flutter," that was impossible to predict with the available empirical tools. The term "nonintegral order" was chosen because the flutter, which involved the coupling of bending and torsion modes, did not occur exclusively at the intersections of the engine order lines and the natural frequency curves of the Campbell diagram (Fig. 7.14). The problem was further exacerbated by the relative supersonic speeds at which the blade tips operated.

At this time (in the early 1960's) there were no applicable aerodynamic theories capable of modeling the complex flowfield in the tip region of the new fanjet geometry, shown schematically in Figs. 7.5 and 7.22. Here, the axial velocity was subsonic, but its vector sum with the wheel speed yielded a supersonic relative speed that placed the leading-edge Mach waves ahead of the leading-edge locus of the blade row. Thus, the relatively simple theory of Lane[54] for supersonic throughflow was inapplicable. Furthermore, the incompressible multiblade theories of Refs. 32–34 were inappropriate for compressible flow and were far too complicated for routine computations on existing computer hardware; hence, they could not be used even for trend studies.

This section will review the first published work to provide a means for identifying the phenomenon and for predicting its behavior.[8] Although the initial paper relied on unsteady aerodynamic theories for isolated airfoils in an incompressible flow, the fundamental principle was sound, and its later use with the supersonic cascade theory of Verdon[11,12] and subsequent aerodynamic theories was shown to be accurate as a predictive design tool by engine manufacturers.[10,107,108]

System Mode Shapes

The vibratory mode shapes that can exist on a rotor consisting of a flexible blade-disk-shroud system are well known to structural dynamicists in the turbomachinery field and a detailed discussion of these modes is beyond the scope and purpose of this section (cf. Chapter 15 of Ref. 53). Although both concentric and diametric modes can occur, the latter are the only system modes of interest. These diametric modes are characterized by node lines lying along the diameters of the wheel and having a constant angular spacing. Thus, for example, a two-nodal-diameter mode would have two node lines intersecting normally at the center of the disk and a three-nodal-diameter mode would have three node lines intersecting at the

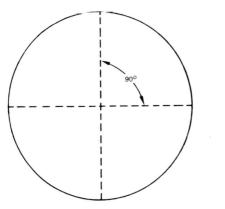

a) Two-nodal diameter pattern.

b) Three-nodal diameter pattern.

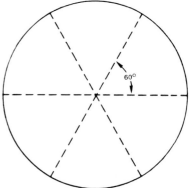

Fig. 7.53 Typical diametric node configuration.

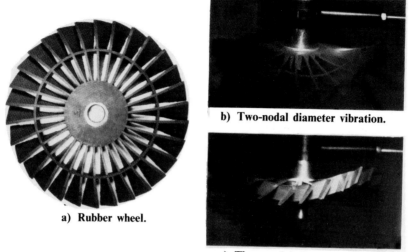

b) Two-nodal diameter vibration.

a) Rubber wheel.

c) Three-nodal diameter vibration.

Fig. 7.54   Rubber wheel deformation.

disk center with an angular spacing of 60 deg between adjacent node lines (see Fig. 7.53). These diametric modes are the physical embodiment of the eigensolutions of the system and it can be shown, using standard structural dynamics techniques, that the system frequency for each mode is primarily a function of the physical distribution of the system mass and stiffness and is only slightly affected by the rotation of the system. Thus, the system frequencies do not necessarily coincide with integral multiples of the rotor speed and in fact, such coincidences of frequency are avoided for the lower frequencies if possible.

A graphic depiction of these coupled disk modes can be found in the "rubber wheel" experiment performed by Stargardter[9] in which a flexible multiblade rotor, with integral part-span ring, was spun over a range of rotational speeds and subjected to integral-order excitations with air jets. The deformation mode shapes were exaggerated relative to a "real" rotor, but left no doubt about the physics of the problem and the key role played by the part-span shroud in coupling the bending and torsion modes. Figure 7.54, taken from the work that led to the paper, shows the flexible wheel in plan view and two other edgewise views of two- and three-nodal diameter vibrations. Of necessity, these are integral-order modes because of the use of an excitation source that was fixed in space. However, they differ from the nonintegral-order flutter modes only in that they are stationary in space, while the nonintegral modes are traveling waves.

In the early 1960's, a number of instances of nonintegral order vibrations at high stress occurred in both engine and test rig compressor rotors. The stress levels reached in a number of these cases were sufficiently high to severely limit the safe operating range of the compressor. Attempts to relate these vibrations to the stall flutter phenomenon or to rotating stall failed, largely because the vibrations often occurred on or near the engine operating line. Subsequent analysis of these cases revealed that the observed frequencies of these instabilities correlated well with the predicted frequencies of the coupled blade-disk-shroud motion described previously.

The initial object of the 1967 analysis was to explore the underlying mechanism of this instability and to show that, under certain conditions of airflow and rotor geometry, this coupled oscillation was capable of extracting energy from the airstream in sufficient quantities to produce an unstable vibratory motion. A further objective was to make the analysis sufficiently general to permit its use with advanced aerodynamic and/or structural dynamic theories and, ultimately, to provide the designer with a tool for flutter-free operation. Reference is first made to Fig. 7.6, which shows the coordinate system of the airfoil under consideration. Once again use will be made of Eqs. (7.19) and (7.20) to represent the complex, time-dependent unsteady lift and moment per unit span, which are repeated here for completeness,

$$L = L_R + iL_I = \pi \rho b^3 \omega^2 (A_h h + A_\alpha \alpha) \tag{7.225}$$

$$M = M_R + iM_I = \pi \rho b^4 \omega^2 (B_h h + B_\alpha \alpha) \tag{7.226}$$

In these equations, the quantities A_h, A_α, B_h, and B_α represent the lift due to bending, the lift due to pitch, the moment due to bending, and the moment due to pitch, respectively. They may be taken from any valid aerodynamic theory from the incompressible isolated airfoil results of Theodorsen[13] to the supersonic cascade results of Verdon.[12] At present, though, the development will use the notation A_h, A_α, B_h, B_α and consequently will be completely general.

It is well known from unsteady aerodynamic theory that the forces and moments acting on an oscillating airfoil are not in phase with the motions producing these forces and moments. A convenient representation of this phenomenon is obtained on writing the unsteady coefficients in complex form as $A_h = A_{hR} + iA_{hI}$, etc., and the time-dependent displacements as

$$h = h_R + ih_I = \bar{h} e^{i\omega t} = \bar{h} \cos \omega t + i\bar{h} \sin \omega t$$

$$\alpha = \alpha_R + i\alpha_I = \bar{\alpha} e^{i(\omega t + \theta)} = \bar{\alpha} \cos(\omega t + \theta) + i\bar{\alpha} \sin(\omega t + \theta) \tag{7.227}$$

where, in general, it has been assumed that the torsional motion leads the bending motion by a phase angle θ. In this equation, $h = h'/b$ is the dimensionless bending displacement (cf. Fig. 7.6) and $\bar{h}$ and $\bar{\alpha}$ the dimensionless amplitudes of the motion in bending and torsion, respectively.

Two-Dimensional Work per Cycle

The differential work done by the aerodynamic forces and moments in the course of this motion is obtained by computing the product of the in-phase components of force and differential vertical displacement and of

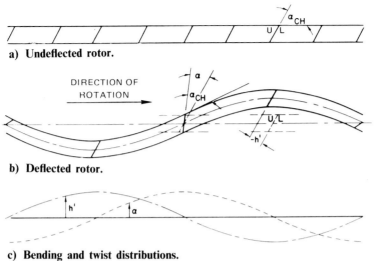

a) **Undeflected rotor.**

b) **Deflected rotor.**

c) **Bending and twist distributions.**

Fig. 7.55   Torsion and bending motions caused by coupled blade-disk-shroud interaction (U = upper surface, L = lower surface).

moment and differential twist. Accordingly, the work done per cycle of motion in each mode is obtained by integrating the differential work in each mode over one cycle. The total work done per cycle of coupled motion is given by the sum

$$W_{\text{TOT}} = -b \oint L_R \, dh_R + \oint M_R \, d\alpha_R \qquad (7.228)$$

where the minus sign is required because L and h are defined to be positive in opposite directions. It is important to note that, in Eq. (7.228), positive work implies instability since these equations represent work done by the air forces on the system.

To compute these integrals, L_R and M_R are obtained from Eqs. (7.225) and (7.226), the real parts of Eqs. (7.227) are differentiated, and these quantities are substituted into Eq. (7.228) to yield

$$W_{\text{TOT}} = -\pi\rho b^4 \omega^2 \{ \bar{h} \oint [A_{hR}\bar{h} \cos\omega t - A_{hI}\bar{h} \sin\omega t$$

$$+ A_{\alpha R}\bar{\alpha} \cos(\omega t + \theta) - A_{\alpha I}\bar{\alpha} \sin(\omega t + \theta)] \sin\omega t \, d(\omega t)$$

$$+ \bar{\alpha} \oint [B_{hR}\bar{h} \cos\omega t - B_{hI}\bar{h} \sin\omega t + B_{\alpha R}\bar{\alpha} \cos(\omega t + \theta)$$

$$- B_{\alpha I}\bar{\alpha} \sin(\omega t + \theta)] \sin(\omega t + \theta) \, d(\omega t) \} \qquad (7.229)$$

The line integrals over one cycle of motion are equivalent to an integration over the range $0 \leqslant \omega t \leqslant 2\pi$; after the indicated integrations in Eq. (7.229) are performed and the equation is simplified, the total work done on the system is given by

$$W_{\text{TOT}} = \pi^2 \rho b^4 \omega^2 \{ A_{hI} \bar{h}^2 + [(A_{\alpha R} - B_{hR}) \sin\theta + (A_{\alpha I} + B_{hI}) \cos\theta] \bar{\alpha}\bar{h} + B_{\alpha I} \bar{\alpha}^2 \}$$

$$(7.230)$$

where the quantities A_{hI} and $B_{\alpha I}$ represent the damping in bending and the damping in pitch, respectively. For an isolated airfoil oscillating at zero incidence in an incompressible flow, both of these damping terms will be negative and hence will contribute to the stability of the system. For cases involving aerodynamic inputs other than those for an isolated airfoil at zero incidence in incompressible flow, the situation may be considerably altered.

The sign of the cross-coupling term in Eq. (7.230) (the term enclosed by the square brackets and multiplied by the product $\bar{\alpha}\bar{h}$) is strongly dependent on the phase angle θ between the motions. In the usual classical flutter analysis, the phase angle remains an unknown until the end of the calculation, at which time it may be evaluated as an output quantity. For the configuration presently under consideration, however, the physical constraint of the structure on the mode shape fixes θ to be a specific input quantity, as will be shown below. This quantity within the square brackets is dominant in specifying regions of unstable operation.

It is shown in Ref. 8 that, for the specific case analyzed therein, the bending motion leads the torsion motion by 90 deg, so $\theta = -\pi/2$. An intuitive demonstration of this result is presented in Fig. 7.55. Here, it is shown that at the nodal points in the disk or shroud rim the blade will experience maximum twist with no normal displacement, whereas at antinodes in the rim the blade will have no twist, but will experience maximum normal displacement.

The value of the phase angle, $\theta = -\pi/2$, may now be substituted into Eq. (7.230) and the resulting two-dimensional aerodynamic work per cycle at each spanwise station reduces to

$$W_{\text{TOT}} = \pi^2 \rho b^2 U^2 [k^2 A_{hI} \bar{h}^2 - (k^2 A_{\alpha R} - k^2 B_{hR}) \bar{\alpha}\bar{h} + k^2 B_{\alpha I} \bar{\alpha}^2] \qquad (7.231)$$

where $k = b\omega/U$ is the reduced frequency parameter. (The use of the combinations $k^2 A_{hI}$, $k^2 A_{\alpha R}$, $k^2 B_{hR}$, and $k^2 B_{\alpha I}$ in the original 1967 paper[8] was dictated by convenience and by the availability of the aerodynamic coefficients in this form.)

Equation (7.231) is an expression for the two-dimensional aerodynamic work per cycle at any arbitrary span station, say at radius r, as measured from the engine centerline. The work done on the entire blade is obtained by integrating Eq. (7.231) over the blade span, from the root at $r = r_0$ to the tip at $r = r_T$; after normalizing the deformations $\bar{h}$ and $\bar{\alpha}$ with respect to the tip bending deflection $\bar{h}_T$, this gives

$$\frac{W}{\bar{h}_T^2} = \pi^2 \int_{r_0}^{r_T} \rho b^2 U^2 \left[k^2 A_{hI} \left(\frac{\bar{h}}{\bar{h}_T} \right)^2 \right.$$

$$\left. - (k^2 A_{\alpha R} - k^2 B_{hR}) \frac{\bar{\alpha}}{\bar{h}_T} \frac{\bar{h}}{\bar{h}_T} + k^2 B_{\alpha I} \left(\frac{\bar{\alpha}}{\bar{h}_T} \right)^2 \right] dr \qquad (7.232)$$

Stability Analysis Using Isolated Airfoil Theory

It was stated earlier that the stability of the system was related to the algebraic sign of the work expression; i.e., positive aerodynamic work implies instability and negative aerodynamic work implies stability. As shown in Eq. (7.232), the aerodynamic work done on the system is a direct function of the squares and products of the oscillatory amplitudes which are ordinarily obtained from a numerical solution of the characteristic equation for the vibratory system. It is well known from elementary vibration theory that the results of such a calculation are given in the form of relative amplitudes rather than absolute amplitudes. Therefore, the aerodynamic work can be calculated only on a relative basis and in its present form it cannot be used to predict either the absolute stability level of a particular configuration or the relative stability levels between two configurations. Since one of the objects of the original analysis was to devise a prediction technique that would permit the evaluation of alternative rotor designs from the standpoint of system stability, the theoretical development was necessarily extended to overcome this deficiency.

In addition to the relative amplitudes of motion, the structural dynamics solution provides average kinetic energy of vibration of the entire blade-disk system, based on the relative amplitudes of motion. It is shown in Ref. 8 that for a simple, linear, spring-mass-dashpot system, the ratio of damping work per cycle to average kinetic energy is proportional to the logarithmic decrement of the system, which is independent of the absolute amplitudes of the system. A comparable ratio of the aerodynamic work done per cycle on the entire blade-disk-shroud system to the average kinetic energy of vibration of the system may be made and this will be equal to the logarithmic decrement of the system. First, however, the normalized aerodynamic work per cycle obtained for one blade in Eq. (7.232) must be multiplied by the number of blades on the entire disk n and then by the proportionality factor, $\frac{1}{4}$, from Eq. (26) in Ref. 8. The result is

$$\delta = -\frac{nW/\bar{h}_T^2}{4\bar{K}_E/\bar{h}_T^2} \qquad (7.233)$$

where $\bar{K}_E/\bar{h}_T^2$ is the average kinetic energy of the system, also normalized with respect ot $\bar{h}_T^2$, and where a positive value of δ represents stable operation. Equation (7.233) yields an absolute measure of system stability that is independent of relative amplitudes. Hence, the results obtained may

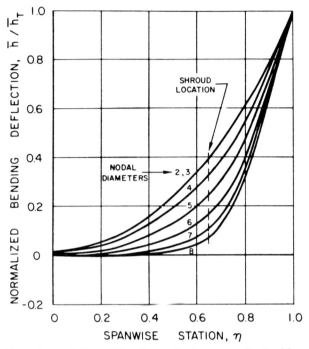

Fig. 7.56 Spanwise variation of bending deflection normalized with respect to tip bending deflection for each nodal diameter.

be used to evaluate both the absolute stability of a particular configuration and the relative stability between two or more configurations.

To illustrate the use of this theory, the stability characteristics of a typical rotor were investigated. The input quantities for use in Eq. (7.232) were obtained both from experiments and analytical studies conducted on an actual rotor at a given rpm. These data consisted of: (1) the geometric parameters for the configuration, α_{CH}, b, r_0, r_T, and n; (2) the steady-state aerodynamic parameters ρ and U; and (3) for each prescribed disk nodal diameter pattern and blade mode, the relative amplitudes of the blade deformation components $\bar{h}$ and $\bar{\alpha}$, the average system kinetic energy $\bar{K}_E$, and the system natural frequency ω_0. For each nodal diameter, both $\bar{h}$ and $\bar{\alpha}$ were then normalized with respect to the tip bending deflection $\bar{h}_T$ and are plotted in Figs. 7.56 and 7.57 as functions of the dimensionless spanwise variable η, defined by

$$\eta = \frac{r - r_0}{r_T - r_0} \tag{7.234}$$

In both of these figures, the shroud location at $\eta = 0.653$ is indicated by short tic marks on each curve.

The dynamic system chosen for analysis in Ref. 8 consisted of a blade-disk-shroud configuration in which the blade oscillates in its first bending and first torsion modes and the overall system vibration modes encompass 2–8 nodal diameters. Figures 7.56 and 7.57 contain the spanwise variations of bending deflection and blade twist, both normalized with respect to the bending deflection at the blade tip for the specific nodal diameter under consideration. Figure 7.56 shows that, as the number of nodal diameters increases, the bending mode shape undergoes a consistent change in which the deformation of the tip region, outboard of the shroud, increases relative to the inboard region.

In Fig. 7.57, however, it is seen that the torsional content of the vibration (relative to the tip bending) first increases and then decreases with the increasing number of nodal diameters. This suggests a variable amount of coupling between bending and torsion as the system mode changes from one nodal diameter to another. Furthermore, the major change in twist distribution for each curve occurs over the portion of the blade inboard of the deformed shroud, which imposes a twisting moment on the blade as a result of this deformation. Hence, the presence of a deformed shroud produces the coupling between blade twist and bending and the degree of this coupling is modified by the diametric modal pattern.

A very revealing and informative plot is shown in Fig. 7.58, in which only the tip value of the normalized deformation ratio $(\bar{\alpha}/\bar{h}_T)_T$, has been plotted as a function of frequency. This figure indicates a very strong variation of $(\bar{\alpha}/\bar{h}_T)_T$ with the number of nodal diameters (i.e., with natural frequency); it is relatively small at both small and large nodal diameters

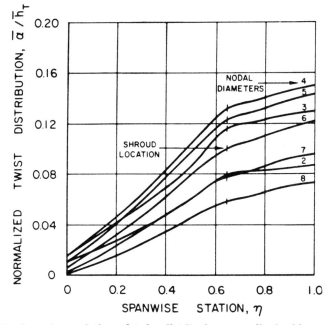

Fig. 7.57 Spanwise variation of twist distribution normalized with respect to tip bending deflection for each nodal diameter.

and reaches a maximum value at approximately 4 or 5 nodal diameters. Thus, it appears that torsion-bending coupling is a maximum for intermediate nodal diameters, with a predominantly bending motion occurring at either extreme. This increase in coupling for intermediate nodal diameters may be regarded as either a relative increase in blade twist or a relative decrease in blade bending. The absolute deformations are unimportant since the stability equations (7.232) or (7.233) are expressed solely in terms of the normalized deformations $\bar{h}/\bar{h}_T$ and $\bar{\alpha}/\bar{h}_T$.

The stability of the system was originally determined using Eqs. (7.232) and (7.233) and employing unsteady isolated airfoil theory. The logarithmic decrement δ was calculated for each nodal diameter (2–8) at the resonant frequencies appropriate for each case. Results of these stability calculations are found in Fig. 7.59. The natural frequency in each case is denoted by the circled point. System stability is indicated by positive values of δ and instability is indicated by negative values of δ. It is seen from this figure that the system is stable for the 2, 6, 7, and 8 nodal diameter modes and is unstable for the 3, 4, and 5 nodal diameter modes, with minimum stability (i.e., maximum instability) occurring at 4 nodal diameters. A comparison of Fig. 7.59 with Fig. 7.58 reveals a rather strong correlation between maximum system instability and maximum torsion-bending coupling, represented in Fig. 7.58 by the maximum values of $(\bar{\alpha}/\bar{h}_T)_T$ at 4 nodal diameters. Similar results were obtained for a number of rotor configurations that were analyzed using these procedures. Thus, it was tentatively concluded that the greater the degree of coupling between torsion and bending in a shrouded rotor, the greater the likelihood of a flutter instability.

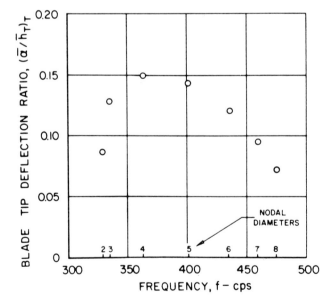

Fig. 7.58 Variation of blade tip deflection ratio with frequency.

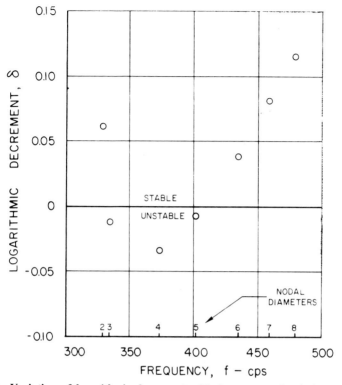

Fig. 7.59   Variation of logarithmic decrement with frequency using isolated airfoil theory.

Comparison between Theory and Experiment

The theoretical procedure for an isolated airfoil at zero incidence described in a previous section was modified slightly (by the engine development groups) for use in evaluating various rotor configurations. An iterative procedure was developed that produced the rotor parameter values for the neutrally stable condition, $\delta = 0$, in which the incompressible isolated airfoil theory of Theodorsen[13] was used at low speeds and the supersonic isolated airfoil theory of Garrick and Rubinow[29] was used at high speeds. A number of rotors were considered in this study and, in each case, values of blade tip deflection ratio $(\bar{\alpha}/\bar{h}_T)_T$ and reduced velocity at the blade tip $(U/b\omega_t)_T$ were obtained for the condition of zero logarithmic decrement, $\delta = 0$. The locus of points so obtained yielded a narrow band of scattered points through which a faired curve was drawn. This formed a single zero-damping stability boundary valid for all configurations, as shown in Fig. 7.60. It is interesting to note that the theoretical curve of Fig. 7.60 may also be confirmed by interpolating Figs. 7.58 and 7.59 to $\delta = 0$, through fictitious curves passed through the circled eigensolutions. This was done for several configurations in the report[109] that formed the basis for the 1967 paper.[8]

In Fig. 7.60, the region beneath and to the left of the curve represents stable operation and the region above and to the right of the curve represents unstable operation. Superimposed on this curve are the results of a number of engine and rotor tests. The solid circular symbols represent configurations that fluttered and the open triangular symbols represent stable configurations. It is seen that the engine experience and the theoretical prediction were in good agreement for these early cases. Furthermore, two of the unstable configurations were redesigned to yield stable rotors, as indicated by the arrows connecting two pairs of points. Hence, it was felt that the theory had some merit in predicting as well as explaining the occurrence of coupled blade-disk-shroud flutter. However, it was also felt that the agreement achieved through the use of incompressible, isolated airfoil theory on a high-speed, multiblade system was tenuous at best and a major effort was continued by a number of researchers to extend the aerodynamic model to include more realistic effects. Before these modern developments are discussed, a brief review of the energy distribution among the modes will be made and a portion of a parametric study of coupling effects will be examined.

Spanwise Variations in the Damping and Coupling Terms

In the previous section on stability analysis, the major emphasis was placed on the variation in stability parameter with changes in frequency (i.e., number of nodal diameters). In this section, some brief considerations

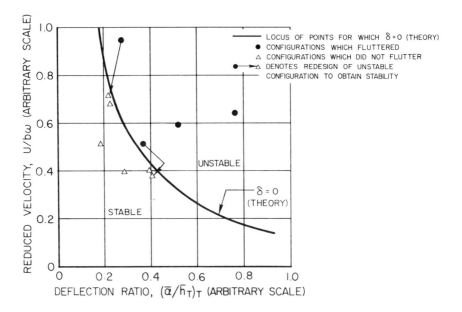

Fig. 7.60 Stability boundary—comparison between theory and experiment.

will be given to the individual terms in the work equation and their spanwise variations as they are affected by changes in frequency.

For convenience the expression for the work per cycle at each spanwise station, Eq. (7.231), will be rewritten as

$$W_{\text{TOT}} = W_B + W_C + W_P \tag{7.235}$$

where

$$W_B = \pi^2 \rho b^2 U^2 k^2 A_{hI} \bar{h}^2 \tag{7.236}$$

is the local two-dimensional work due to bending,

$$W_C = -\pi^2 \rho b^2 U^2 (k^2 A_{\alpha R} - k^2 B_{hR}) \bar{\alpha} \bar{h} \tag{7.237}$$

is the local two-dimensional work due to coupling and

$$W_P = \pi^2 \rho b^2 U^2 k^2 B_{\alpha I} \bar{\alpha}^2 \tag{7.238}$$

is the local two-dimensional work due to pitch. Before these quantities can be usefully investigated, they must be normalized in such a way that meaningful comparisons between various cases may be made. In the course of this study, it was found that the tip value of the local work due to bending,

$$W_{BT} = \pi^2 (\rho b^2 U^2 k^2 A_{hI} \bar{h}^2)_{\text{TIP}} \tag{7.239}$$

was negative and nonzero for all values of frequency in all cases considered. (It will be recalled that negative work implies stability, since this represents work done by the air on the blade.) Therefore, in view of the negative definite behavior of W_{BT}, it was decided to normalize all three local work terms, W_B, W_C, and W_P, with respect to this tip value of W_B for the specific number of nodal diameters being considered. The normalized quantities are

$$\lambda_B = \frac{W_B}{W_{BT}} = \frac{\rho b^2 U^2 k^2 A_{hI} \bar{h}^2}{(\rho b^2 U^2 k^2 A_{hI} \bar{h}^2)_{\text{TIP}}} \tag{7.240}$$

$$\lambda_C = \frac{W_C}{W_{BT}} = -\frac{\rho b^2 U^2 (k^2 A_{aR} - k^2 B_{hR}) \bar{\alpha} \bar{h}}{(\rho b^2 U^2 k^2 A_{hI} \bar{h}^2)_{\text{TIP}}} \tag{7.241}$$

$$\lambda_P = \frac{W_p}{W_{BT}} = \frac{\rho b^2 U^2 k^2 B_{\alpha I} \bar{\alpha}^2}{(\rho b^2 U^2 k^2 A_{hI} \bar{h}^2)_{\text{TIP}}} \tag{7.242}$$

It was stated earlier that the quantity W_{BT} was always a negative, nonzero number and, hence, implied a stable condition for this parameter at the blade tip. By definition the value of λ_B at the tip will be $+1.0$, since at this point W_B is divided by itself. Hence, a positive value of any of the normalized work terms will indicate a stable tendency (in contrast to a negative value of the absolute work parameter) and a negative value of any of the normalized work terms will indicate an unstable tendency.

From Fig. 7.59, the most unstable condition was a 4 nodal diameter vibration, while the most stable was an 8 nodal diameter vibration. The quantities λ_B, λ_C, and λ_P are plotted vs dimensionless span station for these two conditions in Fig. 7.61. An examination of this plot provides a great deal of insight into the mechanism involved in this flutter phenomenon. In both panels, the distributed work is negligible inboard of the midspan and is significant only over the outer one-third of the span. The normalized work in bending λ_B is always positive (stabilizing) and reaches its normal value of 1.0 at the tip. In both instances, the normalized coupling work λ_C is negative (destabilizing) and the normalized work in pitch λ_P is of second order and can be ignored. For the 4 nodal diameter case, the bending and coupling work both reach approximately the same value at the tip, but the distribution over the span is such that the destabilizing integrated work in coupling is greater in magnitude than the stabilizing integrated work in bending. This leads to a net instability of the system of this mode. (A similar situation, with less obvious differences between bending and coupling distributions, is found for the less unstable 3 nodal diameter mode in Fig. 31 of Ref. 109.) In contrast, the spanwise distributions for the 8 nodal diameter mode show that the stabilizing work in bending is significantly greater in magnitude than the coupling work over the entire span and, accordingly, the system is stable at this condition.

Use of Advanced Aerodynamic Theories and Typical Mode Shapes

The results in Fig. 7.61 were for a so-called first family spanwise mode. More typical higher-order family modes, having additional circumferential

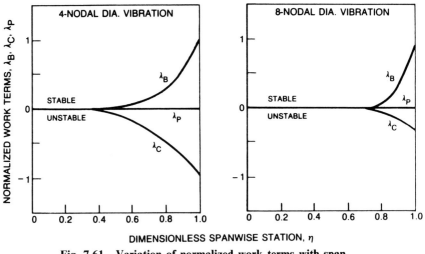

Fig. 7.61 Variation of normalized work terms with span.

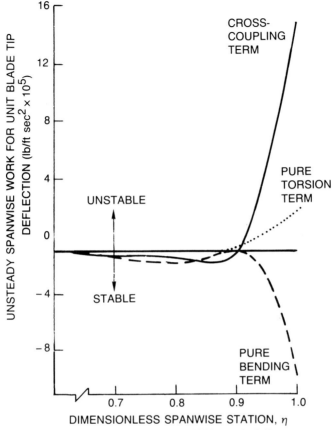

Fig. 7.62   Fan unsteady work components at 100% speed, 4 diameter, second family mode (from Ref. 107).

nodes lines, were examined independently by Mikolajczak et al.[10] and by Halliwell[107] for state-of-the-art compressors of the mid-1970's. Halliwell used the aerodynamic theory of Nagashima and Whitehead[110] and obtained the spanwise distribution of work per cycle shown in Fig. 7.62 for the second family modes. This confirms the distributions of work shown in Fig. 7.61. Note that Halliwell's results are dimensional and are inverted relative to Fig. 7.61. He further computed the integrated work per cycle for each component of the stability equation; the results are shown in Fig. 7.63 as a function of nodal diameter number, again for the second family mode. Here again, the coupling term is destabilizing, and opposite to the bending term. An overall system instability for the 4, 5, and 6 nodal diameter forward traveling waves was predicted and was confirmed as 4 nodal diameters for the test compressor.

The work by Mikolajczak et al.[10] concentrated on the overall aerodynamic damping of several compressor designs. This was preceded by the cascade study of Snyder and Commerford,[55] who also examined the typical compressor designs. The work by Mikolajczak and his coauthors employed several aerodynamic theories, depending on the local aerodynamic conditions. For the supersonic relative flow (with subsonic axial Mach number), use was made of Verdon[11] and for subsonic flows Smith's[41] theory was used. In addition, cambered thin airfoils were treated using an extension of Whitehead's early work[34] or the analysis of Sisto and Ni.[111]

Three radial modes were examined in this work, which concentrated on two compressor rotors. Rotor A was designed specifically to be susceptible to an unstalled supersonic flutter in its second radial mode. It experienced flutter at 13,290 rpm in its second mode with a 4 nodal diameter vibrational pattern. The predicted aerodynamic damping for the three modes, plotted as a function of reduced frequency, is shown in Fig. 7.64. This clearly

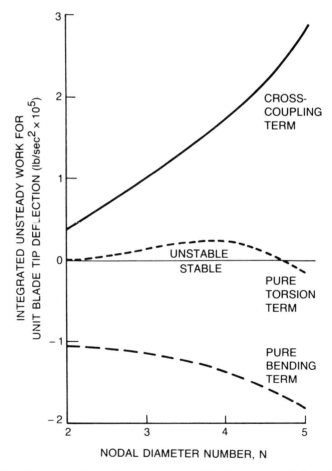

Fig. 7.63 Fan unsteady work components at 100% speed, variation with second family mode number (from Ref. 107).

shows the second mode to be the least stable mode, although the 5 and 6 nodal diameter patterns appear to be theoretically more unstable than the 4 nodal diameter pattern. The sensitivity of predicted stability to changes in rpm for rotor A is shown in the prediction of Fig. 7.65 for the second radial mode. Rotor B (a NASA 1800 ft/s design) was specifically designed to be flutter free over its performance range. It was tested successfully with no flutter up to 12,464 rpm. The predicted aerodynamic damping for the first three radial modes was positive for all nodal diameters, as shown in Fig. 7.66. In a summary of these and several other rotor designs, the Mikola-jczak paper shows that the use of the analytical prediction techniques described here were consistently conservative (at least up to the date of publication) and generally capable of predicting the correct flutter mode when it occurred. This is shown in Fig. 7.67, where the predicted minimum aerodynamic damping for each of several rotors is plotted horizontally for the first three radial modes. The tabulation at the right of this figure briefly describes each rotor and indicates the observed flutter mode when it occurred. It should be noted that the estimated mechanical damping of a typical rotor system (the sum of material and frictional damping) was approximately 0.03. Thus, it can be seen from Fig. 7.67 that, whenever flutter was observed, the analysis predicted a level of negative aerodynamic damping that was comparable to this expected level of mechanical damping of the rotor.

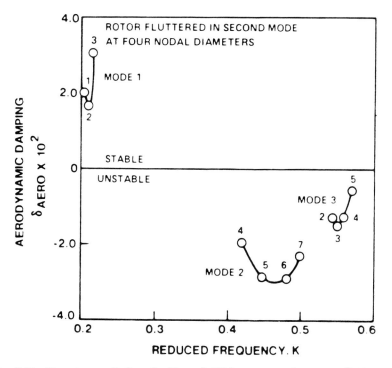

Fig. 7.64   Damping predictions for Pratt & Whitney research rotor at flutter speed.

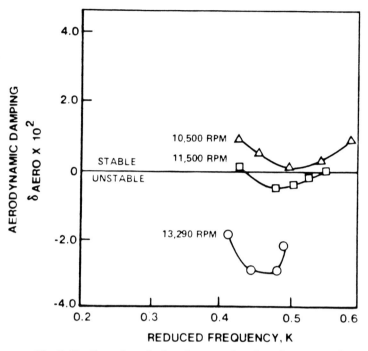

Fig. 7.65   Second mode damping as a function of rotor speed.

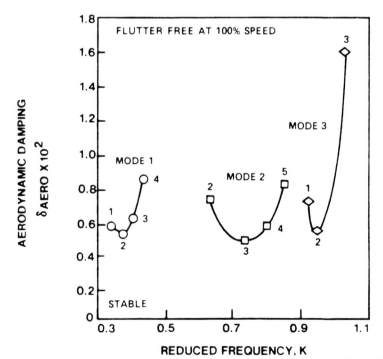

Fig. 7.66   Damping prediction for NASA 1800 ft/s rotor at 100% speed.

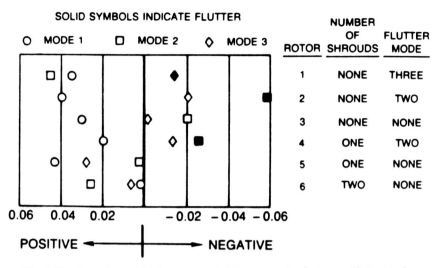

Fig. 7.67   Damping analysis as an unstalled supersonic flutter prediction tool.

Effect of Shroud Location

In the original 1966 report,[109] a parametric study was made of the effect of part-span shroud location on system stability. The results already discussed were for the standard configuration having a part span shroud at the 65.3% span station. Additional locations of 50, 60.5, 69.1, and 80% span were also examined. As before, this study was constrained by the use of isolated flat-plate aerodynamic theory, but it is believed that the physical principles involved are sound and that the relative changes in predicted stability are correct.

The most important finding of this study was that the stability of the system decreases rapidly as the shroud is moved outboard. This is shown in Fig. 7.68 in which a number of normally stable conditions for the basic configuration became quite unstable as the shroud was moved outboard. This unstable tendency is caused by the associated increase in coupling between the torsion and bending modes, shown in Fig. 7.69, in which a number of curves of the normalized twist distribution have been plotted as a function of spanwise station for three values of nodal diameter number (4, 6, and 8) and five shroud locations. For each nodal diameter, curves have been plotted for the shroud location at the 50, 60.5, 65.3, 69.1, and 80% span station. The position of the shroud has been indicated by a short tic mark on each curve.

The primary effect to be noted in Fig. 7.69 is the increase in coupling (i.e., the increase in normalized twist value) within each nodal diameter plot as the shroud location is moved outboard. This increase in coupling is caused by two factors. First, as the shroud is moved outboard, a larger portion of the blade (the inboard portion) is subjected to the oscillating twisting moment of the shroud at resonance in the system mode vibration

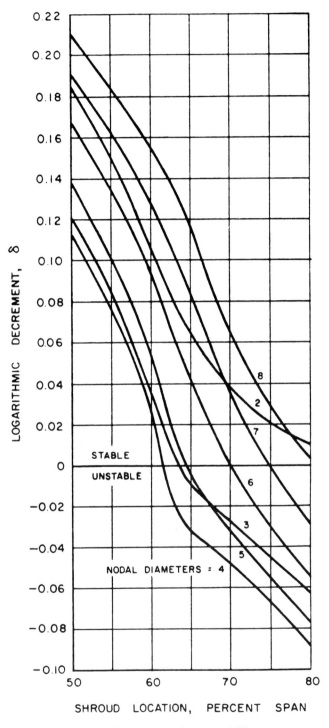

Fig. 7.68 Effect of shroud location on stability parameter.

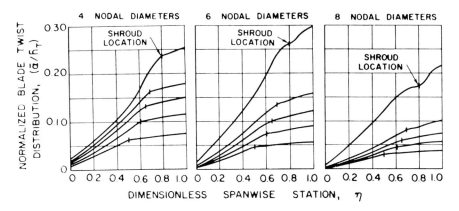

Fig. 7.69   Effect of shroud location on normalized twist distribution.

and a smaller portion of the blade (the remainder of the blade outboard of the shroud) is driven at an off-resonance condition. (Actually, for the 6 and 8 nodal diameter vibrations with an 80% shroud position, it appears that the outboard portion may also be at or near a resonance condition, but it is felt that this is an isolated phenomenon and is not important in general.) Second, as the shroud is moved outboard, the portion of the blade outboard of the shroud becomes stiffer in bending; consequently, the tip bending deflection, which is the normalizing factor in the denominator of $\bar{\alpha}/\bar{h}_T$, becomes relatively smaller and, therefore, the entire level of the curve is raised. In effect, the torsional motion has increased at the expense of the bending motion.

Finally, it is obvious that, as the shroud is moved inboard, the system stability increases—at least for the type of coupled flutter instability being considered here. However, another effect of moving the shroud inboard is to increase the cantilever length of the blade outboard of the shroud, which reduces both the cantilever bending and cantilever torison frequencies of this part of the blade. This reduction in frequency at a fixed value of resultant velocity into the stage produces an increase in reduced velocity $U/b\omega$ and the blade system outboard of the shroud may become susceptible to either a torsional stall flutter, a bending buffet, or both. Clearly, in any design procedure, a compromise must be made between a flutter-free configuration relative to the coupled flutter phenomenon and a flutter-free configuration relative to either torsional stall flutter or bending buffet.

Interblade Phase Angle

In most of the early work described above, the interblade phase angle was necessarily a quantity fixed to the configuration under examination and not usually subjected to parametric scrutiny in a sensitivity analysis. By virtue of its definition, it was inextricably tied to the fixed number of blades in the rotor n and to the number of disturbances over the rotor circumference N by the formula

$$\sigma = \frac{-2\pi}{n/N} = -2\pi\left(\frac{N}{n}\right) \tag{7.243}$$

where the minus sign is associated with a backward traveling wave relative to the rotor and where N is also the number of nodal diameters. Note that, in general, the forward traveling wave is associated with system instability.[109]

Physically, the interblade phase angle is a measure of the phase lag or lead of adjacent blades and has been the subject of several experimental studies at low subsonic speeds of its important effect on aerodynamic damping,[90,93] primarily in cascade. An analysis of the effect of varying σ for a single-degree-of-freedom pitching motion of a supersonic cascade of thin blades operating in a subsonic axial flow was presented by Verdon and McCune.[12] It is well known (e.g., Ref. 93) that the stability of an airfoil executing a pure pitching motion depends only on the sign of the imaginary part of the pitching moment. [Note that in this case, with $h = 0$, Eq. (7.231) reduces to the form

$$W_{\text{TOT}} = \pi^2 \rho b^2 U^2 k^2 B_{\alpha I} \,\bar{\alpha}^2 = \pi \bar{M}_I \bar{\alpha} \qquad (7.244)$$

where the last term is a consequence of using Eq. (7.226).] If $\bar{M}_I$ is positive, then the work per cycle is also positive, which represents an unstable condition. Figure 7.25, discussed earlier, is a phase plane plot of the complex moment coefficient, taken from Verdon and McCune[12] for this condition. The open points connected by curves represent increments of $\Delta\sigma = 30$ deg relative to the specified values of $\sigma = 0$, $\pi/2$, π, and $3\pi/2$. Each curve is for a different compressible reduced frequency, $k_c = 2kM/(M^2 - 1)$, from 0.5 to 2.0. Corresponding values of the complex moment coefficient for an isolated flat-plate airfoil in a supersonic flow, computed from the therory of Garrick and Rubinow,[29] are represented by the solid symbols. Two concepts are revealed by this figure. The first, which is obvious, is that isolated airfoil theory is inadequate to predict the extent of the unstable region for this single-degree-of-freedom oscillation. The second, which is also obvious, but which has subtle implications, is that the interblade phase angle has a significant effect on system stability. For the analysis described here, which was an infinite cascade of idealized flat-plate airfoils, the points and their connecting curves represent a continuum of valid and realizable solutions. Thus, any interblade phase angle can be represented by this plot. It will be shown presently, however, that for a rotor with a finite number of blades, a similar closed diagram is generated by the stability analysis of a coupled motion, in which the only valid solutions are for the specific values of interblade phase angle that satisfy Eq. (7.243). Under these circumstances, a variation in σ implies a corresponding change in number of blades, and the effect on stability will be profound.

An illustration of the assertion that the number of blades has a strong effect on system stability through the interblade phase angle is discussed in the papers by Kaza et al.,[112,113] which deal with single-rotation propfans. Both papers deal with thin, flexible, low-aspect-ratio blades susceptible to large, nonlinear deflections in a strongly three-dimensional flow. To further complicate matters, the blades have large sweep and twist that couples blade bending and torsional motions within each blade, and their flexibility and proximity to one another engenders an aerodynamic coupling similar to that caused by part-span supports. In addition, a flexible hub also contributes to the system coupling.

An analysis of the stability of the basic propfan, denoted by SR3C-X2, was performed for an eight-, a four-, and a single-blade configuration. The results are presented in Fig. 7.70 (taken from Ref. 112), which is a root locus plot of the complex eigenvalues, consisting of the real part μ (proportional to damping) and the imaginary part v (proportional to frequency). Thus, in this phase plane plot, the stable/unstable boundary is at $\mu = 0$, with flutter occurring for a positive real part. The single-blade system has a single eigenvalue, located well within the stable region of the phase plane. The four-blade system has four eigensolutions, represented by interblade phase angles 90 deg apart. This configuration is also fully stable. However, the eight-blade system (with eight eigensolutions spaced 360 deg/8 = 45 deg apart) surrounds the two other system solutions and borders on the unstable region for $\sigma = 180$ and 225 deg. In this instance, it would appear that an increase in the number of blades has intensified the interblade coupling, possibly through the cascading effect, and has caused a deterioration of the system stability. (The prediction is shown to be in good agreement with theory in Fig. 15 of Ref. 112.)

Mistuning

The concluding section to this chapter is, in reality, the introduction to an extensive and very important topic that is currently the subject of intense study. It should be obvious to the reader that the analysis of the

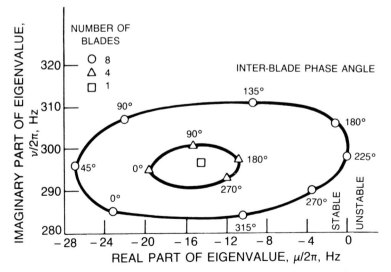

Fig. 7.70   Root locus plot of the mode with least damping: $M = 0.59$, $\Omega = 6080$ rpm, SR3C-X2 rotor (from Ref. 112).

past several pages is based on the ideal situation of a perfectly tuned system in which all blades are identical and vibrate at a common frequency. However, in the real world, there is a statistical spread of the several factors that contribute to the variations about an exact replication of a set of blades from a single ideal design: for example, inaccuracies in dimension, surface contours, and surface finish contribute to variations in frequency and aerodynamic load and response. Clearly, the survey of this subject is well beyond the scope of the present chapter and would constitute a major treatise in itself. It will be the object of this final section to introduce the reader to several basic references to the subject and to focus on a single aspect of "odd-even" mistuning in a propfan study to illustrate one last insight into the physics of the interblade phase angle parameter.

For a general overview of the vibration problems of bladed disks, the reader is referred to the symposium volume of Ref. 114 and to the survey paper[115] that reviews the recent work in this field. Most of the early papers[116-118] concentrated on a distributed mistuning that, in effect, removed interblade phase angle as a major parameter. Under these conditions it is noted by Ewins and Han[117] that it is the blade with the "most mistune" that "is always found to suffer the highest response levels." There

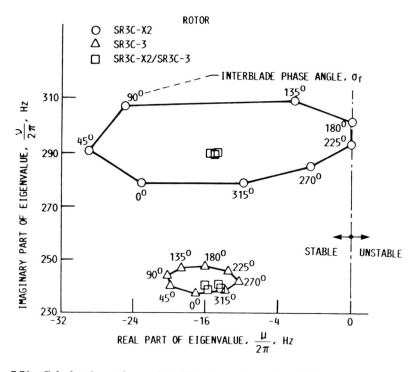

Fig. 7.71 Calculated root locus plot of the lowest damped mode for the SR3C-X2, SR3C-3, and SR3C-X2/SR3C-3 rotors at 6320 rpm and 0.528 freestream Mach number (from Ref. 113).

can be no reference to the interblade phase angle of maximum instability because the system mistuning chosen cannot define such an angle.

There is one recent paper that deals with mistuning in a specialized manner which includes the interblade phase angle and provides an important measure of understanding. This is found in the work of Ref. 113, which is an extension of the propfan work done in Ref. 112. Once again, the SR3C-X2 rotor was analyzed, together with a rotor designated the SR3C-3. Both of these were tuned rotors, as in Ref. 112. In addition, a deliberately mistuned rotor, the SR3C-X2/SR3C-3, was analyzed. It was modeled as an idealized alternately mistuned rotor having four identical blade pairs with two different blades in each pair, one from the SR3C-X2 rotor and one from the SR3C-3 rotor. The analysis yielded the plots shown in Fig. 7.71.

In this case, all three rotors have eight eigensolutions. The SR3C-X2 (circled points) was already discussed in the previous section. It was an unstable rotor with a measured flutter condition that coincided with the real part of the eigenvalues equal to zero. Conversely, the SR3C-3 (triangular points) was a stable rotor during the experiments and yielded eigensolutions comfortably away from the stable/unstable boundary. The aerodynamic coupling of the mistuned rotor (square points) appears to be gone and the eigensolutions are divided into two nested groups, with the high frequency group near the center of the SR3C-X2 eigensolutions and the low-frequency group in the vicinity of the SR3C-3 solutions. It appears that this strong, odd-even mistuning has uncoupled the blade row aerodynamically and has reduced the system to two separate and distinct systems with segregated eigensolutions which behave in much the same manner as the single-blade solutions discussed earlier in Figs. 7.25 and 7.26. It should be obvious that the degree of mistuning will play a crucial role in the phase plane distribution of eigenvalues and in the consequent stability of any real system of blades. This is clearly beyond the scope of this chapter and the reader should consult the references cited for further study and enlightenment.

References

[1]Bisplinghoff, R. L., Ashley, H., and Halfman R. L., *Aeroelasticity*, Addison-Wesley, Reading, MA, 1955.

[2]Pearson, H., "The Aerodynamics of Compressor Blade Vibration," *Proceedings of Fourth Anglo-American Aeronautical Conference*, Sept. 1953, pp. 127–162.

[3]Shannon, J. F., "Vibration Problems in Gas Turbines, Centrifugal and Axial-Flow Compressors," British Aeronautical Research Council, London, R&M 2226, 1945.

[4]Armstrong, E. K. and Stevenson, R. E., "Some Practical Aspects of Compressor Blade Vibration," *Journal of the Royal Aeronautical Society*, Vol. 64, No. 591, March 1960, pp. 117–130.

[5]Sisto, F. and Ni, R. H., Research on the Flutter of Axial-Turbomachine Blading," Stevens Institute of Technology, Hoboken, NJ, TR ME-RT 70004, June 1970.

[6]Carter A. D. S., " Some Preliminary Notes on the Flutter of Axial Compressor Blades," National Gas Turbine Establishment, Memo M.181, Nov. 1953.

[7]Carter, A. D. S. and Kilpatrick, D. A., "Self-Excited Vibration of Axial-Flow Compressor Blades," *Proceedings of the Institution of Mechanical Engineers*, Vol. 171, 1957, pp. 245–281.

[8]Carta, F. O., "Coupled Blade-Disk-Shroud Flutter Instabilities in Turbojet Engine Rotors," *Transactions of ASME, Journal of Engineering for Power*, Ser. A, Vol. 89, No. 3, July 1967, pp. 419–426.

[9]Stargardter, H., "Dynamic Models of Vibrating Rotor States," ASME Paper 66-WA/GT-8, Nov. 1966.

[10]Mikolajczak, A. A., Snyder, L. E., Arnoldi, R. A., and Stargardter, H., "Advances in Fan and Compressor Blade Flutter Analysis and Predictions," *Journal of Aircraft*, Vol. 12, April 1975, pp. 325–332.

[11]Verdon, J. M., "The Unsteady Aerodynamics of a Finite Supersonic Cascade with Subsonic Axial Flow," *Transactions of ASME, Journal of Applied Mechanics*, Ser. E., Vol. 40, No. 3, Sept. 1973, pp. 667–671.

[12]Verdon, J. M. and McCune, J. E., "Unsteady Supersonic Cascade in Subsonic Axial Flow," *AIAA Journal*, Vol. 13, Feb. 1975, pp. 193–201.

[13]Theordorsen, T., "General Theory of Aerodynamic Instability and the Mechanism of Flutter," NACA Rept. 496, 1935.

[14]Schwarz, L., "Berechnung der Druckverteilung einer Harmonisch sich Verformenden Tragfläche in ebener Strömung," *Luftfahrt-Forschung*, Vol. 17, 1940, p. 379.

[15]Smilg, B. and Wasserman, L. S., "Application of Three-Dimensional Flutter Theory to Aircraft Structures," U.S. Air Force, TR 4798, 1942.

[16]Scanlan, R. H. and Rosenbaum, R., *Introduction to the Study of Aircraft Vibration and Flutter*, Macmillan, New York, 1951.

[17]Bisplinghoff, R. L. and Ashley, H., *Principles of Aeroelasticity*, Wiley, New York, 1962.

[18]Pines, S., "An Elementary Explanation of the Flutter Mechanism," *Proceedings of National Specialists Meeting on Dynamics and Aeroelasticity*, Institute of the Aeronautical Sciences, Ft. Worth, TX, Nov. 1958, pp. 52–58.

[19]Den Hartog, J. P., *Mechanical Vibrations*, McGraw-Hill, New York, 1956.

[20]Timoshenko, S., Young, D. H., and Weaver, W., *Vibration Problems in Engineering*, 4th ed., Wiley, New York, 1974.

[21]Reissner, E., "Effect of Finite Span on the Airload Distributions for Oscillating Wings, I—Aerodynamic Theory of Oscillating Wings of Finite Span," NACA TN 1194, 1947.

[22]Söhngen, H., "Die Lösungen der Integralgleichung und deren Anwendung in der Tragflügeltheorie," *Mathematische Zeitschrift*, Vol. 45, 1939, pp. 245–264.

[23]Possio, C., "L'Azione Aerodinamica sul Profilo Oscillante in un Fluido Compressibile a Velocità Iposonora," *L'Aerotecnica*, Vol. XVIII, April 1938 (also British Ministry of Aircraft Production RTP Translation 987).

[24]Dietze, F., "The Air Forces of the Harmonically Vibrating Wing at Subsonic Velocity (Plane Problem)," Pt. I and II, U.S. Air Force Translations F-TS-506-RE and F-TS-948-RE, 1947 (originally, *Luftfahrt-Forschung*, Vol. 16, No. 2, 1939, pp. 84–96).

[25]Fettis, H. E., "An Approximate Method for the Calculation of Nonstationary Air Forces at Subsonic Speeds," U.S. Air Force, Wright Air Development Center, Rept. WADC-52–56, 1952.

[26]Timman, R., van de Vooren, A. I., and Griedanus, J. H., "Aerodynamic Coefficients of an Oscillating Airfoil in Two-Dimensional Subsonic Flow," *Journal of the Aeronautical Sciences*, Vol. 18, Dec. 1951, pp. 797–802 (corrected tabulations appear in *Journal of the Aeronautical Sciences*, Vol. 21, July 1954, pp. 499–500).

[27]Fettis, H. H., "Tables of Lift and Moment Coefficients for an Oscillating Wing Aileron Combination in Two-Dimensional Subsonic Flow," U.S. Air Force, TR 6688, 1951.

[28]Luke, Y. L., "Tables of Coefficients for Compressible Flutter Calculations," U.S. Air Force, TR 6200, 1950.

[29]Garrick, I. E. and Rubinow, S. I., "Flutter and Oscillating Air Force Calculations for an Airfoil in Two-Dimensional Supersonic Flow," NACA Rept. 846, 1946.

[30]Miles, J. W., *The Potential Theory of Unsteady Supersonic Flow*, Cambridge University Press, Cambridge, England, 1959.

[31]Lilley, G. M., "An Investigation of the Flexure-Torsion Flutter Characteristics of Aerofoils in Cascade," College of Aeronautics, Cranfield, England, Rept. 60, 1952.

[32]Sisto, F., "Unsteady Aerodynamic Reactions on Airfoils in Cascade," *Journal of the Aeronautical Sciences*, Vol. 22, May 1955, pp. 297–302.

[33]Lane, R. and Wang, C. T., "A Theoretical Investigation of the Flutter Characteristics of Compressor and Turbine Blade Systems," U.S. Air Force, Wright Air Development Center, Rept. WADC-54–449, 1954.

[34]Whitehead, D. S., "Force and Moment Coefficients for Vibrating Aerofoils in Cascade," British Aeronautical Research Council, London, R&M 3254, 1960.

[35]Whitehead, D. S., "The Aerodynamics of Axial Compressor and Turbine Blade Vibration," Ph.D. Thesis, Cambridge University, Cambridge, England, 1957.

[36]Bromwich, T. J., *An Introduction to the Theory of Infinite Series*, 2nd ed., Macmillan, London, 1955.

[37]Carta, F. O., "Unsteady Aerodynamic Theory of a Staggered Cascade of Oscillating Airfoils in Compressible Flow," United Aircraft Research Laboratories, East Hartford, CT, Rept. R-0582-19, Aug. 1957.

[38]Lane, F. and Friedman, M., "A Theoretical Investigation of Subsonic Oscillatory Blade-Row Aerodynamics," NACA TN 4136, Feb. 1958.

[39]Whitehead, D. S., "Vibration and Sound Generation in a Cascade of Flat Plates in Subsonic Flow," British Aeronautical Research Council, London, R&M 3865, Feb. 1970.

[40]Kaji, S. and Okazaki, T., "Propagation of Sound Waves Through a Blade Row, II: Analysis Based on the Acceleration Potential Method," *Journal of Sound and Vibration*, Vol. 11, March 1970, pp. 355–375.

[41]Smith, S. N., "Discrete Frequency Sound Generation in Axial Flow Turbomachines," British Aeronautical Research Council, London, R&M 3709, 1971.

[42]Runyan, H. L., Woolston, D. S., and Rainey, A. G., "Theoretical and Experimental Investigation of the Effect of Tunnel Walls on the Forces on an Oscillating Airfoil in Two-Dimensional Subsonic Compressible Flow," NACA Rept. 1262, 1956.

[43]Verdon, J. M., Adamczyk, J. J., and Caspar, J. R., "Subsonic Flow Past an Oscillating Cascade with Steady Blade Loading—Basic Formulation," United Aircraft Research Laboratories, East Hartford, CT, Rept. R75-214185-1, April 1975. (*Proceedings of a Symposium on Unsteady Aerodynamics*, sponsored by the University of Arizona and U.S. Air Force Office of Scientific Research, March 1975, Vol. II, pp. 827–851).

[44]Verdon, J. M. and Caspar, J. R., "Subsonic Flow Past an Oscillating Cascade with Finite Mean Flow Deflection," *AIAA Journal*, Vol. 18, May 1980, pp. 540–548.

[45]Verdon, J. M. and Caspar, J. R., "Development of an Unsteady Aerodynamic Analysis for Finite-Deflection Subsonic Cascades," NASA CR 3455, Sept. 1981.

[46]Verdon, J. M. and Caspar, J. R., "Development of a Linear Unsteady Aerodynamic Analysis for Finite-Deflection Subsonic Cascades," *AIAA Journal*, Vol. 20, 1982, pp. 1259–1267.

[47]Verdon, J. M. and Caspar, J. R., "A Linear Aerodynamic Analysis for Unsteady Transonic Cascades," NASA CR 3833, Sept. 1984.

[48]Verdon, J. M. and Caspar, J. R., "A Linearized Unsteady Aerodynamic Analysis for Transonic Cascades," *Journal of Fluid Mechanics*, Vol. 149, Dec. 1984, pp. 403–429.

[49]Verdon, J. M. and Usab, W. J., Jr., "Application of a Linearized Unsteady Aerodynamic Analysis to Standard Cascade Configurations," NASA CR 3940, Jan. 1986.

[50]Caspar, J. R., Hobbs, D. E., and Davis, R. L., "Calculation of Two-Dimensional Potential Cascade Flow Using Finite Area Methods," *AIAA Journal*, Vol. 18, Jan. 1980, pp. 103–109.

[51]Caspar, J. R., "Unconditionally Stable Calculation of Transonic Potential Flow Through Cascades Using an Adaptive Mesh for Shock Capture," *Transactions of ASME, Journal of Engineering for Power*, Vol. 105, No. 3, July 1983, pp. 504–513.

[52]Platzer, M. F. and Carta, F. O. (eds.), *AGARD Manual on Aeroelasticity in Axial Flow Turbomachines*, Vol. 1, "Unsteady Turbomachinery Aerodynamics," AGARDograph 298, Vol. 1, March 1987.

[53]Platzer, M. F. and Carta, F. O. (eds.), *AGARD Manual on Aeroelasticity in Axial Flow Turbomachines*, Vol. 2, "Structural Dynamics and Aeroelasticity in Axial Flow Turbomachines," AGARDograph 298, Vol. 2, June 1988.

[54]Lane, F., "Supersonic Flow Past an Oscillating Cascade With Supersonic Leading-Edge Locus," *Journal of the Aeronautical Sciences*, Vol. 24, Jan. 1957, pp. 65–66.

[55]Snyder, L. E. and Commerford, G. L., "Supersonic Unstalled Flutter in Fan Rotors; Analytical and Experimental Results," *Transactions of ASME, Journal of Engineering for Power*, Vol. 96, No. 4, Oct. 1974, pp. 379–386.

[56]Verdon, J. M., "Further Developments in the Aerodynamic Analysis of Unsteady Supersonic Cascades, Part 1: The Unsteady Pressure Field; Part 2: Aerodynamic Response Predictions," *Transactions of ASME, Journal of Engineering for Power*, Vol. 99, No. 4, Oct. 1977, pp. 509–525.

[57]Verdon, J. M., "Unsteady Aerodynamics for Turbomachinery Aeroelastic Applications," United Technologies Research Center, East Hartford, CT, UTRC Rept. R86-151774-1, June 1986 (prepared for inclusion as a chapter in AIAA Progress in Astronautics and Aeronautics volume 120, "Unsteady Transonic Aerodynamics," edited by D. Nixon, scheduled for publication in 1989).

[58]Kurosaka, M., "On the Unsteady Supersonic Cascade with a Subsonic Leading Edge—An Exact First Order Theory: Parts 1 and 2," *Transactions ASME, Journal of Engineering for Power*, Vol. 96, Jan. 1974, pp. 13–31.

[59]Brix, C. W. and Platzer, M. F., "Theoretical Investigation of Supersonic Flow Past Oscillating Cascades with Subsonic Leading Edge Locus," AIAA Paper 74-14, 1974.

[60]Goldstein, M. E., Braun, W., and Adamczyk, J. J., "Unsteady Flow in a Supersonic Cascade with Strong In-Passage Shocks," *Journal of Fluid Mechanics*, Vol. 83, Pt. 3, 1977, pp. 569–604.

[61]Surampudi, S. P. and Adamczyk, J. J., "Unsteady Transonic Flow over Cascade Blades," *AIAA Journal*, Vol. 24, Feb. 1986, pp. 293–302.

[62]Bratt, J. B. and Scruton, C. "Measurements of Pitching Moment Derivatives for an Aerofoil Oscillating About the Half-Chord Axis," British Aeronautical Research Council, London, R&M 1921, Nov. 1938.

[63]Victory, M., "Flutter at High Incidence," British Aeronautical Research Council, London, R&M 2048, Jan. 1943.

[64]Mendelson, A., "Aerodynamic Hysteresis as a Factor in Critical Flutter Speed of Compressor Blades at Stalling Conditions," *Journal of the Aeronautica Sciences*, Vol. 16, Nov. 1949, pp. 645–652.

[65]Wang, C. T., Vaccaro, R. J., and DeSanto, D. F., "A Practical Approach to the Problem of Stall Flutter, ASME Paper 55-SA-69, June 1955.

[66]Halfman, R. L., Johnson, H. C., and Haley, S. M., "Evaluation of High-Angle-of-Attack Aerodynamic-Derivative Data and Stall-Flutter Prediction Techniques," NACA TN 2533, Nov. 1951.

[67]Baker, J. E., "The Effects of Various Parameters, Including Mach Number, on Propeller-Blade Flutter with Emphasis on Stall Flutter," NACA TN 3357, Jan. 1955.

[68]Brooks, G. W. and Baker, J. E., "An Experimental Investigation of the Effect of Various Parameters, Including Tip Mach Number, on the Flutter of Some Model Helicopter Rotor Blades," NACA TN 4005, Sept. 1958.

[69]Rainey, A. G., "Preliminary Study of Some Factors Which Affect the Stall-Flutter Characteristics of Thin Wings," NACA TN 3622, March 1956.

[70]Regier, A. A., and Rainey, A. G., "Effect of Mean Incidence on Flutter," Paper presented at Seventh Meeting, Structures and Materials Panel, AGARD, March-April 1958.

[71]Rainey, A. G., "Measurement of Aerodynamic Forces for Various Mean Angles of Attack on an Airfoil Oscillating in Pitch and on Two Finite-Span Wings Oscillating in Bending with Emphasis on Damping in the Stall," NACA Rept. 1305, 1957.

[72]Carta, F. O., "Experimental Investigation of the Unsteady Aerodynamic Characteristics of an NACA 0012 Airfoil," United Aircraft Research Lab., East Hartford, CT, Rept. M-1283-1, Aug. 1960.

[73]Liiva, J., Davenport, F. J., Gray, L., and Walton, I. C., *Two-Dimensional Tests of Airfoils Oscillating Near Stall*, Vol. I, "Summary and Evaluation of Results," U.S. Army Aviation Material Laboratories, TR 68-13A, April 1968 (AD 670957).

[74]Carta, F. O., Commerford, G. L., Carlson, R. G., and Blackwell, R. H., "Investigation of Airfoil Dynamic Stall and Its Influence on Helicopter Control Loads," USAAMRDL TR 72-51, Sept. 1972.

[75]Carta, F. O. and Niebanck, C. F., *Prediction of Rotor Instability at High Forward Speeds*, Vol. III, "Stall Flutter," U.S. Army Aviation Material Laboratories, TR 68-18C, Feb. 1969.

[76]Ham, N. D., "Aerodynamic Loading on a Two-Dimensional Airfoil During Dynamic Stall," *AIAA Journal*, Vol. 6, Oct. 1968, pp. 1927–1934.

[77]McCroskey, W. J., and Fisher, R. K., "Detailed Aerodynamic Measurements on a Model Rotor in the Blade Stall Regime," *Journal of the American Helicopter Society*, Vol. 17, Jan. 1972, pp. 20–30.

[78]Carta, F. O., "Effect of Unsteady Pressure Gradient Reduction on Dynamic Stall Delay," *Journal of Aircraft*, Vol. 8, Oct. 1971, pp. 839–841.

[79]Carta, F. O., "Analysis of Oscillatory Pressure Data Including Dynamic Stall Effects," NASA CR-2394, May 1974.

[80]McCroskey, W. J., Carr, L. W., and McAlister, K. W., "Dynamic Stall Measurements on Oscillating Airfoils," AIAA Paper 75-125, 1975.

[81]McCroskey, W. J., McAlister, K. W., and Carr, L. W., "Dynamic Stall Experiments on Oscillating Airfoils," *AIAA Journal*, Vol. 14, Jan. 1976, pp. 57–63.

[82]St. Hilaire, A. O., Carta, F. O., Fink, M. R., and Jepson, W. D., "The Influence of Sweep on the Aerodynamic Loading of an Oscillating NACA 0012 Airfoil, Vol. I—Technical Report," NASA CR 3092, May 1979.

[83]St. Hilaire, A. O. and Carta, F. O., "Analysis of Unswept and Swept Wing Chordwise Pressure Data from an Oscillating NACA 0012 Airfoil Experiment, Vol. I—Technical Report" NASA CR 3567, March 1983.

[84]Lorber, P. F. and Carta, F. O., "Airfoil Dynamic Stall at Constant Pitch Rate and High Reynolds Number," *Journal of Aircraft*, Vol. 25, June 1988, pp. 548–556.

[85]Lorber, P. F. and Carta, F. O., "Unsteady Stall Penetration Experiments at High Reynolds Number," Air Force Office of Scientific Research, AFOSR TR-87-1202, April 1987.

[86]Carta, F. O. and St. Hilaire, A. O., "An Experimental Study on the Aerodynamic Response of a Subsonic Cascade Oscillating Near Stall, United Technologies Research Center, East Hartford, CT, Project SQUID Tech. Rept. UTRC-2-PU, July 1976.

[87]Carta, F. O. and St. Hilaire, A. O., "Experimentally Determined Stability Parameters of a Subsonic Cascade Oscillating Near Stall," *Transactions of ASME, Journal of Engineering for Power*, Vol. 100, No. 1, Jan. 1978, pp. 111–120.

[88]Arnoldi, R. A., Carta, F. O., Ni, R. H., Dalton, W. N., and St. Hilaire, A. O., "Analytical and Experimental Study of Subsonic Stalled Flutter," Air Force Office of Scientific Research, AFOSR-TR-77-0854, July 1977.

[89]Johnson, I. A. and Bullock, R. O., "Aerodynamic Design of Axial Flow Compressors," NASA SP-36, 1965.

[90]Carta. F. O. and St. Hilaire, A. O., "Effect of Interblade Phase Angle and Incidence Angle on Cascade Pitching Stability," *Transactions of ASME, Journal of Engineering for Power*, Vol. 102, April 1980, pp. 391–396.

[91]McAlister, K. W., Carr, L. W., and McCroskey, W. J., "Dynamic Stall Experiments on the NACA 0012 Airfoil," NASA Tech. Paper 1100, Jan. 1978.

[92]Carta, F. O., "An Experimental Investigation of Gapwise Periodicity and Unsteady Aerodynamic Response in an Oscillating Cascade, Vol. I: Experimental and Theoretical Results," NASA CR 3513, Jan. 1982.

[93]Carta, F. O., "Unsteady Aerodynamics and Gapwise Periodicity of Oscillating Cascaded Airfoils," *Transactions of ASME, Journal of Engineering for Power*, Vol. 105, July 1983, pp. 565–574.

[94]Carta, F. O., "An Experimental Investigation of Gapwise Periodicity and Unsteady Aerodynamic Response in an Oscillating Cascade, Vol. II: Data Report," NASA CR 165457, Dec. 1981.

[95]Atassi, H. and Akai, T. J., "Aerodynamic and Aeroelastic Characteristics of Oscillating Loaded Cascades at Low Mach Number," *Transactions of ASME, Journal of Engineering for Power*, Vol. 102, No. 2, April 1980, pp. 344–351.

[96]Bölcs, A. and Fransson, T. H., "Aeroelasticity in Turbomachines: Comparison of Theoretical and Experimental Cascade Results," Lausanne Institute of Technology, Switzerland, Communication du Laboratoire de Thermique Appliquee No. 13, 1986 (two volumes).

[97]Dring, R. P., Joslyn, H. D., and Hardin, L. W., "An Investigation of Axial Compressor Rotor Aerodynamics," *Transactions of ASME, Journal of Engineering for Power*, Vol. 104, 1982, pp. 84–96.

[98]Hardin, L. W., Carta, F. O., and Verdon, J. M., "Unsteady Aerodynamic Measurements on a Rotating Compressor Blade Row at Low Mach Number," *Transactions of ASME, Journal of Turbomachinery*, Vol. 109, Oct. 1987, pp. 499–507.

[99]Triebstein, H., Carstens, V., and Wagener, J., "Unsteady Pressures on a Harmonically Oscillating, Staggered Cascade, Part I: Incompressible Flow; Part II: Compressible Flow," DFVLR, Göttingen, FRG, Repts. DLR-FB 75-57 and DLR-FB 75-58, Oct. 1974 (English translations available from European Space Agency, Paris, as Repts. TT272 and TT273).

[100]Fleeter, S., Novick, A. S., and Riffel, R. E., "Supersonic Unsteady Aerodynamic Phenomena in Airfoil Cascades," Office of Naval Research, Washington, DC, Rept. EDR 8617, Oct. 1975.

[101]Fleeter, S., Novick, A. S., Riffel, R. E., and Caruthers, J. E., "An Experimental Determination of the Unsteady Aerodynamics in a Controlled Oscillating Cascade," *Transactions of ASME, Journal of Engineering for Power* Vol. 99, No. 1, 1977, pp. 88–96.

[102]Fleeter, S., Riffel, R. E., Lindsey, J. H., and Rothrock, M. D., "Time-Variant Translation Mode Aerodynamics of a Classical Transonic Airfoil Cascade," *AGARD Propulsion and Energetics Panel, Symposium on Stresses, Vibrations, Structural Integration, and Engine Integrity*, AGARD CP-248, 1978.

[103]Jay, R. L., Rothrock, M. D., Riffel, R. E., and Sinnet, G. T., "Time-Variant Aerodynamics for Torsional Motion of Large-Turning Airfoil," Final report for Naval Air Systems Command, Contract N00019-79-C-0087, Detroit Diesel Allison Dir, General Motors Corp., Detroit, Rept. 10192, 1979.

[104]Stargardter, H., "Subsonic/Transonic Stall Flutter Study, Final Report," NASA CR 165256, 1979.

[105]Gallus, H. E., Lambertz, J., and Wallman, T., "Blade Row Interaction in an Axial-Flow Subsonic Compressor Stage," *Transactions of ASME, Journal of Engineering for Power*, Vol. 102, Jan. 1980, pp. 169–177.

[106]Szechenyi, E. and Finas, R., "Aeroelasticity Testing in a Straight Cascade Wind Tunnel," *Aeroelasticity in Turbomachines, Proceedings of 2nd International Symposium—Lausanne*, edited by P. Suter, Juris-Verlag-Zurich, Sept. 1980, pp. 143–150.

[107]Halliwell, D. G., "Fan Supersonic Flutter: Prediction and Test Analysis," British Aeronautical Research Council, London, R&M 3789, 1975.

[108]Halliwell, D. G., "Effect of Intake Conditions on Supersonic Unstalled Flutter in Turbofan Engines," *Journal of Aircraft*, Vol. 17, May 1980, pp. 300–304.

[109]Carta, F. O., "A Parametric Study of Coupled Blade-Disk-Shroud Flutter Instabilities in Turbojet Engine Rotors," United Aircraft Research Lab., East Hartford, CT, Rept. E211529-1, 1966.

[110]Nagashima, T. and Whitehead, D. S., "Aerodynamic Forces and Moments for Vibrating Supersonic Cascades of Blades," Cambridge University, Cambridge, England, Rept. CUEDA/A-Turbo/TR59, 1974.

[111]Sisto, F. and Ni, R., "Research on the Flutter of Axial Turbomachine Blading," ONR Technical Report ME-RT 74-008, Stevens Institute of Technology, Hoboken, NJ, 1974.

[112]Kaza, K. R. V., Mehmed, O., Narayanan, G. V., and Murthy, D. V., "Analytical and Experimental Investigation of a Composite Propfan Model," AIAA Paper 87-0738, 1987.

[113]Kaza, K. R. V., Mehmed, O., Williams, M., and Moss, L. A., "Analytical and Experimental Investigation of Mistuning in Propfan Flutter," AIAA Paper 87-0739, 1987.

[114]Ewins, D. J. and Srinivasan, A. V. (eds.), *Vibrations of Bladed Disk Assemblies*, ASME Vibration Conference, 1983 (see also *Transactions of ASME, Journal of Vibration, Acoustics, Stress, and Reliability in Design*, Vol. 106, No. 2, April 1984).

[115]Srinivasan, A. V., "Survey Paper—Vibrations of Bladed Disk Assemblies," *Transactions of ASME, Journal of Vibration, Acoustics, Stress, and Reliability in Design*, Vol. 106, No. 2, April 1984, pp. 165–168.

[116]Whitehead, D. S., "Torsional Flutter of Unstalled Cascade Blades at Zero Deflection," British Aeronautical Research Council, London, R&M 3429, 1964.

[117]Ewins, D. J. and Han, Z. S., "Resonant Vibration Levels of a Mistuned Bladed Disk," *Transactions of ASME, Journal of Vibration, Acoustics, Stress, and Reliability in Design*, Vol. 106, No. 2, April 1984, pp. 211–217.

[118]El-Bayoumy, L. E. and Srinivasan, A. V., "Influence of Mistuning on Rotor Blade Vibrations," *AIAA Journal*, Vol. 13, April 1975, pp. 460–464.

Bibliography

Abramson, H. N., *An Introduction to the Dynamics of Airplanes*, Ronald Press, New York, 1958.

Arnoldi, R. A., Carta, F. O., St. Hilaire, A. O., and Dalton, W. N., "Supersonic Chordwise Bending Flutter in Cascades," Naval Air Systems Command, Rept. PWA-5271, May 31, 1975.

Ashley, H. and Zartarian, G., "Piston Theory—A New Aerodynamic Tool for the Aeroelastician," *Journal of the Aeronautical Sciences*, Vol. 23, Dec. 1956, pp. 1109–1118.

Ashley, H., and Landahl, M., *Aerodynamics of Wings and Bodies*, Addison-Wesley, Reading, MA, 1965, pp. 97, 245–249.

Bell, J. K., "Theoretical Investigation of the Flutter Characteristics of Supersonic Cascades with Subsonic Leading-Edge Locus," Aeronautical Engineering Thesis, Naval Postgraduate School, Monterey, CA, June 1975.

Bratt, J. B., Wight, K. C., and Chinneck, A., "Free Oscillations of an Aerofoil about the Half-Chord Axis at High Incidences, and Pitching Moment Derivatives for Decaying Oscillations," British Aeronautical Research Council, London, R&M 2214, Sept. 1940.

Bratt, J. B. and Wight, K. C., "The Effect of Mean Incidence, Amplitude of Oscillation, Profile and Aspect Ratio on Pitching Moment Derivatives," British Aeronautical Research Council, London, R&M 2064, June 4, 1945.

"Calculation of Aerodynamic Forces and Moments on Airfoils Vibrating in Cascade," Northern Research and Engineering, Cambridge, MA, Rept. 1084-1 (prepared for Pratt & Whitney Aircraft), June 1974.

Cambell, W., "Protection of Steam Turbine Disc Wheels from Axial Vibration," ASME Paper 1920, May 1924.

Campbell, W. and Heckman, W. C., "Tangential Vibration of Steam Turbine Buckets," ASME Paper 1925, May 1924.

Carter, A. D. S., "A Theoretical Investigation of the Factors Affecting Stalling Flutter of Compressor Blades," National Gas Turbine Establishment, British Aeronautical Research Council, London, Tech. Rept. R.172, C.P. 265, April 1955.

Carter, A. D. S., Kilpatrick, D. A., Moss, C. E., and Ritchie, J. "An Experimental Investigation of the Blade Vibratory Stresses in a Single-Stage Compressor," National Gas Turbine Establishment, British Aeronautical Research Council, London, Rept. R.174, C.P. 266, April 1955.

Cavaille, Y., "Aerodynamic Damping in Turbomachinery," ASME Paper 72-GT-8, March 1972.

Chadwick, W. R., Bell, J. K., and Platzer, M. F., "Analysis of Supersonic Flow Past Oscillating Cascades," *Unsteady Flow in Turbomachinery*, AGARD CP-177, Sept. 1975.

Cottney, D. J. and Ewins, D. J., "Towards the Efficient Vibration Analysis of Shrouded Bladed Disk Assemblies," *Transactions of ASME, Journal of Engineering in Industry*, Vol. 96, No. 3, Aug. 1974, pp. 1054–1059.

Dye, R. C. F. and Henry, T. A. "Vibration Amplitudes of Compressor Blades Resulting from Scatter in Blade Natural Frequencies," ASME Paper 68WA/GT-3, Dec. 1968.

Ehrich, F. F., "A Matrix Solution for the Vibration Modes of Nonuniform Disks," *Journal of Applied Mechanics*, Vol. 23, No. 1, March 1956, pp. 1–7.

Ewins, D. J., "Vibration Characteristics of Bladed Disc Assemblies," *Journal of Mechanical Engineering Science*, Vol. 15, No. 3, 1973, pp. 165–186.

Fanti, R. and Carta, F. O., "Aerodynamic Interaction Effects on a Staggered Cascade of Airfoils Oscillating in Two-Dimensional Compressible Flow," United Aircraft Research Laboratories, East Hartford, CT, Rept. R-0582-17, Aug. 1957.

Fleeter, S., "Fluctuating Lift and Moment Coefficients for Cascaded Airfoils in a Nonuniform Compressible Flow," *Journal of Aircraft*, Vol. 10, Feb. 1973, pp. 93–98.

Fleeter, S., McClure, R. B., Sinnet, G. T., and Holtman, R. L., "The Torsional Flutter Characteristics of a Cantilevered Airfoil Cascade in a Supersonic Inlet Flow Field with a Subsonic Axial Component," AIAA Paper 74-530, June 1974.

Fung, Y. C., *An Introduction to the Theory of Aeroelasticity*, Wiley, New York, 1955.

Garelick, M. S., "Nonsteady Airloads on Dynamically Stalling Two-Dimensional Wings," S.M. Thesis, Massachusetts Institute of Technology, Cambridge, June 1967.

Garrick, I. E. (ed.), *Aerodynamic Flutter, AIAA Selected Reprint Series*, Vol. 5, AIAA, New York, March 1969.

Goldstein, M. E., "Cascade with Subsonic Leading-Edge Locus," *AIAA Journal*, Vol. 13, Aug. 1975, pp. 1117-1118.

Gorelov, D. N., "Lattices of Plates in an Unsteady Supersonic Flow," *Fluid Dynamics*, Faraday Press, Vol. 1, No. 4, July-Aug. 1966, pp. 34-39 (translation of *Izvestiya Akademii Nauk SSSR, Mekhanika Zhidkosti i Gaza*, Vol. 1, No. 4, 1966, pp. 50-58).

Graham, R. W. and Costilow, E. L., "Compressor Stall and Blade Vibration," *Aerodynamic Design of Axial-Flow Compressors*, Vol. III, NACA RM E56B03b, Aug. 1, 1956, Chap. XI, pp. 1-52.

Ham, N. D., "Stall Flutter of Helicopter Rotor Blades: A Special Case of the Dynamic Stall Phenomenon," *Journal of the American Helicopter Society*, Vol. 12, Oct. 1967, pp. 19-21.

Hamamoto, I., "Minute Harmonic Oscillations of a Flat Plate Cascade in Transonic Flow," Paper presented at Japanese 2nd Applied Mathematics and Mechanics General Meeting, 1957.

Hawthorne, W. R., "Flow Through Moving Cascades of Lifting Lines with Fluctuating Lift," *Journal of Mechanical Engineering Science*, Vol. 15, No. 1, 1973, pp. 1-10.

Henderson, R. E. and Horlock, J. H., "An Approximate Analysis of the Unsteady Lift on Airfoils in Cascade," *Transactions of ASME, Journal of Engineering for Power*, Vol. 94, Oct. 1972, pp. 233-240.

Huppert, M. C., Calvert, H. F., and Meyer, A. J., "Experimental Investigation of Rotating Stall and Blade Vibration in the Axial-Flow Compressor of a Turbojet Engine," NACA RM E54A08, April 26, 1954.

Ikui, T., Inoue, M., and Kuromaru, M., "Researches on the Two-Dimensional Retarded Cascade," Pts. 1 and 2, *Bulletin of the Japan Society of Mechanical Engineers*, Vol. 15, No. 84, 1972, pp. 705-720; Parts 3 and 4, *Bulletin of JSME*, Vol. 16, No. 92, Feb. 1973, pp. 252-271.

Isogai, K., "An Experimental Study on the Unsteady Behavior of a Short Bubble on an Airfoil During Dynamic Stall with Special Reference to the Mechanism of the Stall Overshoot Effect," Aeroelastic and Structures Research Lab., Massachusetts Institute of Technology, Cambridge, MA, Rept. ASRL TR 130-2, June 1970.

Kemp, N. H. and Sears, W. R., "Aerodynamic Interference Between Moving Blade Rows, *Journal of the Aeronautical Sciences*, Vol. 20, No. 9, 1953, pp. 585-598.

Kilpatrick, D. A. and Ritchie, J., "Compressor Cascade Flutter Tests—20° Camber Blades, Medium and High Stagger Cascades," National Gas Turbine Establishment, British Aeronautical Research Council, London, Tech. Rept. R.133, C.P. 197, Dec. 1953.

Kilpatrick, D. A. and Ritchie, J., "Compressor Cascade Flutter Tests—Part II. 40° Camber Blades, Low and Medium Stagger Cascades," National Gas Turbine Establishment, British Aeronautical Research Council, London, Tech. Rept. R.163, C.P. 296, Oct. 1954.

Kilpatrick, D. A., and Burrows, R. A., "Aspect-Ratio Effects on Compressor Cascade Blade Flutter," British Aeronautical Research Council, London, R&M 3103, July 1956.

Kilpatick, D. A., Carter, A. D. S., and O'Neill, L., "Blade Vibratory Stresses in a Multistage Axial-Flow Compressor," British Aeronautical Research Council, London, R&M 3181, Jan. 1958.

Kurosaka, M. and Verdon, J. M., "Discussion and Author's Closure—Unsteady Aerodynamics of a Finite Supersonic Cascade with Subsonic Axial Flow," *Transactions of ASME, Journal of Applied Mechanics*, Vol. 41, No. 1, June 1974, pp. 539–541.

Kurosaka, M., "On the Issue of Resonance in an Unsteady Supersonic Cascade," *AIAA Journal*, Vol. 13, Nov. 1975, pp. 1514–1516.

Lane, F., "System Mode Shapes in the Flutter of Compressor Blade Rows," *Journal of the Aeronautical Sciences*, Vol. 23, Jan. 1956, pp. 54–66.

Lang, J. D., "A Model for the Dynamics of a Separation Bubble Used to Analyze Control-Surface Buzz and Dynamic Stall," AIAA Paper 75-867, June 1975.

Lefcort, M. D., "An Investigation into Unsteady Blade Forces in Turbomechanics," ASME Paper 64-WA/GTP-3, Nov. 1964.

Lotz, M. and Raabe, J., "Blade Oscillations in One-Stage Axial Turbomachinery," ASME Paper 68-FE-7, May 1968.

McCroskey, W. J., "Inviscid Flow Field of an Unsteady Airfoil," *AIAA Journal*, Vol. 11, Aug. 1973, pp. 1130–1137.

Miles, J. W., "The Compressible Flow Past an Oscillating Airfoil in a Wind Tunnel," *Journal of the Aeronautical Sciences*, Vol. 23, July 1956, pp. 671–678.

Nagashima, T. and Whitehead, D. S., "Aerodynamic Forces and Moments for Vibrating Supersonic Cascade Blades," Dept. of Engineering, University of Cambridge, Cambridge, England, Rept. CUED/A-Turbo/TR59, 1974.

Namba, M., "Subsonic Cascade Flutter with Finite Mean Lift," *AIAA Journal*, Vol. 13, May 1975, pp. 586–593.

Naumann, H. and Yeh, H., "Lift and Pressure Fluctuations of a Cambered Airfoil under Periodic Gusts and Applications in Turbomachinery," ASME Paper 72-GT-30, March 1972.

Ni, R. H., "Nonstationary Aerodynamics of Arbitrary Cascades in Compressible Flow," Ph.D. Dissertation, Stevens Institute of Technology, Hoboken, NJ, June 1974.

Ni, R. H. and Sisto, F., "Numerical Computation of Nonstationary Aerodynamics of Flat Plate Cascades in Compressible Flow," ASME Paper 75-GT-5, 1975.

Nishiyama, T. and Kobayashi, H., "Theoretical Analysis for Unsteady Characteristics of Oscillating Cascade Airfoils in Subsonic Flows," *Technology Reports, Tohoku University*, Vol. 38, No. 1, July 1973, pp. 287–314.

Parker, R. and Watson, J. F., "Interaction Effects Between Blade Rows in Turbomachines," *Proceedings of the Institution of Mechanical Engineers*, Vol. 186, 1972, pp. 331–340.

Platzer, M. F. and Chalkley, H. G., "Theoretical Investigation of Supersonic Cascade Flutter and Related Interference Problems," AIAA Paper 72-377, 1972.

Platzer, M. F., "Unsteady Flows in Turbomachines—A Review of Current Developments," *Unsteady Aerodynamics*, AGARD CP-227, Sept. 1977.

"Propulsion System Structural Integration and Engine Integrity," *Journal of Aircraft*, Vol. 12, April 1975, pp. 193–436.

Rowe, J. R. and Mendelson, A., "Experimental Investigation of Blade Flutter in an Annular Cascade," NACA TN-3581, Nov. 1955.

Samoilovich, G. S., "Resonance Phenomena in Sub- and Supersonic Flow Through an Aerodynamic Cascade," *Izvestiya Akademii Nauk SSSP, Mekhanika Zhidkosti i Gaza*, Vol. 2, No. 3, May-June 1967, pp. 143–144.

Samoilovich, G. S., "Unsteady Flow Around and Aeroelastic Vibration in Turbomachine Cascades" (translated from Russian), Foreign Technology Division, Wright-Patterson AFB, OH, FTD-HT-23-242-70, Feb. 1971.

Schnittger, J. R., "Single Degree of Freedom Flutter of Compressor Blades in Separated Flow," *Journal of the Aeronautical Sciences*, Vol. 21, Jan. 1954, pp. 27–36.

Schniittger, J. R., "The Stress Problem of Vibrating Compressor Blades," *Journal of Applied Mechanics*, Vol. 22, No. 1, 1955, pp. 57–64.

Shinohara, K., Tanaka, H., and Hanamura, Y., "Stall Flutter of a Thin Aerofoil with Leading Edge Separation," *Bulletin of Japan Society of Mechanical Engineers*, Vol. 17, No. 107, May 1974, pp. 578–586.

Sisto, F., "Flutter of Airfoils in Cascade," Sc.D. Dissertation, Massachusetts Institute of Technology, Cambridge, MA, Aug. 1952.

Sisto, F., "Stall Flutter in Cascades," *Journal of the Aeronautical Sciences*, Vol. 20, Sept. 1953, pp. 598–604.

Sisto, F., "Linearized Theory of Nonstationary Cascades at Fully Stalled or Supercaviated Conditions," *ZAMM*, Vol. 47, Pt. 8, 1967, pp. 531–542.

Sisto, F., "Aeroelasticity of Fans and Compressors," AGARD CP-34, Pt. 1, Sept. 1968, Paper 9.

Sisto, F. and Perumal, P. V. K., "Lift and Moment Predictions for an Oscillating Airfoil with a Moving Separation Point," ASME Paper 74-GT-28, 1974.

Snyder, L. E., "Supersonic Unstalled Torsional Flutter," *Aeroelasticity in Turbomachinery*, edited by S. Fleeter, Project SQUID, Office of Naval Research, June 1973, pp. 164–195.

Stargardter, H., "Optical Determination of Rotating Fan Blade Deflections," *Transactions of ASME, Journal of Engineering for Power*, Vol. 99, No. 2, April 1977, pp. 204–210.

Studer, H. L., "Experimentelle Untersuchunger über Flügelschwingungen," Mitt. aus dem Institute fur Aerod., Eidgenossiche Tech. H. S., Zürich, No. 4/5, 1936 ("Experimental Study of Wing Flutter," British Aeronautical Research Council, London, Oscill. Paper 60).

Tanida, Y., Hatta, K., and Asanuma, T. "Experimental Study on Flutter in Cascading Blades," *Bulletin of Japan Society of Mechanical Engineers (JSME)*, Vol. 6, No. 24, 1963, pp. 736–743.

Tanida, Y. and Okazaki T., "Stall Flutter in Cascade, Parts 1 and 2," *Bulletin of JSME*, Vol. 6, No. 24, 1963, pp. 744–757.

Tobias, S. A. and Arnold, R. N., "The Influence of Dynamical Imperfection on the Vibration of Rotating Disks," *Proceedings of the Institution of Mechanical Engineers*, Vol. 171, 1957, pp. 669–690.

Unsteady Phenomena in Turbomachinery, AGARD CP-177, Sept. 1975.

Verdon, J. M., "The Unsteady Supersonic Flow Downstream of an Oscillating Airfoil," *Unsteady Flows in Jet Engines*, edited by F. O. Carta, Project SQUID, Office of Naval Research, July 1974, pp. 237–254.

Verdon, J. M., "The Unsteady Flow Downstream of an Airfoil Oscillating in a Supersonic Stream," *Journal of Aircraft*, Vol. 12, July 1974, pp. 999–1001.

Verdon, J. M., "Comment on the Issue of Resonance in an Unsteady Supersonic Cascade," *AIAA Journal*, Vol. 13, Nov. 1975, pp. 1542–1543.

Whitehead, D. S., "Aerodynamic Aspects of Blade Vibration," *Proceedings of the Institution of Mechanical Engineers*, Vol. 180, Pt. 3I, 1965–1966, pp. 49–60.

Whitehead, D. S., "Effect of Mistuning on the Vibration of Turbomachine Blades Induced by Wakes," *Journal of Mechanical Engineering Science*, Vol. 8, No. 1, 1966, pp. 15–21.

Wieselberger, C., "Electric Measurement of Forces by Varying an Inductivity," Forsch. Ber. 266, 1935 (see p. 452 of Fung).

Woods, L. C., "Aerodynamics Forces on an Oscillating Airfoil Fitted with a Spoiler," *Proceedings of the Royal Society of London*, Ser. A., Vol. 239, 1957, pp. 328–337.

Yashima, S. and Tanaka, H., "Stall Flutter in Cascade with Leading Edge Separation, Part 1: Theory, Part 2: Experiment," *Transactions of Japanese Society of Mechanical Engineers*, Vol. 40, No. 340, Dec. 1974, pp. 3349–3375.

INDEX